Handbook of
Disease Causing Microbes

Handbook of Disease Causing Microbes

Jaishree Paul

Ph.D

School of Life Sciences

Jawaharlal Nehru University

New Delhi, India

CRC Taylor & Francis

Taylor & Francis Group

Boca Raton London New York

CRC is an imprint of the Taylor & Francis Group, an informa business

Ane Books Pvt. Ltd.

New Delhi, India

Handbook of Disease Causing Microbes
Jaishree Paul

First Edition : **2016**

Published by

Ane Books Pvt. Ltd.

4821 Parwana Bhawan, 1st Floor
24Ansari Road, Darya Ganj, New Delhi - 110 002, India
Tel: +91 (011) 2327 6843-44, 2324 6385
Fax: +91 (011) 2327 6863
e-mail: anebooks@vsnl.net
Website: www.anebooks.com

For

CRC Press
Taylor & Francis Group
6000 Broken Sound Parkway, NW, Suite 300
Boca Raton, FL 33487 U.S.A.
Tel : 561 998 2541
Fax : 561 997 7249 or 561 998 2559
Web : www.taylorandfrancis.com

For distribution in rest of the world other than the Indian sub-continent and Africa.

ISBN : 978-14-3986-678-8

British Library Cataloguing in Publication Data
A catalogue record for this book is available from the British Library

Printed at : Gopsons Papers Limited

Preface

With respect to diseases caused by various microorganisms, we live in a global community in which the health of developed and developing nations is intertwined. Infectious diseases can spread rapidly, making global surveillance and control of emerging infections vital to world health. Migration of the organisms to distant continents is taking place at a faster pace due to wind currents, natural catastrophes and through migratory birds and travellers. Global decline of public health systems serves to render the world more susceptible to infectious diseases.

The book provides readers knowledge on the diversity of disease causing microbes including bacteria, viruses, fungi and protozoans- the major diseases they cause and the environmental factors that influence the occurrence, transmission and spread of the microbes along with the epidemiological factors determining the occurrence of disease. Although variety of excellent texts are available providing in depth coverage and analysis of each of infectious microorganisms individually however a book that deals the topics comprehensively at a single place is rarely found. In assembling the present book, I have tried to draw an appropriate balance between the description of the pathogenic organism, mode of action and transmission, pathogenesis, epidemiology, treatment and mode of prevention of the disease. The book attempted to cover a broad spectrum of diseases infecting different organs of human and describes the microorganisms involved in the process. Diseases caused by parasites and helminths are also included in the text since the infective stages of parasites as well as helminths are microscopic in nature and need to be distinguished from bacterial infections by appropriate diagnosis. Secondly, their presence in the body causes damage to the host resembling that is caused by other microbes. The reader would also find the details to better understand the factors contributing to the emerging and reemerging diseases, microbial resistances and the available control measures.

The focus of this book has been entirely on human infections. The book is planned and designed at a level appropriate for postgraduate students, researchers in Microbiology, Pathology, Infectious disease specialists, Public health experts who could understand and appreciate the relevance and interrelatedness of these topics.

The book is divided into eight chapters. The first chapter deals with an introduction to the infectious diseases covering the spectrum of diseases including neglected tropical diseases, emerging and reemerging diseases and the factors contributing to it. The chapter also includes information sources available globally including India for reporting of epidemics and global surveillance net work for highly infectious diseases. Next seven chapters give the glimpses of human diseases caused by microbes.

Chapter 2 gives a description of normal skin flora and their mode of colonization. Among the bacterial infections, hair follicle infection, atopic dermatitis, scalded skin syndrome, pyoderma and lyme disease are described in length. Viral infections described here are Human Papilloma virus, Herpes virus, Herpes simplex virus, chicken pox and mumps viruses. Athelet's foot, Jock Itch and ringworm are among the fungal infections.

Chapter 3 deals with the respiratory tract infections caused by bacteria, viruses and fungi. Among bacterial infections, Pharyngitis, Diphtheria, Otis media and sinusitis are described in the lower respiratory tract infections. Important diseases covered in the upper respiratory tract infections are bacterial pneumonia, pertussis, tuberculosis and legionellosis.

Chapter 4 broadly deals with the gastrointestinal tract infections of upper and lower alimentary tract. Various pathogens transmitted by fecal oral route, their pathogenesis, epidemiology, treatment and prevention are described in detail. These pathogens are either bacteria, fungi, viruses or parasites.

Chapter 5 focuses on two sections. The first section deals with urinary tract infections caused by microbes leading to bladder and kidney infections. The second section deals with the sexually transmitted diseases, their mode of transmission, pathogenesis, epidemiology, treatment and prevention.

Chapter 6 provides a description of the blood and lymphatic infections. It includes the bacterial infections of blood vascular system, lymph nodes and spleen; viral diseases of the lymphoid system leading to Burkitt's lymphoma, HIV, Chikungunya, dengue fever; protozoal infections like Chagas disease, toxoplasmosis, malaria, leishmaniasis, schistosomiasis, lymphatic filariasis.

Chapter 7 is dedicated to the pathogens that are responsible for serious diseases of our nervous system. Microbial infections like meningitis caused by different types of pathogens, including viruses, bacteria, fungi and protozoa are discussed here. Among other diseases, the chapter also includes the description of the microorganisms causing Tetanus, Botulism, leprosy, Poliomyelitis and human rabies.

Chapter 8 deals with the health-care associated infections. The principal microorganisms involved in nosocomial infections and their specific sites of infection are discussed in detail. The mode of acquiring antibiotic resistance by some of the gram negative bacteria and how these strains become part of the microbiota of hospital personnel and patients leading to a reservoir of resistant strains that initiates a chain of transmission.

I hope that this book will receive attention and appreciation from students, teachers and health care professionals. I believe that the information provided in this book will take us one step forward in our awareness to respond and prevent the outbreaks of infectious diseases in future.

Jaishree Paul

Foreword

The publication of the "Handbook of Disease causing Microorganisms" could not have been better planned with the current threat of an Ebola outbreak in West Africa spinning out of control and posing perhaps the greatest contemporary threat to humankind. The other 'bug' making international news these days is Salmonella enteritidis, which has affected 156 people in England and is the cause of the National outbreak of Salmonella reportedly linked to an Oriental restaurant in East Leigh. But to get a full and frightening sense of the scale of seriousness of the problem of infectious diseases outbreaks in humans, a visit to the Google Global outbreaks site (www.outbreaks.globalincidentmap.com) is illuminating from where it becomes readily apparent that several threats of viral, bacterial and parasitic pathogens prevail over the world at any given time. All that is required is a little bit of carelessness, coupled with the right ecological and epidemiological conditions and the right climatic conditions to create the "perfect microbial storm" manifest in the form of large unmanageable outbreaks. An example of how pervasive, intrusive and destructive outbreaks are is explicit from the Haitian outbreak of cholera which started in October 2010 and since has virtually decimated the Caribbean country which had not seen cholera for the past several decades. This brings to sharp focus on how vulnerable the world, especially the poor and underprepared, are today from such similar global threats.

It is against this grim background that the comprehensive "Handbook of Disease causing Microorganisms" written by Dr. Jaishree Paul is published. The nice thing about this book is that a conscious effort has been made to comprehensively deal with the existing scores of pathogens in a systematic way based on diseases caused at specific human anatomical sites. The book starts with a general overview of the spectrum of infectious diseases and causative agents with subchapters on transmission, community and emerging and reemerging diseases. The author then describes diseases resulting in skin infections, respiratory tract infections, gastrointestinal tract infections, urinary and genital tract infections, blood and lymphatic system infections and nervous system infections. The book is a one-stop shop for a wealth of information on the etiologic agents and the diseases they cause and a departure from books that focus on a single disease, its etiology and pathogenesis.

The handbook is clearly a product of some careful research and hard work. I am sure the book will find a great niche among post graduate and post-doctoral students and will be of profound use to researchers in this area of work. The handbook will also serve as a ready reference for public health officials and epidemiologist working at the field. The author deserves to be congratulated for undertaking this initiative and for accomplishing it so well.

G. Balakrish Nair
(Ph.D, FNA, FASc, FNASc, FTWAS, FAAM)
Executive Director
Translational Health Science and Technology Institute
Gurgaon 122016
Haryana, India

UC San Diego
SCHOOL OF MEDICINE

ANE Books Publishing

Forwarding Letter for "Handbook on Disease Causing Microbes" by Dr. Jaishree Paul
July 29, 2014

Dear Publishers:

I am pleased to write a forwarding letter for Dr. Jaishree Paul's "Handbook on Disease Causing Microbes". I have known Dr. Paul for more than 15 years through our shared research interest on the pathogenesis of amebiasis. Dr. Paul has made many contributions as a Professor at the School of Life Sciences, Jawaharlal Nehru University both through her research and teaching. Her research program has been well funded to study the role of retrotransposons in Entamoeba histolytica, the epidemiology of amoebaisis in NE India, important studies on intestional microbiota in Inflammatory Bowel Disease as well as rapid diagnostic detection of parasites.

Dr. Jaishree now brings her strong scientific background in Microbiology, Biochemistry, and Genetics to her lengthy teaching experience in the important field of Clinical Microbiology as the lead for Microbiology courses for both M.Phil and M.Sc. students at Jawaharlal Nehru University. In the constantly changing world of emerging infections, a new focus on human diseases organized by site of infection is very welcome. This new book will be appreciated by students and health care professionals at all levels.

Sincerely,

Sharon L. Reed, MD

Sharon L. Reed, M.D.
Professor of Pathology and Medicine
Director, Microbiology Laboratory
University of California, San Diego School of Medicine
Email: slreed@ucsd.edu
Telephone: 858-822-2808

Acknowledgements

This book is dedicated to my beloved parents who gave me the best thing in life–life itself

Primary acknowledgements go to many dedicated scientists who have discovered the principles of microbiology. The scientific literature acknowledges individual contributions, but there are plenty of well-endowed investigators whom I could not give adequate attention. In addition, I wish to thank all those who granted permission to adapt figures and data for reproduction that had been previously published. I would also like to acknowledge the staff of Ane Books for their patience and editorial guidance in making this venture possible. I extend sincere thanks to all my colleagues, research scholars and my students in JNU for providing me with an excellent environment to carry out this project smoothly.

Finally, I am grateful to my husband and children who have always shown full understanding for the priorities and extended invaluable support throughout the writing process of this book. I acknowledge my heartfelt thanks to them.

Jaishree Paul

Contents

Preface *v*
Foreword *vii*
Acknowledgements *xi*

1. INTRODUCTION TO INFECTIOUS DISEASES **1–58**

1.1 Spectrum of Infectious Diseases and Their Causative Organisms 2
1.2 Neglected Tropical Diseases 5
1.2.1 Selected Neglected Diseases 6
1.3 Recognition of an Infectious Disease in a Population 11
1.3.1 Occurrence of Infectious Diseases 12
1.4 Mode of Transmission of Pathogens 15
1.5 Host Community 17
1.6 Emerging and Re-emerging Diseases 20
1.6.1 Historical Perspectives 20
1.6.2 Factors Contributing to the Emergence of Infectious Diseases 31
1.6.3 Dynamics, Morbidity and Mortality Rates 31
1.6.4 Deliberately Emerging Infections 37
1.6.5 Resources on Emerging Diseases and Infections 39

2. SKIN INFECTIONS **59–86**

2.1 Normal Skin Flora 59
2.1.1 Pathological Manifestations 60
2.2 Bacterial Infections—Pathogenesis, Epidemiology, Treatment and Prevention 63
2.2.1 Infections of the Hair Follicles-Cause Folliculitis, Furuncles and Carbuncles 63
2.2.2 Atopic Dermatitis (AD) 64
2.2.3 Scalded Skin Syndrome 66
2.2.4 Pyoderma 67
2.2.5 Lyme Disease 68
2.3 Viral Infections–Pathogenesis, Epidemiology, Treatment and Prevention 71
2.3.1 Human Papilloma Virus (HPV) 71
2.3.2 Human Herpesviruses (HHV) 74
2.3.2.1 Herpes Simplex Virus 74
2.3.2.2 HHV-6 76
2.3.3 Varicella (Chicken pox) 77
2.3.4 Mumps Virus 79

2.4 Fungal Infections 82
2.4.1 Athlete's Foot (*Tinea Pedis*) 82
2.4.2 Jock Itch (*Tinea Cruris*) 83
2.4.3 Ringworm (*Tinea Corporis*) 83
2.4.4 Tinea Versicolor 83

3. RESPIRATORY TRACT INFECTIONS **87–132**
3.1 Upper Respiratory Tract Infections 87
3.1.1 Normal Upper Respiratory Tract Flora and mode of Colonization 87
3.1.2 Bacterial Infections-Pathogenesis, Epidemiology, Treatment and Prevention 91
3.1.2.1 Streptococcal Pharyngitis: Pathogenesis, Epidemiology and Treatment 91
3.1.2.2 Diphtheria 94
3.1.2.3 Otitis Media (OM) and Sinusitis 98
3.2 Viral Infections of Upper Respiratory Tract-Pathogenesis, Epidemiology, Treatment and Prevention 100
3.2.1 Common Cold 100
3.3 Lower Respiratory Tract Infections 102
3.3.1 Normal Lower Respiratory Tract Flora and Mode of Colonization 102
3.3.2 Bacterial infections-Pathogenesis, Epidemiology, Treatment and Prevention 103
3.3.2.1 Bacterial Pneumonia 103
3.3.2.2 Pertussis Caused by Bordetella Pertussis 110
3.3.2.3 Tuberculosis 113
3.3.2.4 Legionellosis 117
3.3.3 Viral Infections of the Lower Respiratory Tract 119
3.3.4 Fungal Infections of the Lower Respiratory Tract 122
3.3.4.1 Coccidioidomycosis (Valley Fever) 122
3.3.4.2 Histoplasmosis 124
3.3.4.3 Pneumocystis Pneumonia (PCP) 126

4. GASTROINTESTINAL TRACT INFECTIONS **133–194**
4.1 Upper Alimentary Tract Infections 133
4.1.1 Oral Bacterial Infections- Pathogenesis, Epidemiology, Treatment and Prevention 133
4.1.1.1 Dental Plaque 134
4.1.1.2 Peridontitis/Gingivitis 135
4.1.2 Viral Diseases of the Mouth and Salivary Glands-Pathogenesis, Epidemiology, Treatment and Prevention 136
4.1.2.1 Herpes Simplex Virus Infection 136
4.1.2.2 Mumps Virus 139
4.1.3 Infections of Stomach-causative Agent, Pathogenesis, Epidemiology, Treatment and Prevention 141
4.2 Lower Alimentary Tract Infections 145
4.2.1 Infectious Intestinal Diseases 145

4.2.2 Viral Infections of the lower intestinal tract-Pathogenesis, Epidemiology, Treatment and prevention 164
4.2.2.1 Rotavirus 164
4.2.2.2 Enteric Adenoviruses 166
4.2.2.3 Norwalk Virus and Other Caliciviruses 167
4.2.3 Fungal Infections-Pathogenesis, Epidemiology, Treatment and Prevention 167
4.2.4 Parasitic Infections 168
4.2.4.1 Intestinal Helminthes and Nematodes:Pathogenesis, Epidemiology, Treatment and Prevention 168
4.2.4.2. Intestinal Protozoan Parasites 172

5. URINARY AND GENITAL TRACT INFECTIONS 195–220
5.1 Urinary Tract Infections 195
5.1.1 Bladder Infections 195
5.1.2 Kidney Infections 197
5.2 Genital Tract Infections 198
5.2.1 Sexually Transmitted Infections (STI) 198
5.2.1.1 Chlamydia 198
5.2.1.2 Gonorrhoea 200
5.2.1.3 Genital Warts 202
5.2.1.4 Genital Herpes 204
5.2.1.5 Trichomonas 205
5.2.1.6 Syphilis 208
5.2.1.7 HIV 211
5.2.2 Other Infections 215
5.2.2.1 Candidiasis 215
5.2.2.2 Bacterial Vaginosis 217

6. BLOOD AND LYMPHATIC INFECTIONS 221–280
6.1 Bacterial Infections of the Blood Vascular System 221
6.1.1 Bacterial Endocarditis 221
6.1.2 Septicemia or "Blood Poisoning" 222
6.1.3 Bacterial Diseases Involving Lymph Nodes and Spleen 225
6.1.3.1 Tularemia 225
6.1.3.2 Brucellosis 227
6.1.3.3 Plague 229
6.2 Viral Diseases of the Lymphoid System-Burkitt's Lymphoma, Mononucleosis, HIV Infection, Chikungunya Virus, Dengue Virus, Ebola Virus 234
6.2.1 Burkitt's Lymphoma 234
6.2.2 Mononucleosis 235
6.2.3 HIV Infection 236
6.2.4 Chikungunya Fever 236
6.2.5 Dengue Virus 239
6.2.6 Ebola Virus 243

6.3 Protozoal Diseases of the Blood and Lymphatic System 246
6.3.1 Chagas Disease 246
6.3.2 Toxoplasmosis 250
6.3.3 Malaria 254
6.3.4 Leishmaniasis 261
6.3.5 Babesiosis 265
6.3.6 Schistosomiasis 265
6.3.7 Lymphatic Filariasis 269

7. NERVOUS SYSTEM INFECTIONS 281–318

7.1 Bacterial Infections of the Nervous System 281
7.1.1 Meningitis–Mode of Action, Laboratory Diagnosis and Prevention 281
7.1.2 Mycobacterium Leprae 285
7.1.3 Tetanus 289
7.1.4 Botulism 292
7.2 Viral Infections of the Nervous System 294
7.2.1 Encephalitis 295
7.2.2 Poliomyletis–Pathogenesis, Epidemiology and Prevention 297
7.2.3 Human rabies–Pathogenesis, Symptoms, Diagnosis, Transmission and Epidemiology, Treatment and Prevention 301
7.3 Fungal infections of the Central Nervous system 305
7.3.1 Cryptococcal Meningoencephalitis—Pathogenesis, Symptoms, Diagnosis, Epidemiology, Treatment and Prevention 305
7.4 Protozoal Infection 308
7.4.1 African Sleeping Sickness- Causal Organism, Mode of Action, Symptoms, Diagnosis, Pathogenesis, Epidemiology, Treatment and Prevention 308
7.4.2 Amebic Meningoencephalitis 313
7.4.3 Granulomatous Amebic Encephalitis (GAE) 314

8. NOSOCOMIAL INFECTIONS 319–330

8.1 Ventilator-Associated Pneumonia (VAP) 320
8.2 Bloodstream Infection 323
8.3 Hospital Acquired Urinary Tract Infection (HAUTI) 324
8.4 Other Antibiotic-Resistant Nosocomial Infections 326

1

Introduction to Infectious Diseases

Chapter Outline

1.1 Spectrum of Infectious diseases and their causative organisms
1.2 Neglected Tropical diseases
1.3 Recognition of an infectious disease in a population
1.4 Mode of transmission of the pathogen
1.5 Host community
1.6 Emerging and reemerging diseases

INTRODUCTION

The growth of medical Microbiology began in the eighteenth century. It started with the learning that microbes are causative agents to many human diseases and lead to breakthroughs in their treatments thereby preventing diseases that once took millions of lives. Vaccines were developed against some of the most infectious diseases. The discovery of antimicrobial compounds made it possible for the first time, to cure infectious diseases. Eradication of small pox was first such global success. Till late 1970, the human victory over disease-causing microbes seemed to be on a sound footing. However, microbes constantly presented newer challenges to medical microbiologists. Acquired immunodeficiency syndrome (AIDS), a new viral disease emerged. A short time later, new bacterial diseases such as Legionnaire's and Lyme diseases were documented. Tuberculosis reappeared in developed countries. An epidemic of cholera, a disease previously seen mainly in Europe and Asia, swept through South America. Large and widespread outbreaks of food borne disease began to emerge with a high frequency. Medication failed in some patients infected with bacteria that were resistant to many antibiotics. These occurrences renewed efforts in studying microbial pathogenesis of human diseases.

1.1 Spectrum of Infectious Diseases and Their Causative Organisms

Diseases caused by bacteria, viruses, fungi and other parasites are major causes of death, disability and eventually lead to social and economic disruption for millions of people. Though safe and effective measures to cure these infections are available today but these remain inaccessible to economically weak sections of society. This leads to huge losses to the communities in productivity income besides incurring high cost to cure the diseases. Rate of infections remains high in developing countries, where co-infection is also very common. The impact is most severe among low income group people, who are left out with limited financial resources and thus remain deprived of integrated health care system. As a result, millions of people in these countries remain vulnerable to infectious diseases. The global awareness programs and strong health care system for such diseases need to be extended to a vast section of people in the less developed countries. Over 9.5 million people die every year due to infectious diseases – mostly living in developing countries.

Children are particularly vulnerable to infectious diseases. In children below the age group of five, pneumonia, diarrhea and malaria are as leading causes of death. Cerebral malaria causes permanent mental impairment. Infectious diseases are also of great concern to the health of adults since it may cause disability, bad quality of life, reduces productivity or sometime in death of the earning member.

As a person gets infected with one infectious disease become more susceptible to other diseases as well. Classical examples are HIV/AIDS co-infection with tuberculosis or malaria co-infection with other neglected diseases. Timely interventions can cure and save patients from suffering since most of the infectious diseases are preventable and treatable.

By definition, a disease is an illness that is characterized usually by two of the criteria- recognizing the etiologic agent, typical signs and symptoms of a disease. Most diseases follow a characteristic pattern as shown in Fig. 1.1. As the disease sets in due to an exposure to an infectious agent, pathological changes are noticed and progression towards clinical manifestation occurs till medical interventions are done. Because, not all infections result in disease therefore, conducting a outbreak investigation requires familiarity with a natural history of a disease in order to identify who are infected people.

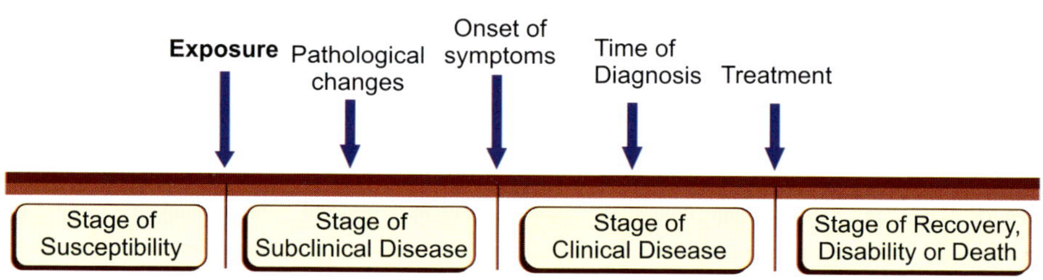

Fig. 1.1: Stages towards progression of a disease

The infectious disease spectrum can be divided into three classes of infection- 1) infection not very frequent 2) infection results in clinical disease that is rarely fatal, 3) infection results in severe disease that is usually fatal.

Infectious Disease Killers

The World Health Organization collects information on global deaths by International Classification of Disease (ICD) code categories. The following table lists the top infectious disease killers which caused more than 100,000 estimated deaths in 2002. 1993 data is included for comparison (Table 1)

Table 1.1: Worldwide mortality due to infectious diseases ["The World Health Report (Annex Table 2)]" 2004[1]. (Reproduced with permission)

Rank	Cause of death	Deaths 2002	Percentage of all deaths	Deaths 1993	1993 Rank
N/A	All infectious diseases	14.7 million	25.9%	16.4 million	32.2%
1	Lower respiratory infections	3.9 million	6.9%	4.1 million	1
2	HIV/AIDS	2.8 million	4.9%	0.7 million	7
3	Diarrheal diseases	1.8 million	3.2%	3.0 million	2
4	Tuberculosis (TB)	1.6 million	2.7%	2.7 million	3
5	Malaria	1.3 million	2.2%	2.0 million	4
6	Measles	0.6 million	1.1%	1.1 million	5
7	Pertussis	0.29 million	0.5%	0.36 million	7
8	Tetanus	0.21 million	0.4%	0.15 million	12
9	Meningitis	0.17 million	0.3%	0.25 million	8
10	Syphilis	0.16 million	0.3%	0.19 million	11
11	Hepatitis B	0.10 million	0.2%	0.93 million	6
12-17	Tropical diseases (6)*	0.13 million	0.2%	0.53 million	9, 10, 16-18

Note: Other causes of death include maternal and perinatal conditions (5.2%), nutritional deficiencies (0.9%),noncommunicable conditions (58.8%), and injuries (9.1%).

*Tropical diseases include Chagas disease, dengue fever, lymphatic filariasis, leishmaniasis,onchocerciasis, schistosomiasis and trypanosomiasis.

The top three single agent/disease killers are HIV/AIDS, TB and malaria. While the deaths due to nearly every disease have decreased however deaths due to HIV/AIDS have increased fourfold. Major childhood diseases that are of great concern are pertussis, poliomyelitis, diphtheria, measles and tetanus. Children also make up a large percentage of lower respiratory and diarrheal deaths.

Indian Scenario

Table 1.2: New infectious diseases recognized in india since 1973.
http://www.medindia.net/health_statistics/general/newdiseases.asp (reproduced with permission)

Year	Agent	Type	Disease
1973	Rotavirus	Virus	Major cause of Infantile Diarrhoea worldwide
1975	Parvovirus B19	Virus	Aplastic crisis in chronic haemolyticanaemia
1976	*Cryptosporidium parvum*	Parasite	Acute and chronic diarrhoea
1977	Ebola virus	Virus	Ebola haemorrhagic fever
1977	*Legionella pneumophila*	Bacterium	Legionnaires' Disease
1977	Hantaan virus	Virus	Haemorrhagic fever with Renal Syndrome (HFRS)
1977	*Campylobacter jejuni*	Bacterium	Enteric pathoger distributed globally
1980	Human T-Lymphotrompic virus 1(HTLV1)	Virus	T-cell lymphoma leukaemia
1981	Toxin-producing stains of *Staphylococcus aureus*	Bacterium	Toxic shock syndrome
1982	*Escherichia Coli 0157:H7*	Bacterium	Haemorrhagic colitis; haemolyticuraemic syndrome
1982	*Borrelia burgdorferi*	Bacterium	Lyme disease
1982	HTLV-2	Virus	Hairy cell leukaemia
1983	Human Immunodeficiency Virus	Virus	Acquired Immuno- deficiency Syndrome (AIDS)
1983	*Helicobacter pylori*	Bacterium	Peptic Ulcer disease
1985	*Enterocytozoon bieneusi*	Parasite	Persistent diarrhoea
1986	*Cyclospara cayetanensis*	Parasite	Persistent diarrhoea
1986	BSE agent?	Non-Conventional Agent	Bovine Spongiform encephalopathy in cattle
1988	Human herpesvirus 6 (HHV-6)	Virus	Exanthumsubitum
1988	Hepatitis E virus	Virus	Enterically transmitted non-A, non-B virus
1989	*Ehrlichia chaffeensis*	Bacterium	Human ehrlichiosis
1989	Hepatitis C virus	Virus	Parentally transmitted non-A,non-B liver hepatitis

Contd...

Year	Agent	Type	Disease
1991	Guanarito virus	Virus	Venezuelan haemorrhagic fever
1991	*Encephalitozoon hellem*	Parasite	Conjunctivitus, disseminated disease
1991	New Species of Babesia	Parasite	Atypical babesiosis
1992	*Vibrio Cholerae O139*	Bacterium	New strain associated with epidemic cholera
1992	*Bartonella henselae*	Bacterium	Cat-scratch disease; bacillary angiomatosis
1993	Sin Nambre virus	Virus	Hantavirus pulmonary syndrome
1993	*Encephalitozoon cuniculi*	Parasite	Disseminated disease
1994	Sabia virus	Virus	Brazilian haemorrhagic fever
1995	Human herpesvirus 8 (HHV-8)	Virus	Associated with Kaposi's sarcoma in AIDS patients

1.2 Neglected Tropical Diseases

Nearly one billion people spread over 150 countries are infected with neglected tropical diseases (NTDs). These diseases are caused by both parasites and bacteria but somehow did not get much attention compared to more popular diseases like malaria, TB, and HIV/AIDS. The neglected diseases have devastating acute and chronic effects on health. Besides the high rate of infectivity of NTDs, they also kill more than a half-million annually. In addition, the loss due to economic, psychosocial and physical damage sustained due to these neglected diseases is the equivalent of 57 million disability-adjusted life years (DALYs) per year.

Some of the most common neglected diseases are: soil-transmitted helminthes, schistosomiasis, lymphatic filariasis (elephantiasis), trachoma, onchocerciasis, leishmaniasis, chagas disease, Human African trypansomiasis, guinea worm (dracunculiasis), buruli ulcer and dengue.

The toll of the neglected diseases on individuals, communities, and societies extends beyond physical suffering to the economic burden of medical expenses and lost productivity. Thus hampers the economic development of people where it is most needed. Moreover, affected individuals often suffer from multiple diseases, compounding the personal burden of disease and disability.

- Neglected diseases can be prevented and treated through effective interventions.
- Often these diseases leave children and adults weak and malnourished. For example, intestinal worms are a major cause of malnutrition, stunting and impaired development among children.
- Diseases such as lymphatic filariasis, trachoma, intestinal parasites, leprosy and onchocerciasis cause chronic disability and stigma for millions. Many are spread through insects or contact with contaminated water.

- Neglected diseases are mentioned within the framework of the Millennium Development Goals under MDG 6: to combat HIV/AIDS, malaria, and other diseases. Though the MDG does not cite official targets for neglected diseases, international organizations, governments and corporations are working to reduce their toll.

Nearly three out of four countries are burdened with at least one or more neglected disease (Fig. 1.2 and Fig. 1.3). Among those countries, 28 are struggling with six or eight neglected diseases. Africa is believed to bear approximately 90 percent of the global burden of neglected diseases[2].

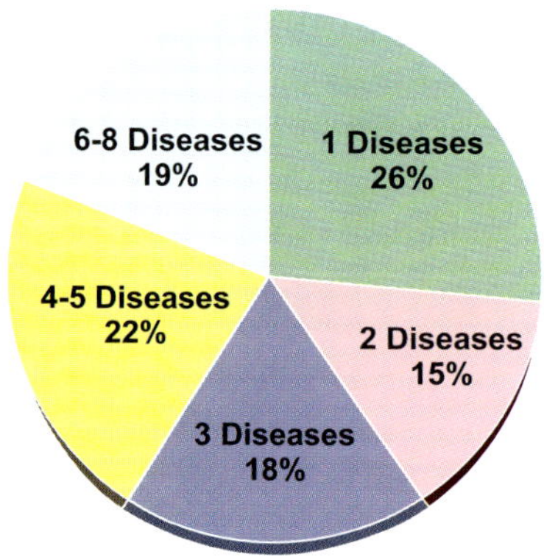

Fig. 1.2: Countries in tropical regions with one or multiple neglected disease world health organization. 2009. Neglected tropical diseases: hidden successes, emerging opportunities. Geneva: who. Available from: whqlibdoc.Who. Int/publications/2009/9789241598705_eng.Pdf] (reproduced with permission)

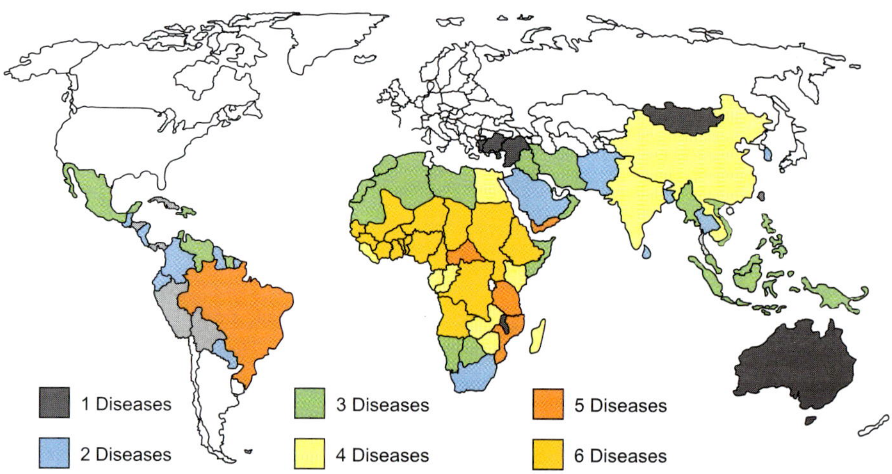

Fig. 1.3: Countries burdened with neglected tropical diseases

1.2.1 Selected Neglected Diseases

Soil-transmitted helminthes (STH) – *or infectious intestinal worms* – It is the most prevalent disease among the neglected diseases. Infections can lead to anemia, vitamin A deficiency, stunting, malnutrition, impaired development, and intestinal obstruction. About 1.6 billion people are at risk throughout the world for soil-transmitted helminthes, which can be treated through de-worming interventions.[3]

Helminthes are spread in areas where there is poor sanitation. The absence of sanitation systems allows parasite eggs, found in feces of infected people, to seep into land and water sources, facilitating the broader infection of a community. 60 percent of those infected with helminthes are children. The three most common STHs are:

- **Ascaris (round worm)** – cause of the most common worm infection; 1.2 billion people live with roundworm infection and it causes 60,000 deaths annually, mostly among children.
- **Trichuris (whipworm)** – affects 795 million people; between 20 to 25 percent live in Africa.
- **Necatoror Ancylostoma (hookworm)** – affects 740 million people of which about one-third live in Africa.

Success story in protecting children from soil-transmitted helminthes and schistosomiasis[4]

In 2001, more than 70% of school-aged children in Cambodia were infected with soil-transmitted helminthes. With the support of international donors, the Cambodian Government now provides preventive medication to all children attending school, with the goal of ensuring that all children are worm-free. This program's success was achieved by training teachers to deliver pills and treatment kits to students twice each year.

Treatment for schistosomiasis was also provided, leading to country-wide elimination of this disease. Cambodia's plan serves as an example for other nations burdened by these diseases.

Schistosomiasis – or Snail Fever – is caused by a fluke, a parasitic worm that is transmitted by freshwater snails. The lack of sanitation systems spurs transmission of the worm. The worms develop in fresh water snails and penetrate the skin, often during swimming or bathing in contaminated waters. The worms enter blood vessels, where they produce eggs. The eggs travel to various organs of the body, causing fever, cough, muscle aches, and impaired growth and performance in school. Long-term infection can result in life-threatening conditions of the bladder (e.g., bladder cancer), kidneys, liver, and lungs[5]. More than 650 million people are at risk of schistosomiasis and an estimated 200 million people live with the disease worldwide. It causes the largest amount of socioeconomic damage after malaria. Nearly 90 percent of the burden of the disease lies in Africa alone.

Water can be treated with iodine or chlorine to remove parasites, or permitted to sit for 48 to 72 hours before use. Communities should be taught to avoid bathing in, washing or drinking contaminated water.

Lymphatic filariasis (LF) – or Elephantiasis – is a crippling, mosquito-borne infection characterized by swollen limbs and breasts, genital damage and thickened, hardened skin. Globally, it is the second leading cause of permanent disability. Anti-parasitic drugs administered annually for at least five years can break the cycle of transmission. More than 1.3 billion people are at risk of infection with 120 million infected people. Many of those infected are permanently disabled. More than one-third of the disease burden is concentrated in sub-Saharan Africa, with a majority of the remaining cases in India, Southeast Asia and Brazil[6].

Trachoma, a bacterial infection of the eye, can result in blindness after multiple infections. It is transmissible from person to person by direct contact or contact with contaminated clothing; flies also help to spread trachoma. Trachoma has become the leading cause of preventable blindness due to its ease of transmission. Repeated infections slowly cause scarring of the upper eyelid that can lead to total blindness during adulthood.

Eight million people have been left visually impaired due to trachoma infection. Women are disproportionately affected; their likelihood of blindness is triple that of men. This may be because mothers and children often pass the infection back and forth. Trachoma occurs most often in densely populated areas with inadequate access to clean water sources. With an estimated 84 million people infected and 150 million requiring treatment, trachoma can now be controlled with a strategy called SAFE. Forty percent of the disease burden is concentrated in Africa only.

Onchocerciasis – or River Blindness – is characterized by intense itching, disfiguring skin lesions and blindness. Infection with onchocerciasis can reduce life expectancy by 15 years; about 6 million of the 37 million people living with the disease become blind. More than 99 percent of those infected live in Africa, but it is also prevalent in parts of Mexico and Venzuela, Guatemala, Brazil, Colombia, Ecuador and Yemen. It has been particularly problematic in Western Africa, where infection was widespread among people seeking fertile land for farming.

The disease is caused by a microscopic worm, *Onchocerca volvulus*, which is carried by blackflies that swarm around natural and artificial waterways. When the fly bites a person, it releases microfilariae (larvae) into the bloodstream of the host. Once in the host, the larvae mature into worms that produce more larvae. When another fly bites the infected person, it picks up the larvae and passes the infection to another host. People remain infectious for 10-15 years if not treated.

Economic losses have been incurred as farmers were forced to move away from fertile but blackfly-infested areas. Onchocerciasis interventions include treatment of water, protecting skin with clothing and drug treatment.

Leishmaniasis – is an infection of the skin and mucous membranes (subcutaneous or mucosal leishmaniasis) that results in disfiguring lesions. In visceral leishmaniasis (kala-azar), the disease spreads to the internal organs, causing fever, anemia, and loss of strength and energy; if left untreated, the disease may cause death.

The disease is caused by infectious protozoa (unicellular parasites) that are transmitted from person to person via sandfly bites. Interventions have focused on ridding regions of sandflies by spraying insecticides in infested homes and places where sandflies breed. Using fine-mesh screens and bed nets for protection during the night help offer protection in homes. Up to two million new infections occur each year and 12 million people are living with the disease. About 350 million people are believed to be at risk for leishmaniasis.[7] Co-infection with HIV is not uncommon in southern Europe and is gaining ground in East Africa and Asia. It is the second-leading cause of parasite-related deaths – malaria is the only parasitic infection that causes more deaths. Treatments for leishmaniasis are expensive and must be administered in a health clinic, making treatment difficult for many to obtain.[7]

Chagas disease – or American trypanosomiasis – affects eight million people, mostly children. Acute chagas disease symptoms includc fever, swelling of the lymph nodes, a lingering feeling of weakness and illness, and enlargement of the liver or spleen. Though between 70% and 80% of people experience no symptoms, the remaining portion have chronic debilitating effects in adulthood such as damage to the heart, esophagus and colon.

Chagas disease is caused by Trypansoma cruzi, protozoa that are transmitted through the bite of triatomine bugs, blood transfusion, or to a fetus by an infected mother. The disease is prevalent in Paraguay, Argentina, Brazil, Chile and Uruguay. Control measures have focused on killing the insects responsible for much of the transmission. Vaccine development is unlikely, as the disease stimulates an autoimmune response that would interfere with a vaccine, but some medications are available[8].

Leprosy – is a bacterial disease that in 2008 affected 212,802 people in sub-Saharan Africa, India, Nepal and Brazil. The disease is transmitted through contact with droplets from the nose and mouth of an infected person who is not receiving treatment. Leprosy affects the nervous system, skin, limbs and eyes. It can be treated using multidrug therapy, though disability or disfigurement may be permanent.

Transmission elimination efforts have succeeded in eliminating disease transmission in 116 of 122 countries where it was endemic. As a result, prevalence has decreased 90% between 1985 and 2006. More than 14 million people have been cured with multi-drug therapy and deformities have been prevented in 2-3 million of those affected. Although between 1 and 2 million people have received treatment to cure their active disease, they remain permanently disabled or disfigured[9].

Human African trypanosomiasis – or African sleeping sickness affects 70,000 people in sub-Saharan Africa, causing physical and neurological damage. During outbreaks, this disease has higher mortality rates than AIDS. It is caused by Trypanosoma brucei rhodesiense and Trypanosoma brucei gambiense, protozoa transmitted by the Tsetse fly.

Symptoms include fever, anemia, headache, enlarged lymph nodes and spleen, and joint and tissue pain10. Trypanosoma brucei rhodesiense infections can be rapid (within days or weeks) and severe – acting on the central nervous system, which may lead to mental deterioration, coma, or death. Trypanosoma brucei gambiense infections may also progress over several years and the impact on health is less severe. Several treatments are available, depending on disease progression.

Dracunculiasis – or guinea worm disease – meaning "affliction with little dragons," is caused by a parasitic roundworm called Dracunculus medinensis. In 2008, the global burden of drancunculiasis was less than 5,000. Most infections in 2008 were concentrated in six endemic countries: Ethiopia, Ghana, Mali, Niger, Nigeria and Sudan. There is an international effort underway to eradicate this disease.

Guinea worm victims are often disabled for two or three months, with symptoms including nausea, vomiting, fever, diarrhea, difficulty breathing and sometimes hives. The infection is characterized by a blister that forms on the lower extremities of the host, usually on the foot. The area around the blister is typically itchy and causes a burning sensation.

Infection occurs when a person ingests water contaminated with aquatic crustaceans or water fleas that carry the Guinea worm larvae. Once ingested, the larvae move through the body to the abdominal cavity and mature into worms. The worms grow to reach two to three feet in length after one year and produce larvae. The worms release their larvae through the blister when it is immersed in water, where the larvae are ingested by aquatic crustaceans or water fleas.

Guinea worm disease often afflicts women who are exposed to contaminated water sources, often when washing clothes or collecting water. The worm also takes a toll on societies by causing absenteeism among infected school-aged children. Interventions include behavioral health education and the provision of safe, filtered drinking water. Parts of the Middle East and India that have implemented these interventions have successfully eliminated the disease.

Buruli ulcer – is an ulcer on the skin, often on the hands and feet, that occurs most commonly in children under 15 inhabiting wet, tropical regions. The number of people infected is unknown, though new infections have been increasing since 19809. The disease causes disfiguring skin ulcers that may limit joint motion and cause disability.

The infectious agent is *Mycobacterium ulcerans*. Though reports of *M. ulcerans* have appeared in more than 30 countries, the heaviest burden lies in Africa, particularly West Africa. The bacteria typically enter the host through cuts, wounds or bites of insects that thrive in infected waters. Infection may cause permanent scarring of tissue.

Outbreaks of buruli ulcer are most common after flooding, which may be the result of deforestation or construction of dams and irrigation systems. Frequent use of ponds or other small, stagnant water sources exposes populations to the infectious agent: Access to household sources of clean water would reduce opportunities for infection[11].

Dengue – is an infection prevalent in tropical and sub-tropical climates that is carried by Aedes aegypti mosquitos and caused by one of four closely-related Flaviviridae viruses12. The symptoms are similar to those of influenza, though are often more severe. Life-threatening complications can ensue if the infection is not addressed in a timely manner. A serious complication may occur if it is dengue hemorrhagic fever, a condition with a fatality rate greater than 20 percent among those who do not receive care. However, it can be reduced to 1 percent for those who receive proper medical care. The rate of new dengue infections has increased steadily over the last few decades as the process of globalization is becoming more prevalent. The number of countries with epidemics increased between 1970 and 1995. Currently, roughly 2.5 billion people are at risk of dengue and an estimated 50 million are infected annually. Dengue is now epidemic in more than 100 countries. It is believed that the mosquito that carries the virus spread to all continents largely via traded commodities arriving with stagnant water, which provides suitable breeding conditions. Dengue hemorrhagic fever is more prevalent in Asia, though more than 20,000 cases have been reported in the Americas in 2007 alone.

The increasing global disease burden, the epidemic nature of the disease and the ability of the mosquito to thrive in densely-populated urban environments has made dengue a growing concern in the global health community.

1.3 Recognition of an Infectious Disease in a Population

This quotation from Darwin's Variation in Animals and Plants Under Domestication illustrates not only how well developed the contagion school was before Pasteur and Koch's demonstration of bacteria as the etiologic agents of disease, but also how infectious disease has played a secondary role in evolutionary biology.

"A particle of small-pox matter, so minute as to be borne by the wind, must multiply itself many thousandfold in a person thus inoculated; and so with the contagious matter of scarlet fever. It has recently been ascertained that a minute portion of the mucous discharge from an animal infected with rinderpest, if placed in the blood of a healthy ox, increases so fast that in a short space of time 'the whole mass of blood, weighing many pounds, is infected, and every small particle of the blood contains enough poison to give, within less than forty-eight hours, the disease to another animal."

http://science. education.nih.gov/supplements/nih1/diseases (Reproduced with permission)

The terms "infection" and "disease" are not synonymous. An infection results when a pathogen invades and begins to grow within a host. Disease results when tissue function is impaired as a consequence of the invasion and growth of a pathogen. Human body has defence mechanisms to prevent infection and when this mechanism fails to prevent onset of the disease, infection sets in. Some infectious agents are easily transmitted and they are referred as contagious, but they are not very likely to cause disease that is, they are not so virulent. The polio virus is an example: It probably infects most people who contract it, but only about 5 to 10 percent of those infected actually develop clinical disease. Other infectious agents are highly virulent, but not terribly contagious. The terror surrounding Ebola hemorrhagic fever is based on the virulence of the virus which has recorded 50 to 90 percent fatality rate among those infected; however, the virus itself is not transmitted easily by casual contact but through body fluids. The infectious agents both contagious and highly virulent are difficult to control.

In order to cause disease, pathogens must meet the following criteria:

1. able to enter the host body,
2. adhere to specific host cells,
3. invade and colonize host tissues,
4. damage the host tissues
5. capable of multiplying inside the host

Entrance to the host typically occurs through natural openings of the body such as the mouth, eyes, or genital openings, or through wounds due to which the damaged skin get exposed to pathogens. Only some pathogens can grow at the initial entry site, mostly they must invade areas of the body by attaching to specific host cells. The pathogen may adopt different mechanism for multiplying inside the host. Some pathogens multiply between host cells or within body fluids, while others such as viruses and some bacterial species enter the host cells and grow there. Although the growth of pathogens may be enough to cause tissue damage in some cases, whereas in other cases, the pathogen is capable of producing toxins or destructive enzymes. For example, *Corynebacterium diphtheriae*, the bacterium that causes diphtheria, grows only on nasal and throat surfaces. However, the toxin it produces is distributed to other tissues by the circulatory system. This toxin is capable of damaging heart, liver, and nerve tissues. *Streptococcus pyogenes*, the infectious agent associated with several diseases including strep throat and "flesh-eating disease". The organism produces several enzymes that work in a synergistic way to break down barriers between epithelial cells and remove fibrin clots thus helping the bacteria to invade tissues.

1.3.1 Occurrence of Infectious Diseases

Epidemiology

Epidemiology is the study of the occurrence of disease in populations. Epidemiologists are concerned not only with infectious diseases, but also with noninfectious diseases such as cancer and atherosclerosis, and with environmental diseases such as lead or arsenic poisoning etc. These epidemiologists work to achieve the following targets:

1. To prevent or minimize the impact of diseases in the population.
2. To identify unusually high incidences of a particular disease,
3. To determine the effectiveness of a vaccine, and
4. Calculating the cost effectiveness of various means of controlling disease transmission.
5. To track down the cause of a "new" disease,
6. To determine its reservoir and mode of transmission, and
7. Help to organize and mobilize various health care workers to bring the disease under control.

Acquisition of Disease

1. **Sources of infectious disease**
   a. **Sources through which one may acquire an infectious disease include:**
      - inanimate reservoirs
      - sick individuals
      - asymptomatic carriers
      - other species

2. **Reservoir**
   a. **Source of disease:**
      - A *reservoir* is simply the source of an infecting microorganism.
      - An infectious disease spreads from a reservoir to individual.
      - Includes inanimates, animates (such as animals), or humans
3. **Carrier**
   a. **Intraspecific reservoir:**
      - In case of human disease, a carrier is defined as a person who is infected and carries a communicable disease for transmission to another uninfected person. This carrier serves as an "intraspecific" reservoir
   b. **Asymptomatic carriers:**
      - An individual who is asymptomatic or carrying an infection without showing any symptoms of the disease may also be a potential carrier of the disease.
4. **Zoonosis**
   a. **Animal reservoir:**
      - A zoonosis is a communicable disease which is transmitted from a non-human animal to a human.
      - Here the non-human animal is the reservoir.
   b. **Uni- or bi-directional:**
      - *Zoonotic* reservoirs can be either unidirectional e.g., transmitting only from non-humans to humans; or bi-directional i.e., back and forth between non-human animals and humans.
      - The kind of replication the microorganism does might differ between its multiple hosts *e.g.,* Plasmodium spp.
   c. The passage of pathogens between populations can range from rare to common (e.g., emerging infectious diseases vs., Lyme disease). About 150 zoonotic human disease are known. Some of the important ones include:
   d. Anthrax, brucellosis, bubonic plague, cat-scratch fever, chagas' disease, hydatid cyst, influenza, leptospirosis, Lyme disease, malaria, pneumonic plague, psittacosis, Q fever, rabies, ringworm, Rocky Mountain spotted fever, salmonellosis, tapeworm, trichinosis, toxoplasmosis, tularemia, typhus fever, western equine encephalitis, yellow fever.
5. **Endogenous infection**
   - An *endogenous infection* is one which is caused by an opportunistic pathogen from an individual's own normal microbiota.
   - An individual faces this consequence when the immune system is weak when an opportunist being deposited in a location other than that in which it typically benignly resides.
6. **Portals of entry**
   a. **Pathogen route of entry:** A *portal of entry* is the route a pathogen takes to enter a host. Typical routes include:
      - mucous membranes
      - skin
      - parenteral

b. Typical portals of entry:
- Each pathogen takes a specific and preferred portal of entry and exit.
- Many pathogens are not able to cause disease if their usual portal of entry is artificially bypassed.

7. Mucous membrane portal of entry

a. *Mucous membranes* line or cover:
- respiratory tract
- gastrointestinal tract
- genitourinary tract
- eyes

b. The gastrointestinal and respiratory tracts are particularly common *mucous membrane portals of entry*.

c. Microbes gain access to these *mucous membranes* during the following activities:
- during inspiration
- upon ingestion
- touching the eyes
- touching the genitourinary tract

8. Skin portal of entry
- Normally very impenetrable to most microorganisms, the *skin* can serve as an efficient portal of entry for many pathogens only when broken.
- Some pathogens are capable of entering through otherwise intact *skin* either via natural pores or by actually boring through, however.
- Examples of pathogens which can enter through intact skin include the hookwarm larvae of *Necator americanus* and some fungi.

9. Parenteral portal of entry
- A parenteral portal of entry is defined as a means of entry where direct deposit of pathogens takes place into body cavities or wounds.
- portal of entry through a break in skin is often considered as a parenteral portal of entry.

10. Portals of exit
- Respiratory infections mostly utilize the mouth and nose as *portals of exit*.
- Gastrointestinal infections tend to *exit* in feces or in saliva
- Sexually transmitted diseases tend to have *portals of exit* at the urethra or genital region.
- Blood-borne diseases tend to *exit via*:
  - ♦ Vectors like arthropods
  - ♦ Needles (hyperdermic syringe)
  - ♦ bleeding

A more general *portal of exit* can be seen when an infected animal is butchered or an infected person undergoes surgery.

1.4 Mode of Transmission of Pathogens

Epidemiologists follow transmission of a disease by correlating geographical, seasonal, and age-group incidence of a disease with possible modes of transmission. A disease limited to a restricted geographical location may suggest a particular vector; malaria for example, is transmitted by a mosquito found mainly in tropical regions.

Different pathogens have different modes of transmission, which are usually related to the habitats of the organisms in the body. For instance, respiratory pathogens are generally airborne, whereas intestinal pathogens are spread by food and water. If the pathogen has to survive, it must undergo transmission from one host to another. Even environmental factors may play a role in survival of pathogen, and such variables as weather patterns may influence exposure to pathogen. For example, California encephalitis, caused by a group of Bunyaviruses occurs primarily during the summer and fall months, and disappears every winter, in a predictable cyclic pattern. The virus is transmitted from mosquitoes which, of course, are prevalent only during summer months. Disappearance of the insect vector cause the disease to disappear until the vector reappears and retransmits the virus in the summer months. The process of transmission involves following steps:

Movement Between Environments

Transmission is the transfer of a disease-causing microorganism from one environment to another, particularly from an external environment to a susceptible individual. Technically it is explained as a process where the pathogen is transferred from a portal of exit to a portal of entry.

1. **Contact transmission**
    - direct contact (person-to-person)
    - indirect contact
    - droplets

Person-to-Person Transmission

Direct contact transmission is a physical form of transmission. For example, direct contact transmission includes: touching, kissing or sexual intercourse

Prevention

Best way to prevent *direct contact transmission* is through the use of barriers such as:

- Use of gloves or masks
- Use of condoms
- Use of a barrier to avoid physical contact

Indirect Contact Transmission [Fomite]

Indirect contact transmission is a form of contact transmission where transmission occurs from a reservoir via inanimate objects. These objects are generally known as fomites. Fomites are basically almost any surface an infected individual (or reservoir) can touch. Upon touching the surface, residue of a contagious pathogen may get deposited on the surface. Some of the inanimates are referred to as vehicles like food, air and liquids. Typically it is more difficult to avoid indirect contact transmission than it is to avoid direct contact transmission. The transmitted pathogen should have the ability to survive passage on a fomite.

Prevention

- avoiding contact with fomites
- avoiding contact of hands with mucous membranes while handling fomites
- the use of barriers when handling fomites
- disinfecting fomites before handling

2. **Airborne Transmission**

   Transmission of pathogens occurs via aerosol. Airborne transmission is a form of vehicle transmission. This type of transmission occurs via droplets for example mucous droplets which remain airborne and may be associated with dust particles. It has been observed that airborne transmission is a problem within airliners due to reduced turnover of cabin air and therefore air recycling is increased. Therefore, the potential pathogenic organisms get an opportunity to remain viable and circulate repeatedly thus may infect passengers.

3. **Foodborne Transmission**

   Foodborne transmission is a form of vehicle transmission. Any number of pathogens that are found in food and not killed during processing may be transmitted via food. Generally pathogens that undergo *foodborne transmission* are either being transmitted from the:

   - food handlers
   - flora that are originally associated with the food product. *e.g. Salmonella* spp. tends to be part of the normal flora of chickens and consequently associated with chicken products.

4. **Waterborne Transmission**

   Waterborne transmission is a form of vehicle transmission. Generally, this is via sewage (i.e., fecal) contaminated water supplies. Gastrointestinal pathogens that are present in feces may get transmitted through contaminated water if it enters drinking water supply for some reason. Transmission of waterborne diseases mainly takes place through fecal oral route.

Portals of Entry to Nervous System

Four common routes of entry: The brain typically is fairly resistant to bacterial infection. However, a pathogen may take advantage of four common portals of entry to the nervous system. The organism needs to display specialized adaptation mechanism to enter through these portals. They are: (*a*) parental, (*b*) via the blood, (*c*) via the lymphatic systems and (*d*) peripheral nerve axons.

The figure below (Fig. 1.4) depicts where the interventions can be made in the cycle of infection.

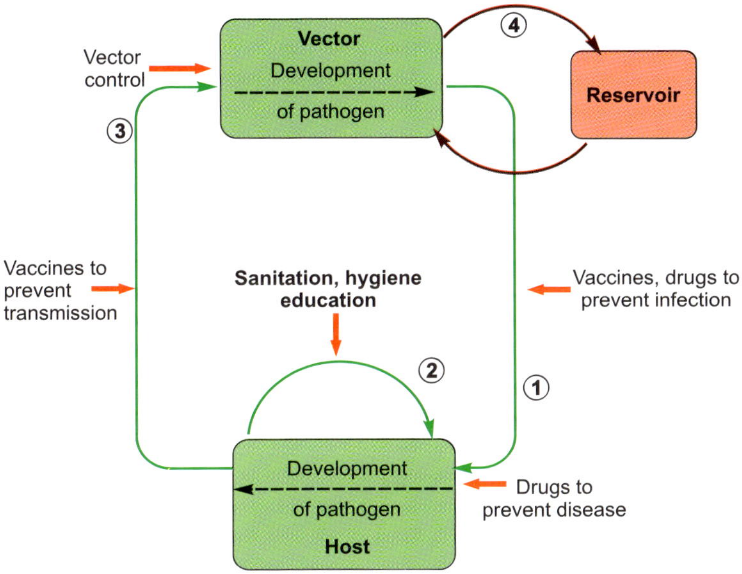

Fig. 1.4: The black arrows illustrate a generalized infectious cycle; the shaded arrows indicate points where infectious diseases can be prevented. ❶ A host is infected by the reservoir or a vector for the pathogen. This individual may infect ❷ other hosts in a population or ❸ new vectors. ❹ The pathogen may cycle between the vector and a reservoir.

Source: [http://www.ncbi.nlm.nih.gov/books/NBK20370/#ui-ncbiinpagenav-heading-14] (reproduced with permission)

1.5 Host Community

The colonization of a susceptible, unimmunized host by a parasite may first lead to an explosive infection and epidemic. As the host population develops resistance, however, the spread of the parasite is checked, and eventually a balance is reached in which host and parasite are in equilibrium. In an extreme case, failure to reach equilibrium could result in death and eventual extinction of the host species. If the pathogen has no other host, then the extinction of the host could also result in the extinction of the pathogen. Thus, the evolutionary success of the pathogen is best measured by its ability to establish a balanced equilibrium with the host, rather than by pathogen's ability to destroy the host. In effect, the host and parasite are affecting each other's evolution; that is the host and the parasites are coevolving.

Coevolution of a Host and a Parasite

Host-parasite co evolution is a special case of co evolution which exerts selective pressure on each other for their coexistence. The evolution of parasites is not independent of the evolution of their hosts, who defend themselves. Generally, parasites have an advantage in this arms race due to their shorter generation time and greater numbers. Co-evolution leads to extensive local adaptation. That is, parasites in a particular area infect hosts from that area more efficiently than they infect hosts from another geographically distinct population. Local adaptation is a general phenomenon found in most host-parasite relationships.

For organisms with short generation times, host-parasite coevolution can be observed in a relatively small time periods. Therefore, experiments can be conducted to study evolutionary change in real-time under both field and laboratory conditions. Host-parasite co evolution is a ubiquitous phenomenon of potential importance to all living organisms, including humans. Many medically relevant diseases like malaria, AIDS and influenza are caused by coevolving parasites. Therefore detailed understanding of the coevolutionary adaptations between parasite 'attack strategy' and host immune system may result in the development of novel medications and vaccines.

During the process of host-parasite co evolution, reciprocal genetic changes may take place leading to changes in allele frequencies within populations. These changes may be determined by three main types of selection dynamics:

Negative Frequency Dependent Selection

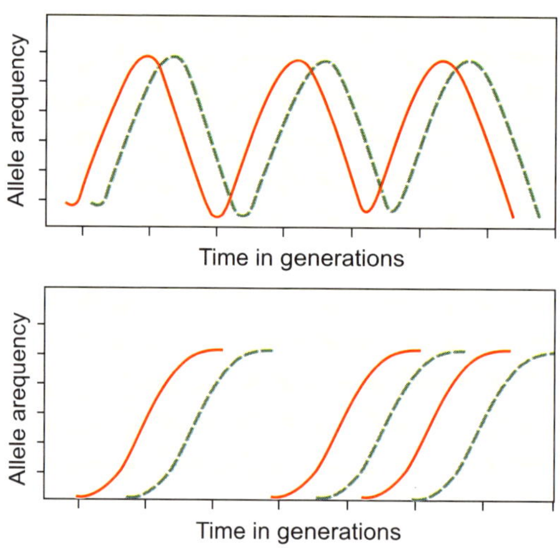

A
Negative frequency dependent selection in the coevolution of hosts and parasites; black curve: host; dotted curve: parasite

B
Directional selection: recurrent selective sweeps in the coevolution of hosts and parasites; black curve: host; dotted curve: parasite

Fig. 1.5A and B: (***Source***: wikipedia.org/wiki/Host–parasite_coevolution)

An allele is subject to negative frequency dependent selection if a rare allelic variant has a selective advantage (Fig.1.5A). For example, if the parasite adapts to the most common host genotype, then the parasite is capable of infecting a large number of hosts. In turn, a rare host genotype will be then favored by selection. This will increase its frequency and eventually it becomes common. Subsequently the parasite should adapt to the former infrequent genotype[13]. Coevolution determined by negative frequency dependent selection is rather rapid and occurring across few generations. This negative frequency dependent selection favors uncommon alleles. It maintains high genetic diversity by favoring uncommon alleles.

The phenomenon of overdominance occurs when the heterozygote phenotype has a fitness advantage over both homozygotes. One example is sickle cell anemia. It is due to a mutation in the hemoglobin gene leading to sickle shape formation of red blood cells, causing clotting of blood vessels, restricted blood flow and reduced oxygen transport. Interestingly, the mutation also confers resistance to malaria, caused by Plasmodium parasites. These parasites are passed off in red blood cells after transmission to humans by mosquitoes. Hence, homozygote and heterozygote genotypes for the sickle-cell disease allele show malaria resistance. The homozygote suffers from severe disease phenotype. The alternative homozygote, which does not carry the sickle cell disease allele, is susceptible to Plasmodium. As a result, the heterozygote genotype is selectively favored in areas with a high incidence of malaria.

Directional Selection

If an allele provides a fitness benefit, its frequency will increase within a population - selection is directional or positive (Fig. 1.5B). Selective sweeps are one form of directional selection, where the increase in frequency will eventually lead to the fixation of the advantageous allele. The process is considered to be slower in comparison to negative frequency dependent selection. A competition takes place between the repeated origin and fixation of new parasite virulence and host defense traits[14] This mode of interactions also take place between unicellular organisms and viruses due to large population sizes, short generation times, often haploid genomes and horizontal gene transfer, which increase the probability of beneficial mutations arising and spreading through populations.

Experiments Under Controlled Lab Conditions

In order to understand the importance of particular selective factors on evolution and to assess the dynamics and consequences of co evolution, elegant laboratory-based experiments have been designed. These experiments have helped to test confounding environmental influences on one test factor at a time which can be varied between treatments. Therefore, this approach may allow distinction of cause and effect. Such evolution experiments require host systems with short generation times like bacteria or invertebrates. It is quite challenging to relate the laboratory results performed in a defined environment to the complex situation encountered in nature.

Creation of a Model System

Potamopyrgus Antipodarum and its Trematode The New Zealand freshwater snail *Potamopyrgus antipodarum* and its different trematode parasites represent a rather special model system. Populations of *P. antipodarum* consist of asexual clones and sexual individuals. This has been explored to study the evolution and advantages of sexual reproduction. Long term laboratory experiments and field studies revealed that a correlation exists between the presence of parasites and the frequency of sexual individuals within the different populations. This result is consistent with the Red Queen hypothesis that sexual reproduction is favoured during host-parasite coevolution [15]

Tribolium Castaneum* and *Nosema Whitei As many other arthropods *Tribolium castaneum*, the red flour beetle,(Fig. 1.6) is a host for the microsporidia *Nosema whitei*. Interestingly, it kills its host during transmission, thus the host's lifespan is important for the parasite's success. In turn, the fitness of the parasite depends on a trade-off between transmission (spore load) and virulence: A higher virulence would increase the potential for the production of more offspring, but a higher spore load would affect the host's lifespan and therefore the transmission rate. This finding has been supported by coevolutionary experiments, which revealed the decrease of virulence, a constant transmission potential and an increase in the host's lifespan over a period of time[16].Further experiments demonstrated a higher recombination rate in the host during coevolutionary interactions, which may be selectively advantageous because it should increase diversity of host genotype[17].

Fig. 1.6: *Tribolium castaneum*

1.6 Emerging and Re-emerging Diseases

1.6.1 Historical Perspectives

Epidemics and infectious diseases have been documented throughout history. In ancient Greece and Egypt accounts describe epidemics of smallpox, Leprosy, Tuberculosis, meningococcal infections and diphtheria[18a]. The morbidity and mortality of infectious diseases profoundly shaped politics, commerce and culture. Interestingly, epidemics did not spare anybody. At times political upheavals exasperated the spread of disease. The Spartan wars caused massive dislocation of Greeks into Athens triggering the Athens epidemic of 430-427 B.C. that killed up to one half of the population of ancient Athens. Several modern epidemiologists have hypothesized on the causative agent. Langmuir et al. (1985)[18b] has described 'The Thucydides syndrome', a new hypothesis for the cause of the plague of Athens favor a combined influenza and toxin- producing Staphylococcus epidemic, while Morrens and Chu suggest Rift Valley Fever[19]. From the earliest times, man has sought to understand the natural forces and risk factors affecting the patterns of illness and death in society.

Between mid-century and 1992, there was a consensus that the battle against infectious diseases had been won, and the Surgeon General announced that it was time to close the book. Finally, to attract attention at the international level, the World Health Organization (WHO), designated April 7 as World Health Day. A conference was held in 1997 where WHO declared the theme as 'Emerging Infectious Diseases – Global Alert, Global Response'. It was concluded that in a global village, no nation is immuned[20]. In addition to the voices of scientists, elected officials, and the public health community, the popular press gave extensive coverage to the new and unexpected danger, especially when four events of the 1990s were highlighted that captured attention worldwide. These infectious diseases discussed were:

1. Experience with human immunodeficiency virus/acquired immunodeficiency syndrome,
2. The return of cholera to the Americas in 1991,
3. The plague outbreak in India in 1994, and
4. The emergence of Ebola in Zaire in 1995

This created awareness in people about return of epidemics and realized that how vulnerable the world population is due to population growth, unplanned urbanization, antimicrobial resistance, poverty, societal change, and rapid mass movement of people. Simultaneously, the epidemiologists also witnessed the changes in the virulence pattern of some pathogenic organisms. For example, increasing virulence of dengue fever with dengue hemorrhagic fever and dengue shock syndrome was observed. Therefore, the epidemiologists started disproving the theory of the evolution toward commensalism, and the discovery of the microbial origin of peptic ulcer due to a commensal bacterium *Helicobacter pylori* demonstrated the changing pattern of infectious diseases. Then the Institute of Medicine coined the term 'emerging and reemerging diseases' to explain that the world had entered an era in which the vulnerability to epidemics would be greater globally than before. The World Health Organization instituted rapid response systems to monitor and contain disease outbreaks. Much research efforts were made to develop new strategies to fight against reemerging microbes. These mechanisms were tested by severe acute respiratory syndrome in 2003[21a]. Important emerging and re emerging diseases are described below (Fig. 1.7)[21b].

A. HIV/AIDS

HIV/AIDS was first described by Krause in the scientific literature in June 1981[22a]. The disease first emerged more than thirty years back as a pandemic disease however, it is now considered as a fundamental matrix disease. Nearly 40 million people throughout the world are currently infected with HIV, of which two third of affected population live in sub-Saharan Africa. In 2004, three million people died from AIDS and nearly five million adults and children became infected with HIV. Nearly 90 percent of new infections occur in the developing world (Fig. 1.8A and Fig. 1.8B).

When HIV/AIDS first emerged, facts were not clear regarding the period between first infection and its manifestation. Now with the current understanding, it has been measured that a patient with AIDS developes advanced HIV disease within a period of ten years. Therefore, it can be presumed that those patients who were first identified in 1981 had likely been infected with HIV for many years before developing symptoms of the advanced disease.

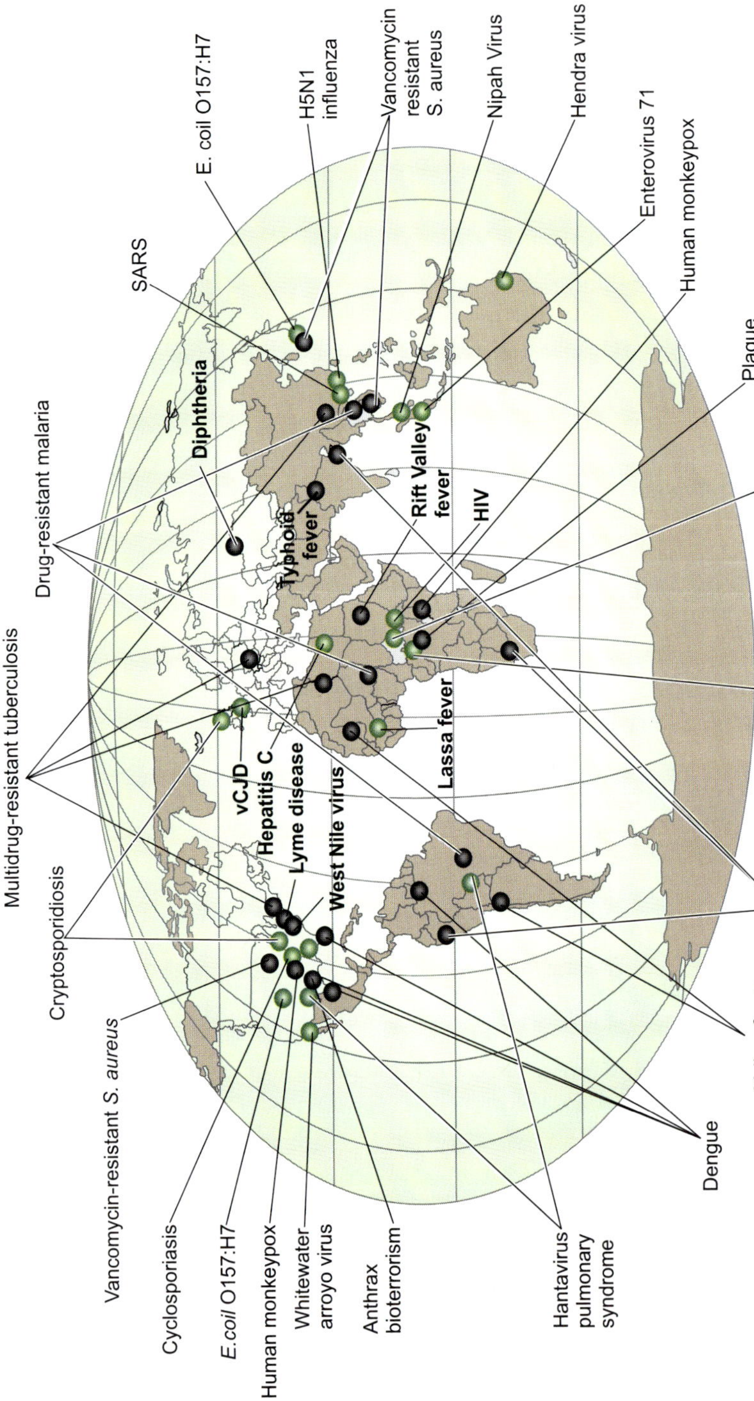

Fig. 1.7: Important emerging and re-emerging diseases across the globe (adapted from fauci A.S. (2001)[21B])

Despite the current concentration of disease burden in sub-Saharan Africa, Asia likely will be the next epicenter of this discasc. Currently about five million people in India are infected with HIV, but that is less than 1 percent of the adult population. India has a total population of more than a billion therefore if the disease is not contained in time, an enormous impact is expected. In the United States, almost one million AIDS have been diagnosed with 525,000 deaths. About a million people have been recorded living with HIV infection, half of whom are untested, untreated, or both. Most of the new infections occur in people under the age of 25. The rate of infection was found to be significantly higher among minorities, particularly African Americans and Hispanics.

Achievements in AIDS Research

Despite the huge challenge in front of the scientific community engaged in HIV/AIDS research, great achievement has been made in the field of Biomedical research on HIV by developing potent antiretroviral drugs that have prolonged and improved the lives of HIV-infected individuals. Since then AIDS research model serves as an ideal example for using in other infectious diseases of global importance. More than 20 drugs have been approved by the Food and Drug Administration for the treatment of HIV/AIDS. This has led to a dramatic decline in AIDS deaths in the USA since the mid-1990s. A suitable combination of antiretroviral therapy (ART) was introduced. However, the irony is, as fewer people die and a consistent 40,000 people become infected each year, the number of people living with AIDS continues to increase.

HIV/AIDS is a totally preventable disease. Different modalities can be adopted on a routine basis for prevention:

1. Education and behavioral changes
2. Treatment of drug users,
3. Distribution of condoms, clean syringes,
4. Use of disposable needles and syringes
5. Use of topical microbicides,
6. Antiretroviral therapy, and abstinence

A characteristic feature of this infection is that so far not a single case has been reported where the viral load from an infected individual has approached to zero even after drug therapy. Lymphocyte cultures from people who clinically are doing well still yield the virus. Even for people on drug therapy for eight to nine years with undetectable viral loads, the virus can be cultured from their lymphocytes. To address the many hurdles in HIV vaccine development, the Global HIV/AIDS Vaccine Enterprise was proposed in 2003 and endorsed by the G8 summit in June 2004.

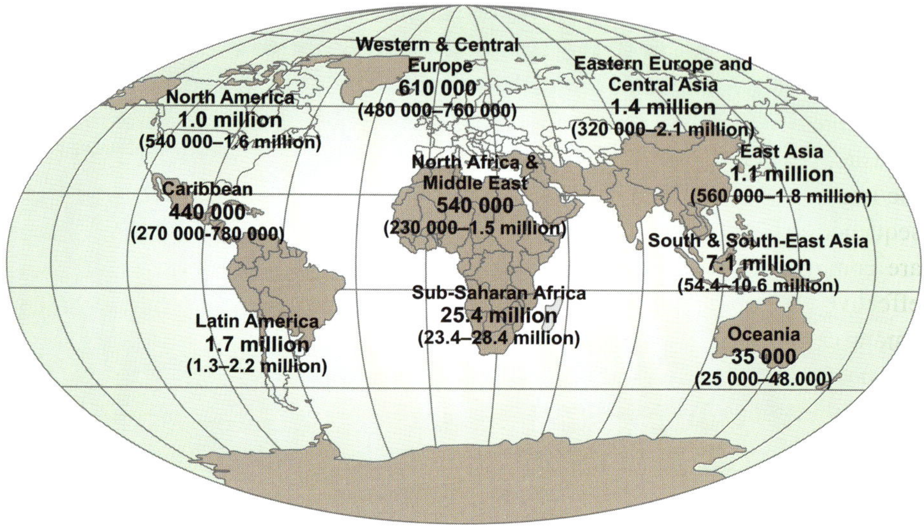

Fig. 1.8A: Adults and children estimated to be living with HIV as of the End of 2004 (Total: 39.4 million) (Map: UNAIDS/WHO)[22] (reproduced with permission)

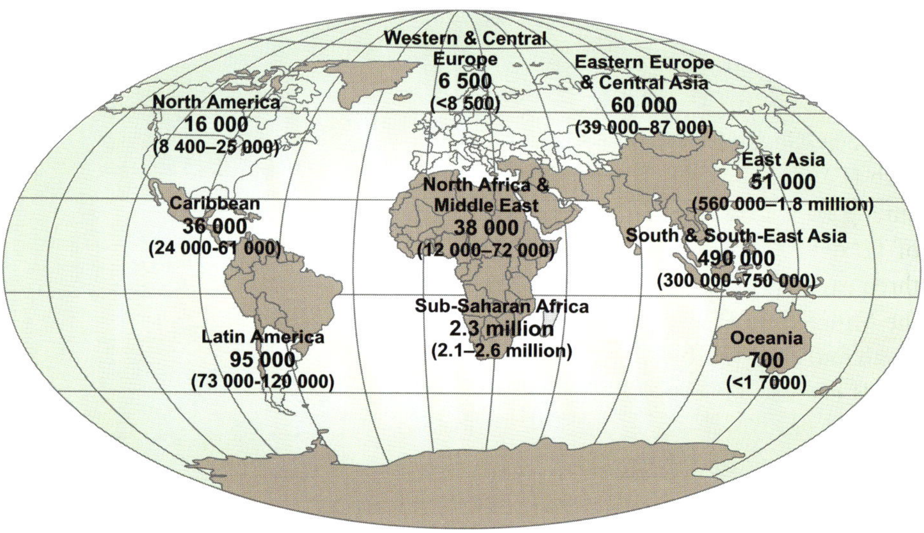

Fig. 1.8B: Estimated adult and child deaths from aids during 2004[22] (Total: 3.1 [2.8–3.5] million) (Map: UNAIDS/WHO) (reproduced with permission)

Development of an HIV vaccine remains one of the most difficult challenges confronting biomedical research today. Strategically, vaccines that are designed based on recently transmitted viruses hold the best hope of inducing relevant immune responses against currently circulating strains.

B. Malaria

The prevalence of malaria is very high in developing countries. It is a public health problem today in more than 100 countries inhabited by nearly 2,400 million people which covers 40 percent of the world's population. Death rate almost touches to one million people every year. Every 30 seconds, a child dies of malaria. Over the past few years, considerable progress has been made in malaria research. Scientists now have completed the genomic sequence of the most virulent malaria-causing parasite, *Plasmodium falciparum*, as well as that of *Anopheles gambiae*, one of the important mosquito species that carries the parasite. With the human genome sequence also available, now the genomes of the host, the vector, and the microbe are completely sequenced. This information has given a boost in the design of effective drugs and vaccines. More recently, a trial of a malaria vaccine conducted among children in Mozambique showed 30 percent efficacy in preventing infection and nearly 60 percent efficacy in preventing severe disease. The malaria vaccine is not as efficacious as most commonly used childhood vaccines however it has the potential for saving millions of lives.

C. Tuberculosis

It is considered as another major killer, causing the deaths of about two million people each year [who.int/mediacentre/factsheets/fs104/en/index.html]. About one third of the world's population is infected with the mycobacterium that causes tuberculosis, and almost four million have the active disease at any time. This also includes 300,000 new cases of multiple-drug-resistant tuberculosis that develop each year. Tuberculosis is especially prevalent among patients with HIV/AIDS. It has been estimated that about 46 percent of individuals in the developing world with HIV are co-infected with tuberculosis, and around 13 percent of the deaths among HIV-infected individuals are from disseminated tuberculosis. Over 95% of TB deaths occur in low- and middle-income countries, and it is among the top three causes of death for women aged 15 to 44. The world is on track to achieve the Millennium Development Goal to reverse the spread of TB by 2015. The TB death rate dropped by 41% between 1990 and 2011.

TB occurs in every part of the world. In 2011, the largest number of new TB cases occurred in Asia, accounting for 60% of new cases globally. However, Sub-Saharan Africa carried the greatest proportion of new cases per population with over 260 cases per 100 000 population in 2011. One of the challenges faced in treating patients suffering from TB is due to development of Multidrug-resistant tuberculosis (MDR-TB) bacteria which is a form of TB caused by bacteria that do not respond to, at least, isoniazid and rifampicin, the two most powerful, first-line (or standard) anti-TB drugs.

D. Influenza

Each year citizens in every country confront with seasonal or pandemic influenza. About 250,000 to 300,000 deaths are reported globally due to this disease. Currently most deaths are found to be associated with people from industrialized countries and prevalence was found to be higher among the elderly people over 65 years of age.

Pandemic Influenza

A pandemic disease occurs through exposure to a microbe for which there is no baseline immunity in the population. The influenza A virus has undergone major genetic changes in the last century resulting in global pandemics and large tolls in terms of both disease and deaths. The worst influenza pandemic in history occurred in 1918–1919 designated as "Spanish Flu" There were 40 million deaths throughout the world due to this infamous flu. Unlike the seasonal flu that typically kills the elderly, the 1918 pandemic killed young people as well. Due to lack of any background immunity to the virus, people belonging to different age groups who were healthy otherwise lost their lives as they contracted with the virulent strain of this virus. Influenza virus strains are designated by the composition of their hemagglutinin and neuraminidase proteins. Thereafter, two other influenza A pandemics occurred in 1957 ("Asian influenza") and 1968 ("Hong Kong influenza") and caused significant morbidity and mortality globally.

The major influenza strain circulating since 2005 was an H3N2 strain. Virtually every year, the sequence of the prevalent strain mutates slightly in a process known as drift. A drift from an H3N2 Panama strain to an H3N2 Fujian strain occurred in 2003 to 2004. The change in the influenza strain represents a relatively minor mutation. In case a person is vaccinated or immune to a Panama strain, one would also harbor some degree of cross-reacting, baseline immunity to a Fujian strain also. However, an antigenic shift occurs when an influenza strain emerges that is substantially different from anything to which the population has been previously exposed. The timetables of the three influenza pandemics that occurred in the 20th century are as follows. In 1918, H1N1 first appeared; in 1957 H2N2 emerged, and in 1968, H3N2 was encountered for the first time.

Over the past few years, many more influenza strains have emerged with the capability of infecting humans. The H5N1 strain likely evolved from a few flocks of chickens in Hong Kong where it was first noticed in 1996. It also infected a small number of humans in 1997 and since then it has infected numerous flocks, as well as wild birds throughout Southeast Asia. Hundreds of millions of birds have been infected; more than 100 million chickens have died or have been culled to slow the spread of the H5N1 virus. The greater challenge to the community is when the virus jumps to humans, the problem gets more severe.

Between December 2003 and June 2005, 108 laboratories-confirmed cases of human infection with the H5N1 avian influenza virus and 54 deaths were reported to the World Health Organization. Clearly, the virus is percolating among avian populations in Southeast Asia. Basic research is being conducted to understand the properties of the virulence factors allowing transmission from one animal species to another animal species and the mechanisms of transmission. H5N1 vaccine has been developed based on the virus that is currently circulating in Asia. A clinical trial of this vaccine was initiated in April 2005. Initial results show that the vaccine is safe and immunogenic. The antiviral compound known as Tamiflu, is effective against the H5N1 virus. These efforts are part of the HHS Pandemic Influenza Response and Preparedness Plan Fig. 1.9 [http://www.flu.gov/planningpreparedness/federal/hhspandemicinfluenzaplan.pdf]

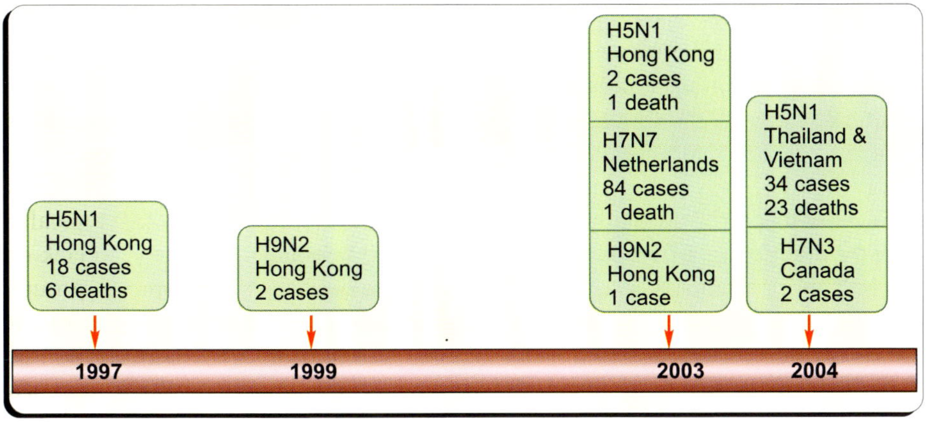

Fig. 1.9: Documented human infections with avian influenza viruses, 1997–2004[23]

Human infection with avian influenza A H7N9 virus emerged in eastern China in February, 2013, and has been associated with exposure to poultry[24]. Cross species poultry-to-person transmission of this new reassortant H7N9 virus is associated with severe pneumonia and multiorgan dysfunction in human beings. Monitoring of the viral evolution and further study of disease pathogenesis will improve disease management, epidemic control, and pandemic preparedness.

E. SARS

In the recent past, the world experienced another newly emerging microbe, a previously unknown coronavirus that caused severe acute respiratory syndrome (SARS). Fortunately, the morbidity and mortality associated with the SARS outbreak were not as high as observed with influenza. The SARS outbreak became an important subject of study from epidemiological point of view with regard to tracking the point source, the spread, and the containment. First case of SARS was detected in Guangdong Province in China. However, this case remained unreported to authorities until it emerged in Hong Kong due to an index case happened to travel from Guangdong to Hong Kong, stayed at the Metropole hotel and infected at least 14 people[25]. Some of these individuals traveled in various parts of the world. Within months another epidemic hit that temporarily transfixed the world and did extraordinary economic damage in Canada, China, Hong Kong, and Vietnam. There were 8,098 reported cases and 774 deaths[26].

The severe morbidity and mortality associated with SARS make it imperative that an effective vaccine be developed to prevent reemergence and epidemics in the future. The new microbe was identified in March 2003 and was rapidly sequenced. This was followed by the development of a vaccine for protection against SARS. In December 2004 a clinical trial of the SARS vaccine began[27] and the trial was carried out at a rapid pace. Such rapid progress could not have occurred 30 to 40 years ago. Thus, as microbes continue to emerge in one hand and the development in technological advances on the other hand, hopefully, keep us one step ahead of our microbial foes.

F. West Nile Virus

SARS is recognized as an emerging infection whereas West Nile virus is a re-emerging infection. It was earlier reported from Africa and the Middle East. West Nile virus (WNV) is a mosquito-borne flavivirus in the Japanese encephalitis antigenic group. Sporadic outbreaks have been reported in humans, horses and birds from the warmer regions of Europe for at least 20 years[28]. In 1999, it landed in Queens, New York, via an unknown route. The infection then moved across the country generating different hot spots each year [29]. In the year 2000, the disease did not spread extensively therefore, people were not deeply concerned however, the disease showed all the parameters of an epidemic in place: the right mosquito, the right microbe, and suitable hosts. In 2001, the hotspots expanded throughout New York State and to Connecticut, Maryland, and Florida. In 2002, hotspots appeared in the Midwest, and farther west in 2003. By 2004, the epidemic had crossed the Rocky Mountains and emerged in California, Arizona, and Colorado. In 2004, 2,470 cases and 88 deaths were reported.[36] West Nile virus also followed the same pattern as seen with other infectious encephalopathies such as St. Louis encephalitis and eastern equine encephalitis. Thus this was then included in the list of reemerging infectious diseases.

The scientific community also moved rapidly to respond to West Nile virus. The scientists could create a chimeric vaccine against West Nile virus. Genes coding for the immunodominant antigens of the West Nile virus, a flavivirus were spliced into the genome of an attenuated yellow fever virus another flavivirus vaccine to yield a chimeric yellow fever/West Nile virus vaccine. Soon the vaccine was subjected to clinical trials with promising early results. Another chimeric vaccine has also been created with an attenuated dengue virus vector. Other West Nile vaccine approaches also are being pursued including a DNA vaccine and recombinant subunit vaccines and other new approaches to therapy are also been tried[30].

Indian Scenario

In India, antibodies against WNV were first detected in humans in Bombay in 1952. Virus activity has been reported in southern, central, and western India. WNV has been isolated in India from *Culexvishnui* mosquitoes in Andhra Pradesh and Tamil Nadu, from *Cx. quinquefasciatus* mosquitoes in Maharashtra, and from humans in Karnataka State. Japanese encephalitis is also one of the major public health problems in Assam, northeast India[31]. WNV is suspected to be the cause of a substantial number of acute encephalitis syndrome cases in this region.

F. Marburg Virus

Marburg virus (MARV) is a member of the family Filoviridae. It is known to cause lethal haemorrhagic fever in humans and non-human primates. Primarily the outbreaks are reported from Central Africa, however these viruses are also considered as potential agents of imported infectious diseases and bioterrorism in non-African countries. This virus is closely related to Ebola virus. This was first detected in 1967 in Marburg, Germany. Carriers were infected people working with monkeys from Uganda resulting in seven deaths. The virus has re-emerged over the years in Kenya, South Africa, Democratic Republic of the Congo and also from Angola. For Marburg outbreaks have been mostly reported from localized regions and have not triggered epidemics throughout the world. The disease gets transmitted when the infected individual is seriously ill. The people at greatest risk of contracting disease are family members, physicians, and nurses in hospitals and other people who happen to come in close contact with infected individuals. Unless the virus is used as an agent of bioterrorism, it is unlikely that the world will experience an epidemic of Marburg because of its restricted transmission[32].

G. Mumps Virus

Mumps virus (MuV) is the etiological agent responsible for an acute viral infection which presents clinically with parotitis accompanied by a low-grade fever, headache, malaise, anorexia, rash, and abdominal pain. Symptoms are normally benign and not life threatening. However, complications develop following MuV infection in approximately 10 percent of cases. Complications may include pancreatitis, orchitis, oophoritis, mastitis, and neurological involvement (meningoencephalitis), which can result in deafness and other severe neurological sequelae[33]. In some of the patients, MuV can invade the central nervous system (CNS) with symptoms of clinical mumps as evidenced by cerebrospinalfluid (CSF) pleocytosis. Symptomatic CNS infection (aseptic meningitis) is much less frequent, occurring in <10 percent of cases, and encephalitis occurs in <1percent of mumps cases[34]. A gender bias has been observed for neurological manifestations, with males disproportionately affected (ca. 3:1 male/female ratio)[35,36]. Routine childhood vaccination with measles-mumps-rubella (MMR) vaccine administered normally between 12-15 months can protect a child from MuV infection. It is observed that the child gets complete protection only when all the doses of the MMR vaccine are taken.

H. Chikungunya: a Potentially Emerging Epidemic

Chikungunya virus is a mosquito-borne emerging pathogen that has a major health impact in humans and causes fever disease, headache, rash, nausea, vomiting, myalgia, and arthralgia. Indigenous to tropical Africa, recent large outbreaks have been reported in parts of South East Asia and several of its neighboring islands in 2005–07 and in Europe in 2007. Chikungunya virus (CHIKV), a mosquito-borne pathogen listed by National Institute of Allergy and Infectious Diseases (NIAID) as a Category C Priority Pathogen that causes Chikungunya fever (CHIKF). In recent times, the disease has spread in several continents like Asia, Africa, and parts of Europe[37,38]. Chikungunya virus in Africa shows the interconnection between the sylvatic cycle and the urban cycle. (Fig. 1.10). CHIKV is an arthropod-borne virus (arbovirus) and is transmitted to humans primarily by *Aedes aegypti*, the infamous yellow fever propagator. CHIKV infection is marked by severe joint pain, contorting its victims into unusual postures[38]. Based on phylogenetic studies, CHIKV isolates have been classified into three genotypes. Based on the gene sequences of an Envelope protein (E1), these genotypes are Asian, East/Central/South African, and West African[39.40,41]. Using phylogenetic models, It has been estimated that the Asian genotype of CHIKV emerged between 50 and

310 years ago, and the West and East African genotypes diverged between 100 and 840 years ago. In the course of time period, CHIKV has incorporated several mutations and has continued to cause epidemics in several regions. In 2005-2006, CHIKV caused an epidemic in India, followed by a sudden explosion of cases in 2007. An estimated 1.3 million people across 13 states were reported to be infected in India[42,43] and CHIKV was also widespread in Malaysia, Sri Lanka, and Indonesia[44].

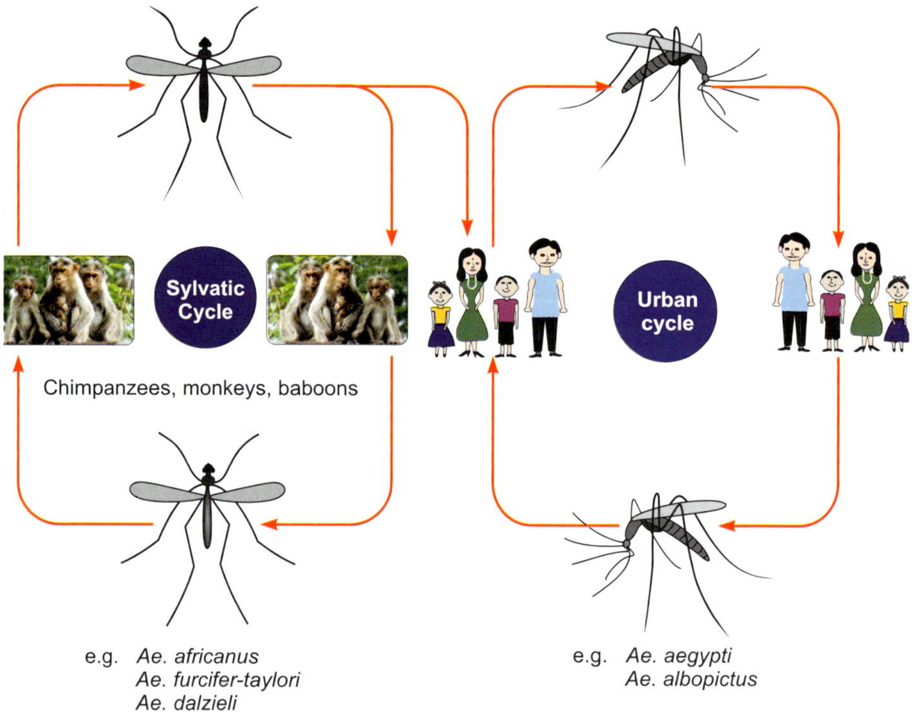

Fig. 1.10: Life cycle of Chikungunya virus in Africa showing the interconnection between the sylvatic cycle on the left and the urban cycle on the right. Particularly in Africa, the virus is maintained in a sylvatic cycle comprising non-human primates and different species of forest-dwelling mosquitoes including *Aedene mosquitoes* (*Ae. africanus, Ae. furcifer-taylori, Ae. dalzieli, etc.,*) and non *Aedene* mosquitoes (Mansonia, Culex, *etc.*)

Source: from Thiboutot et al. (2010)[45]

Indian Scenario

It was assumed that CHIKV had disappeared from India, however, in 2005 the virus reemerged on Reunion Island and in India. Confirmed cases of CHIKV infection have been reported from Delhi, Haryana, Uttar Pradesh, and Rajasthan provinces in northern India, however did not result in large-scale epidemics. *Aedes aegypti* mosquitoes are common vectors for dengue virus and chikungunya virus. In areas where both viruses co-circulate, they can be transmitted together. During a dengue outbreak in Delhi in 2006, 17 of 69 serum samples were positive for chikungunya virus as detected by reverse transcription–PCR. Out of these samples,6 samples were found to be positive for both viruses[46].

Though CHIKV was designated as a potential biological weapon with a high potency of infection, so far no vaccine was available for protection. Live attenuated vaccine trials were carried out in 2000 with little success. Newer approaches such as DNA vaccines appear promising over conventional strategies like live attenuated or inactivated virus and thus call for further investigation. Recent advances such as electroporation delivery and incorporation of adjuvants has boosted DNA vaccine efficacy[47]. Despite the low antibody response to DNA vaccines, it has several other advantages like it mounds both humoral and cellular immune response[48].

1.6.2 Factors Contributing to the Emergence of Infectious Diseases

Several human activities have led to the emergence and spread of new diseases[49]

- Encroachment on wildlife habitats
- The construction of new villages and housing developments in rural areas force animals to live in dense populations. This overcrowding creates opportunities for microbes to mutate and thus leads to emerging infectious diseases of wild life
- *Changes in agriculture:* The introduction of new crops in the fields attracts new crop pests and the microbes they carry to farming communities exposing people to several unfamiliar diseases.
- *Destruction of rain forests:* In many countries, the rain forests are destroyed for building roads and creating settlements and thereby exposing people to insects and other animals harboring unknown microorganisms.
- *Uncontrolled urbanization:* The rapid growth of cities in many developing countries tends to concentrate large numbers of people into crowded areas with poor sanitation. These conditions are favorable for transmission of infectious diseases.
- *Modern transport:* Ships and other cargo carriers often harbor "passengers"who are potential source of infection and may spread diseases to faraway destinations. The rate of transmission further increases through international travel by airplane where an individual carrying a infectious agent can transmit disease to large number of fellow passengers coming from distant places in a shorter period of time.
- *Pollution of the environment:* Changes in the climate such as global warming can cause microorganisms to adapt and create new strains giving them an evolutionary advantage.

1.6.3 Dynamics, Morbidity and Mortality Rates

Predictions that infectious diseases would be eliminated as a major threat to human health have been shattered by emerging and reemerging infections, among them acquired immunodeficiency syndrome (AIDS), hemorrhagic fevers are quite notable. The world has seen marked increases in infections caused by antimicrobial-resistant bacteria, and the resurgence of tuberculosis and malaria. Therefore, it is critical to evaluate the dynamics of emerging and reemerging infections in order to minimize the morbidity and mortality of such infections. This should help to establish policy related to preparedness for infectious threats. The dynamics and determinants of epidemic development have been studied since the nineteenth century for several infectious diseases. For example in case of influenza, prospective epidemics have been predicted using deterministic and stochastic mathematical models. These predictions were quite accurate when compared with actual epidemic data. However some of the models describing and predicting epidemics for diseases that are not transmitted by person to person contact have often yielded mixed results but have been helpful in planning public health responses to epidemics caused by HIV and few other diseases[50].

(*a*) *The Basic Reproduction Number* is often referred as **basic reproductive rate** or **basic reproductive ratio.** This number is defined as a mean number of secondary cases generated from a single infected case in a population with no immunity to the disease and where no interventions have been made to control the infection. It is often denoted as R_0. This metric value decides if the infection has a potential to turn into an epidemic or not in a given population. The roots of the basic reproduction concept has been developed by the efforts of few epidemiologists like Alfred Lotka, Ronald Ross, and others. However the application of this concept was made in epidemiology by George MacDonald in 1952 who constructed population models of the spread of malaria.

$R_0 < 1$, the infection will not spread and eventually die out. When $R_0 > 1$, the infection has a potential to spread in a population. Large values of R_0 indicate the possibility of a major epidemic

The larger the value of R_0, it is more difficult to control the epidemic. One can then determine the proportion of the population that needs to be vaccinated to prevent spread of the infection by $1 - 1/R_0$. The basic reproductive rate is affected by several factors including the duration of infectivity of affected patients, how infectious the organism is, and the number of unaffected people with whom the infected patients come in contact. The table 1.3 below describes the R_0 values, mode of transmission and herd immunity threshold values for different emerging and remerging diseases.

Table 1.3: Values of R_0 of well-known infectious diseases*

Disease	Transmission	R_0	Herd Immunity Threshold
Measles	Airborne	12–18	83–94%
Pertussis	Airborne droplet	12–17	92–94%
Diphtheria	Saliva	6–7	85%
Smallpox	Social contact	5–7	83–85%
Polio	Fecal-oral route	5–7	80–86%
Rubella	Airborne droplet	5–7	80–85%
Mumps	Airborne droplet	4–7	75–86%
HIV/AIDS	Sexual contact	2–5	—
SARS	Airborne droplet	2–5	—
Influenza (1918 pandemic strain)	Airborne droplet	2–3	—

*Adapted from Epidemiology and History of Global Smallpox Eradication (CDC and WHO)

(b) Herd Immunity can be defined as a state where a high percentage of a population is immune to a disease and as a result the disease can not spread further due to lack of new hosts. This concept is also referred to as "community immunity." The threshold value for herd immunity is not constant and therefore varies depending on the disease, its virulence. In order to achieve the desired herd immunity, the percentage of population to be brought under vaccination program may vary from disease to disease. Besides disease prevention, 'community immunity' is also utilized to fight ongoing outbreaks. Most vaccination policies are directed to creating herd immunity. In order to protect children from common diseases, most of the countries practice mass vaccinations in childhood to ensure their safety in case these diseases make an entry in the population at some point in life time. It also gives protection to adults who are previously vaccinated. Establishment of herd immunity is especially important in crowded environment which facilitate the spread of disease, like schools, markets etc.

Herd immunity can be defined as a method to reduce the number of infected individuals in a segment of unimmunized population by immunizing a proportion of population instead of the whole population. The effect of herd immunity can be measured by testing a sample of the population for the presence of a chosen disease against which the population has been vaccinated. Herd effect will also lead to measurable decline in incidence of the disease in the unimmunized segment of the population where the program has been started. Herd effect also applies to other health interventions which reduce the probability of transmission especially if infections get transmitted human to human, directly or via vector. The induced herd immunity of a given vaccine exhibits geographic variation and therefore it depends upon coverage and efficacy of the vaccine. Herd effect is determined by herd immunity as well as the force of transmission of the corresponding infection. It is important to understand this phenomenon in order to design immunization programs more effectively. This will further help at control, elimination or eradication of vaccine preventable infectious diseases[51].

Application – Experts in the field tried to prevent the outbreak of diseases by creating herd immunity, however success rate is not always high. Quite often the pathogen gets mutated and therefore, the body fails to mount an immune response against the original pathogen for which the vaccine was developed. This creates a situation in which an outbreak can occur as majority of the population is vulnerable. In the event of an outbreak of a major disease, agencies like the World Health Organization responds to the situation by dispatching teams within days to figure out the cause of the outbreak and help to develop a vaccine to create herd immunity to stop the outbreak. For some diseases, herd immunity thresholds are as low as 50%, especially when combined with good hygiene. In other instances, up to 90% of the population may need to be vaccinated to create the desired herd immunity. Since few vaccines often lose their efficacy over time, it is mandatory to receive booster doses otherwise people remain vulnerable to an outbreak. Herd immunity played an important role to the eradication of smallpox. This has also contributed largely to eradicate diseases such as polio and diphtheria from developed nations following established vaccination policies.

(*c*) *The Endemic Steady State* is an infectious disease which is said to be endemic, if it can sustain in a population without involving any external inputs. On an average, a person when infects another person or more than one person, the number of people infected will grow exponentially and then it is termed as épidemic'. However, if the number of infected individuals reduces over a time period, then the disease die out. In mathematical terms, that is:

$$R_0 \times S = 1$$

The basic reproduction number (R_0) of the disease is therefore calculated assuming everyone is susceptible, multiplied by the proportion of the population that is actually susceptible (S). Those who are not susceptible are not included in the calculations as they cannot contract the disease. This relation means that the disease is at an endemic steady state, the higher the basic reproduction number, the lower the proportion of the population susceptible for the disease.

Further it is assumed that everyone in the population lives to age L and then dies. If the average age of infection is A, then on average individuals younger than A are susceptible and those older than A are immuned or infectious. Thus the proportion of the population that is susceptible is given by:

$$S = A/L$$

But the mathematical definition of the endemic steady state can be rearranged to give:

$$S = 1/R_0$$

And therefore, since things equal to the same thing are equal to each other:

$$1/R_0 = A/L$$

This provides us with a simple way to estimate the parameter R_0 using easily available data.

$$R_0 = L/A$$

In a population with an exponential age distribution:

For a population with an exponential age distribution, it turns out that

$$R_0 = 1 + L/A$$

The mathematics required to calculate this is a little more complicated than that above, and thus beyond the scope of this article. However, the above term permits to work out the basic reproduction number of a disease given A and L in either type of population distribution.

(*d*) *The Mathematics of Mass Vaccination*, if the proportion of the population that is immune exceeds the herd immunity level for the disease, then the disease can no longer persist in the population. Thus, if this level can be exceeded by vaccination, the disease can be eliminated. An example of this being successfully achieved worldwide is the global eradication of smallpox, with the last wild case in 1977. Currently, the WHO is carrying out a similar campaign of vaccination in an attempt to eradicate polio.

The herd immunity level will be denoted q. Recall that, for a stable state:

$$R_0 \cdot S = 1$$

S will be $(1 - q)$, since q is the proportion of the population that is immune and $q + S$ must equal one (since in this simplified model, everyone is either susceptible or immune). Then:

$$R_0 \cdot (1 - q) = 1$$
$$1 - q = 1/R_0$$
$$q = 1 - 1/R_0$$

If the proportion of immune individuals exceeds this level due to a mass vaccination programme, the disease will die out.

Critical immunisation threshold (denoted q_c). It is the minimum proportion of the population that must be immunised at birth or close to birth so that the infection dies out in the population.

$$q_c = 1 - 1/R_0$$

Outbreaks can occur in a wholly susceptible population when the basic reproductive number (R_0) exceeds one; that is, an infectious individual gives rise, on average, to more than one new infection[52]. In a partially vaccinated population, the effective reproductive number, R must take into account the cases of infectious individuals that may make contact with non vaccinated population. This is the more appropriate measure of outbreak risk. The effective reproductive number can be calculated as $R = R_{0\ pqd} / d_{max}$, where p and q are the probabilities of becoming infected and infectious, respectively, and d is the infectious period[53]. The parameter d_{max} is infectious period estimated as the sample mean of the control individuals.

When a mass vaccination programme cannot exceed the herd immunity: If the vaccine used is insufficiently effective or fails to meet the required coverage, the program may not be able to exceed q_c. Such a program can disturb the balance of the infection without eliminating it and may lead to often unforeseen problems.

If it is supposed that a proportion of the population q (where $q < q_c$) is immunised at birth against an infection with $R_0 > 1$. The vaccination program then changes from R_0 to R_q where

$$R_q = R_0 (1 - q)$$

This change occurs simply because there are now few susceptible individuals in the population who can be infected. R_q is simply R_0 minus those who are immunized.

As a consequence of this lower basic reproduction number, the average age of infection A will also change to some new value A_q in those who have been left unvaccinated. By recalling the relation that linked R_0, A and L. Assuming that life expectancy has not changed, now:

$$R_q = L/Aq$$

$$A_q = L/Rq$$

$$A_q = L/R_0(1 - q)$$

If, $$R_0 = L/A$$

then, $$A_q = L/(L/A)(1 - q)$$

$$A_q = AL/L(1 - q)$$

$$A_q = A/(1 - q)$$

Thus the vaccination program will produce an increase in the average age of infection which appears to be justified. In this situation we can assume that unvaccinated individuals now experience a reduced force of infection due to the presence of the vaccinated group. However, it is important to consider this effect when vaccinating against diseases which increase in severity with age. A vaccination program against such a disease (catching the disease later in life) that does not exceed q_c may cause more deaths and complications than there were before the program was enforced. These unforeseen outcomes of a vaccination program are called perverse effects.

When a mass vaccination program exceeds the herd immunity: If a vaccination program causes the proportion of immune individuals in a population to exceed the critical threshold for a significant length of time, transmission of the infectious disease in that population will gradually come to a halt. This is known as elimination of the infection and is different from eradication.

1. **Elimination:** This can be defined as preventing endemic transmission of an infectious disease by maintaining vaccination coverage to keep the proportion of immune individuals above the critical immunization threshold.
2. **Eradication:** Reduction of infective organisms to zero worldwide. So far, this has only been achieved for smallpox.

Mathematical Modelling

It is used to determine the impact of emerging epidemics. Estimation of overall influenza mortality has been a difficult task because fatal infections are often neither diagnosed nor accurately recorded in hospital records and death certificates, especially in the elderly. Recent epidemiological attempts to obtain improved influenza mortality estimates indicated that influenza mortality may be greater than was previously suspected. This discrepancy mainly arises because influenza deaths are frequently coded under seemingly unrelated categories such as cardiovascular diseases. The same approaches also show that other influenza-like deaths may actually be due to other agents, such as respiratory syncytial virus (RSV), a common childhood virus that in the past decade has emerged as a major cause of adult mortality now[54]. Fig 1.11 represents the dynamics of interaction between a Pathogen and a host.

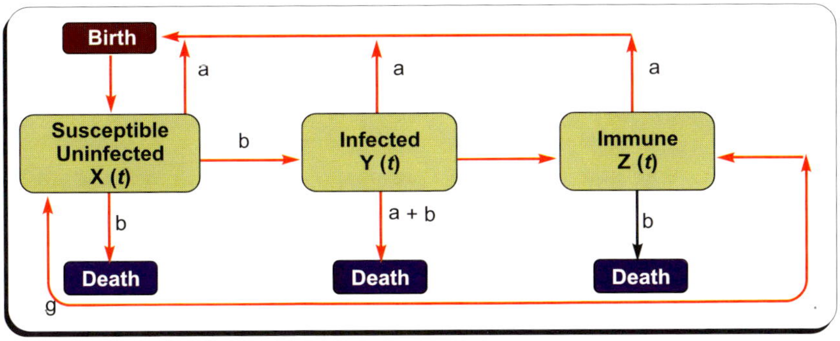

Fig. 1.11: Schematic representation of the flow of hosts between susceptible [X(t)]; infected [Y(t)] and immune [Z(t)] classes, which records the dynamic interaction between a directly transmitted microparasite and its host population. In this diagram the host reproduce at a per capita rate a and die at **a** per capita rate **b**. The infected hosts experience an additional death rate **α**, induced by microparasitic infection. The average duration of stay in the infected and immune classes are denoted by 1/v and 1/**γ** respectively. The transmission coefficient which determines the rate at which new infections arise as a consequence of mixing between the susceptible and infected individuals is defined by **β**

1.6.4 Deliberately Emerging Infections

They are defined as microbes that have been developed by man usually for destructive purpose. The term 'deliberately emerging' refers to both naturally occurring microbial agents such as anthrax and those that are bioengineered microorganisms such as those created by the insertion of genetic virulence factors that produce or aggravate the disease[55]. Deliberately emerging microbes may include microorganisms or toxins produced in a form that would cause maximal harm because of its ease for enhanced infectivity or high level of pathogenicity.

Bioterrorism and Biowarfare – The intentional release and dissemination of *Bacillus anthracis* spores by contaminated U.S. Postal Service mail in the fall of 2001 made public health officials acutely aware of the importance of rapidly and accurately detecting such events. Since this remains an unpredictable event, therefore reliable detection platforms are important to minimize these threats in order to protect the public health.

These platforms must be sensitive and specific and able to accurately detect a variety of pathogens, including modified or previously uncharacterized agents directly from complex samples. Various detection techniques are currently available to identify biological threat agents. Some of the techniques employed are based on tests utilizing biochemical, immunological, nucleic acid, and bioluminescence procedures. Recently, more user friendly and accurate tests have been developed to identify such agents using aptamers, biochips, biosensors, cantilevers, and other innovative technologies. Clinical microbiology laboratories have access to the Laboratory Response Network, which is an organized system designed by the Centers for Disease Control and Prevention (CDC) to structured guidance for the detection, diagnosis, and reporting of biological threat agents (http://www.bt.cdc.gov/lrn/factsheet.asp)

The Chemical and Biological Weapons Non proliferation Program maintains an open-source database containing information on all known incidents worldwide involving chemical, biological, or nuclear materials since write as 1900[56]. This database is searchable, and annual summary reports of activity are available. At the same time, large amounts of biological weapons material remain unaccounted. There is a critical need to develop methods for rapid, accurate detection of biothreat agents. Bioterror agents have been grouped into three categories of risk[57].

Category A Agents

1. Highest priority agents.
2. Risk to national security.
3. Can be easily disseminated or transmitted person-to-person.
4. Cause high mortality and a major public health impact.
5. Might cause public panic and social disruption.
6. Require special action for public health preparedness.

Examples: *Variola major* (smallpox), *Bacillus anthracis* (anthrax), *Yersinia pestis* (plague), *Clostridium botulinum* toxin (botulism), *Francisella tularensis* (tularemia), Hemorrhagic fever (Ebola,Marburg, Lassa viruses).

Category B Agents

1. Foodborne or waterborne pathogens: *Salmonella sps, Shigella dysenteriae, Escherichia coli* O157:H7, *Vibrio cholerae, Cryptosporidium parvum.*
2. Pathogens that are moderately easy to disseminate.
3. Cause moderate morbidity and low mortality.
4. Require special attention for disease surveillance.

Examples: *Coxiella burnetti (Q fever), Brucella species (brucellosis), Burkholderia mallei (glanders), Alphaviruses, Epsilon toxin of Clostridium perfringens, Staphylococcus enterotoxin B.*

Category C Agents

1. Those pathogens that could be engineered for mass dissemination.
2. Ease of production and dissemination.
3. Potential for high morbidity and mortality and major health impact.

Examples: Nipah virus, Hantaviruses, Tickborne hemorrhagic fever viruses, Tickborne Encephalitis viruses, Yellow fever virus, Multidrug-resistant *Mycobacterium tuberculosis*.

Biosurveillance

Early detection and rapid response to bioterrorism depend on close cooperation between public health authorities and law enforcement bodies. The available detection systems and vaccine stockpiles should be made accessible to state officials as and when required. In 1999, the University of Pittsburgh's Center for Biomedical Informatics deployed the first automated bioterrorism detection system, called RODS (Real-Time Outbreak Disease Surveillance). This detection system is capable of collecting data from many data sources to build up a database and use them to perform signal detection to detect a possible bioterrorism event at the earliest possible moment. The various sources of data generation are clinic data, laboratory data, and data from over-the-counter drug sales[58]. These efforts eventually led to the establishment of the National Retail Data Monitor, a system which collects data from 20,000 retail locations nation-wide. In Europe, disease surveillance is organized on the continent-wide scale to track any biological emergency. The system not only monitors infected persons but also attempts to track the origin of the outbreak.

Data which potentially could assist in early detection of a bioterrorism event include many categories of information. Health-related data such as that from hospital computer systems, clinical laboratories, electronic health record systems, medical examiner record-keeping systems, call center computers, and veterinary medical record systems. Other potential sources of information that can be considered include data generated by ranching and feedlot operations, food processors, drinking water systems, school attendance recording etc. It is advisable to collect more than one type of data on a suspected bioterrorism so that the system is less prone to false alarms. Researchers are experimenting with devices to detect the existence of a threat:

- Designing tiny electronic chips that would contain living nerve cells to warn of the presence of bacterial toxins
- Fiber-optic tubes lined with antibodies coupled to light-emitting molecules (identification of specific pathogens, such as anthrax, botulinum, ricin)

1.6.5 Resources on Emerging Diseases and Infections

Association of State and Territorial Directors of Health Promotion and Public Health Education. Addressing Infectious Disease Threats. Current Fact Sheets.

http://www.astdhpphe.org/infectintro.asp

This fact sheet describes what infectious diseases are and also explains both emerging and reemerging infectious diseases, provides prevention suggestions and concludes by describing the critical role for health professionals in communicating disease risk. Hotlines and a table of infectious diseases, agents, and mode of transmission are made available for health professional use.

Centers for Disease Control and Prevention. National Center for Infectious Diseases.

http://www.cdc.gov/ncidod/

This is a valuable link that provides information in infectious disease from the CDC about specific diseases and to the vectors that cause the disease. Both technical information such as clinical features, etiologic agent, reservoir, incidence, sequelae, transmission, risk groups, surveillance, challenges and opportunities, and additional information such as references are given in the extensive list of infectious diseases. Specific links are also available on the site highlighting the important emerging diseases such as SARS and food-related diseases. From this site, travelers can also access prior information on infectious diseases and the surveillance mechanism of a place they plan to visit.

Centers for Disease Control and Prevention: National Center for Infectious Diseases and Emerging Infectious Disease Resource Links

http://www.cdc.gov/ncidod/id_links.htm

This valuable collection of links to resources in emerging infectious disease from the CDC points readers to articles and lectures, journals, newsletters and other publications, to data, to programs and networks, and to agencies, organizations and associations, educational institutions and departmental Web sites.

National Institute of Allergy and Infectious Disease. Division of Microbiology & Infectious Diseases. Emerging Infectious Diseases.

http://www.niaid.nih.gov/dmid/eid/

Many infectious diseases are expected to increase in incidence or change in distribution under conditions in which currently available control measures prove insufficient. Therefore, periodic epidemics of influenza or cholera experienced in the last century provide familiar examples. This site provides fact sheets and brochures about antimicrobial resistance, news releases, staff presentations and papers, reports and articles, research plans and priorities and meetings. Useful links are also offered to the new visitor to the site.

National Institute of Allergy and Infectious Disease. Emerging Infectious Disease Publications.

http://www.niaid.nih.gov/publications/eid.htm

Lists fact sheets and brochures, as well as news releases on such topics as SARS, Dengue Fever, the genetic blueprint of disease-causing microbes, and related topics are available on this site. From this page the reader can find links to various emerging infectious diseases as well.

National Institute of Allergy and Infectious Diseases. Antimicrobial Resistance

http://www.niaid.nih.gov/dmid/antimicrob/default.htm

Increasing use of antimicrobials in humans, animals, and agriculture has resulted in many pathogens developing resistance to the powerful drugs. All major groups of pathogens — viruses, fungi, parasites, and bacteria can become resistant to antimicrobials. This site provides fact sheets and brochures about antimicrobial resistance, news releases, staff presentations and papers, reports and articles, research plans and priorities and meetings.

StatePublicHealth.org

http://www.statepublichealth.org/

This site is supported by The Robert Wood Johnson Foundation, by the State Health Leadership Initiative at the National Governors Association, and by the

Association of State and Territorial Health Officials. This search engine provides information for morc than 20,000 emerging disease documents from the states. One can use either "emerging diseases" or "emerging infectious diseases" as search terms.

Institute of Medicine. Board on Global Health

http://www.iom.edu/board.asp?id=3783

The Board on Global Health (BGH) is concerned with advancing the health of populations worldwide. This involves addressing health issues related to developing countries. It also describes the role of United States in enhancing the global health and addressing health issues that have implications for U.S. health policy. The Board focuses on public health programs for prevention and control of disease and disability. This includes assessment of biomedical knowledge, research and opportunities in reducing behavioral, socioeconomic, and environmental risks to public health. This also provides guidance on ethical issues in public health; and recognition of opportunities to apply scientific knowledge to public policy making.

New York Academy of Medicine. Emerging Infectious Diseases Discussion Group.

http://www.nyas.org/channels/index.asp?channelID=4

The site provides links to upcoming emerging infectious disease events and related events and reports sponsored by the Academy.

ProMED (Program for Monitoring Emerging Diseases)

http://www.fas.org/promed/

Provides links to the ProMED-mail archives and to AHEAD/ILIAD. This is a site that provides surveillance information on animal and zoonotic diseases.

Saint Louis University School of Public Health. Center for the Study of Emerging Diseases.

http://www.emerginginfections.slu.edu/

In addition to providing a high quality list of emerging infections, the site also provides complete educational material, key references including online articles and case studies, official remarks and reports, news and Internet resources. Using this site, visitors can search for information regarding the mode of transmission, organism, and syndrome of a given infectious disease. The site also provides clear image of each organism on each disease mentioned.

Society for Medical Anthropology, Global Health and Emerging Disease Study Group (GHEDSG) (formerly the International Health and Infectious Disease Study Group)

http://www.medanthro.net/research/globalhealth/index.html

Describes the Study Group's activities and discussion lists and provides academic resources and other links of related interest.

U.S. Department of Agriculture Animal and Plant Health Inspection Service Center for Emerging Issues Emerging Disease Notices

http://www.aphis.usda.gov/vs/ceah/cei/notices.htm

Emerging Disease Notices examine critically the occurrences of a new or emerging diseases. Areas covered in an Emerging Disease Notice include a brief review of what is known about the epidemiology of the disease, mode of transmission, reservoir species, and zoonotic potential if any. In addition, the notice also provides information regarding production and trade in livestock and livestock products likely to be affected by the disease and associated risk factors for disease spread.

U.S. Department of Defense. Global Emerging Infections Surveillance and Response System

http://www.geis.fhp.osd.mil/

The DoD Global Emerging Infections Surveillance and Response System (DoD-GEIS) is designed to strengthen the prevention of, surveillance of and response to infectious diseases that are a threat to military personnel and families. It also provides information if the disease presents any risk to U.S. national security. The site is divided into several sections: News about the GEIS system, Infectious Disease News, and Infectious Disease Topics of Interest. Additionally, information about training, global networks, locations and programs are listed.

Commercial Web Sites

Medscape Resource Center - Emerging and Reemerging Infectious Diseases

http://www.medscape.com/pages/editorial/resourcecenters/public/infections/rc-infections.ov

This commercial site provides clinicians with the latest news, reference material, literature selections and continuing education about emerging and reemerging infectious diseases. It is designed to help healthcare professionals to understand the epidemiology of emerging diseases, to identify the signs and symptoms of such diseases, and to aptly consider potentially unusual etiologies in the differential diagnosis of patients. The major section is titled, "Focus On" and guides clinicians to important emerging infectious diseases such as SARS and avian influenza. One can get an overview of current EIDs, recent news, general information and the latest updatesfrom the literature. Also valuable are the CME programs, some available as Webcasts and others through the Medscape Personal Professor, and "Conference Coverage" sections. While on the site readers can also sign up for newsletters on a variety of specialty areas such as "Infectious Diseases" and "Public Health & Prevention."

Resources for Selected Emerging Infectious Diseases

Hardin MD: Infectious Diseases/Bacterial Diseases/Microbiology

http://www.lib.uiowa.edu/hardin/md/micro.html

Hardin MD is a list of directory sites. The Infectious Diseases / Bacterial Diseases/ Microbiology site points the reader to large, medium length and short lists of sites on specific emerging diseases. Of particular interest to readers is the large number of images included on the listed sites. Only the most important and constantly updated directory sites are listed.

1. Avian Influenza

Centers for Disease Control and Prevention. Basic Information About Avian Influenza (Bird Flu)

http://www.cdc.gov/flu/avian/facts.htm

Describes in detail what the flu is and how to protect from getting the flu, or if one is suffering from it, how to prevent spreading the flu. The section for health professionals offers the following links: Vaccination, Clinical description and diagnosis, Background, Training, Infection control, Treatment, Patient education, References and resources, and Surveillance. Content is available in five additional languages including Spanish, Vietnamese, Chinese and Japanese.

World Health Organization. Avian Influenza

http://www.who.int/health_topics/avian_influenza/en/

This page provides links to descriptions of activities, reports, news and events, as well as contacts and cooperating partners in the various WHO programs and offices working on this topic. Also shown are links to related web sites and topics.

2. Gastrointestinal Illnesses

Hepatitis C
Hepatitis Information Network: HepNet-Hepatitis C

http://www.hepnet.com/hepc.html

The site gives information on news, events, and information for physicians and patients. The home page lists New Additions and provides links to Epidemiology, Natural History of the disease, Management and Future Directions. Links to different categories of Hepatitis A, B, D, and E are also given.

Centers for Disease Control and Prevention. National Center for Infectious Diseases. Viral Hepatitis C.

http://www.cdc.gov/ncidod/diseases/hepatitis/c/index.htm

The liver disease caused by the hepatitis C virus (HCV) can be detected from the blood samples of affected individuals. HCV is spread by contact with the blood of an infected person. In addition to a fact sheet about Hepatitis C and a good FAQ, visitors to the site can find recommendations and can get on-line training regarding preventive measures. The site is also available in Spanish. Freely available information for the public and for clinicians and other health care providers is also made available. One can register for online training using the Training link.

3. Lyme Disease

Centers for Disease Control and Prevention. Division of Vector-Borne Infectious Diseases. Lyme Disease

http://www.cdc.gov/ncidod/dvbid/lyme/

Describes the history of the disease and also the vector (tick responsible for transmitting the disease), the risk of getting the disease, provides answers to questions, describes the tick ecology and also provides information for the clinician for diagnosis. Prevention and control suggestions are also discussed. Links to the scientific literature, to the epidemiology of the disease and to a PBS documentary give both professionals and the public information on Lyme disease.

Lyme Disease Network (LymeNet)

http://www.lymenet.org/

The Lyme Disease Network, is a non-profit foundation dedicated to public education of the prevention and treatment of Lyme disease and other tick-borne illnesses. The network maintains Support groups, an online library, legal resources, medical abstracts, a newsletter, books, pictures and additional information.

4. Norovirus

National Center for Infectious Diseases (CDC). Respiratory and Enteric Viruses Branch. Viral Gastroenteritis

http://www.cdc.gov/ncidod/dvrd/revb/gastro/faq.htm

The above web site describes various aspects of viral gastroenteritis like causal organism, symptoms, contagion, and treatment.

Centers for Disease Control and Prevention.Respiratory and Enteric Viruses Branch: Norovirus

http://www.cdc.gov/ncidod/dvrd/revb/gastro/norovirus.htm

Noroviruses (genus Norovirus, family Caliciviridae) are a group of related, single-stranded RNA, nonenveloped viruses that cause acute gastroenteritis in humans. Norovirus was recently approved as the official genus name for the group of viruses provisionally described as Norwalk-like viruses (NLV). One can find Norovirus Q&A, Norovirus Technical Fact Sheet, and information about Norovirus and foodhandlers. The Q&A link sends the reader detail information on what a norovirus is, what the symptoms are, how serious the norovirus disease is, alternate names for the illness caused by the virus, infection process, contagiousness of the virus, and treatment. For the health care practitioner, similar information is presented from the clinician standpoint. Appropriate information for food handlers is also available.

5. Tick Relapsing Fever (Ornithodoros Turicata)

Available from: URL: http://www.cdc.gov/ncidod/EID/vol9no9/03-0280.htm

The site describes the history of tick-borne relapsing fever, the vector that causes it and the study that lead to the publication of this article.

Washington State Department of Health. Tick-borne Relapsing Fever

http://www.doh.wa.gov/ehp/ts/Zoo/TBRF.html

The above state Health Department document describes "relapsing fever" and what it is, how it is transmitted, its symptoms, diagnosis and treatment. Of particular interest to the public is the information on how to avoid getting relapsing fever, importantly measures to control rodents and ticks.

6. Tuberculosis

Centers for Disease Control and Prevention. Health Topic: Tuberculosis

http://www.cdc.gov/tb/

CDC has prepared sections of interest to the public, to travelers, workers and to clinicians and epidemiologists in the following site links: Questions & Answers about TB, Travelers' Health Information on Tuberculosis, how to prevent an attack of Tuberculosis in the workplace, Tuberculosis surveillance reports, and Tuberculosis vaccinations.

7. West Nile Virus

Centers for Disease Control and Prevention. Division of Vector-Borne Infectious Diseases.

http://www.cdc.gov/ncidod/dvbid/westnile/index.htm

One can learn about ways to protect the family and the community from West Nile virus. Publications and presentations from the 2003 WNV National Conference are also accessible from this page. Links are also available for the state and local government sites with information on West Nile Virus.

Washington State Department of Health. West Nile Virus in Washington

http://www.doh.wa.gov/ehp/ts/Zoo/WNV/WNV.html

The Department of Health Web site is very helpful with respect to this disease. Clinical information for health care providers, environmental health information for state and local agencies and other organizations, useful information for the public and maps, statistics, and news releases is also available. Information on mosquito repellants and on mosquitoes is available in two brochures.

8. SARS

Centers for Disease Control and Prevention. Severe Acute Respiratory Syndrome (SARS)

http://www.cdc.gov/ncidod/sars/

The CDC has collected information for specific groups and settings such as the general public, clinicians, health departments, clinical laboratories and the legal community. It presents information on community containment (quarantine), clinical evaluation and diagnosis, exposure management and infection control. Additional information such as updated guidance and news is offered as well as Hotlines for reporting cases of SARS.

CBC News Online SARS: Timeline

http://www.cbc.ca/news/background/sars/timeline.html

Describes the onset of SARS from its discovery since February 15, 2003 and traces further events in China and Canada to June 11, 2003.

9. Food Issues and Food Security

FoodNet: Foodborne Diseases Active Surveillance Network (CDC's Emerging Infections Program): a program of the Foodborne & Diarrheal Disease Branch

http://www.cdc.gov/foodnet/

Food Net is a collaborative project of the CDC, ten EIP sites (California, Colorado, Connecticut, Georgia, New York, Maryland, Minnesota, Oregon, Tennessee and New Mexico), the U.S. Department of Agriculture (USDA) (http://www.usda.gov/), and the Food and Drug Administration (FDA) (http://vm.cfsan.fda.gov/). The project consists of active surveillance for foodborne diseases and related epidemiologic studies designed to help public health officials better understand the epidemiology of foodborne diseases in the United States. It provides a network for responding to new and emerging foodborne diseases of national importance. This program also monitors the burden of foodborne diseases, and help in identifying the sources of specific foodborne diseases. One may find information on the diseases and pathogens under surveillance, reports and studies, and conferences, related links and links to the Food Net member site.

Iowa State University. College of Veterinary Medicine. Center for Food Security and Public Health

http://www.vetmed.iastate.edu/services/institutes/cfsph/

The mission of the CFSPH is to increase national preparedness for accidental or intentional introduction of disease agents that threaten food production or public health. The Center is involved in three major activities: Awareness Education, The Public Health Auxiliary Support Team for Emergency Response (PHASTER), and providing MPH education. It is one of the CDC Specialty Centers in Public Health Preparedness. Find disease fact sheets and also collects information on agroterrorism via training sessions. The fact sheets include the World Organization for Animal Health enlisting the diseases that are either not present in the US or that are reportable diseases and also include information regarding number of reportable agents.

International Web Sites

AFRIMS: Global Emerging Infections System (GEIS)

http://www.afrims.org/geis.html

The mission of this site is to support the Department of Defense's GEIS program by enhancing the infectious disease surveillance and response components of the international public health infrastructure in Asia. Visitors to the site can read about

GEIS current Projects, about outbreak investigations, collaborators, GIS Disease Mapping, training activities and to links to related content.

Asia-Pacific Economic Cooperation, Regional Health Threats

http://www.apec.org/apec/apec_groups/som_committee_on_economic/working_groups/health/apec_information_on.html

The site describes emerging infectious health threats such as SARS and Avian Flu in the Asia-Pacific Rim region. It also looks at these public health problems from the economic standpoint, on the affected economies. APEC health ministers provide statements on emerging disease issues of importance to their countries. Helpful documents such as press releases, self-assessment questionnaires for officials, prevention and control documents and articles from epidemiological journals are also linked from the sidebar.

Emerging/Reemerging Diseases (PAHO)

http://www.paho.org/english/ad/dpc/cd/eid-eer.htm

This PAHO site provides access to EID Weekly Updates: Emerging and Reemerging Infectious Diseases, Region of the Americas and to various emerging and/or reemerging diseases such as Severe Acute Respiratory Syndrome (SARS), Anthrax, Dengue, Malaria, Tuberculosis and West Nile Virus. In addition to these, readers may also find "Situation Reports," "Incidence/Mortality/Case Fatality" documents, "Guidelines and Manuals," "Resistance Surveillance," "Epidemiological Surveillance Systems (ESS)," and "Country Profiles."

Emerging, Re-emerging and Other Communicable Diseases - DDC - WHO/AFRO (African Region)

http://www.afro.who.int/emc/

The mission of this agency is to provide technical support to the countries of the African region in order to reduce morbidity, disability and mortality due to epidemic diseases through the improvement of epidemic preparedness, detection and response. Of some interest are the outbreak control measures at the community level for Cholera, Meningitis, Plague, Dysentery, Hemorrhagic Fever, and Yellow Fever. Outbreak control measures cover such topics as Case finding and management, contact and family management, environmental sanitation, personal hygiene and health education.

Health & Borders: Comparing global to international & local visions of health and disease.

http://caribou.cc.trincoll.edu/~jtrostle/Health & Borders/main.html

The main point of this site is that "communicable diseases do not stop at borders." Some information on new and re-emerging diseases, on "classic" international diseases such as malaria, AIDS, cholera, diarrheal diseases and respiratory infections, on refugees and their health, global health organizations and a case study of cholera in Ecuador. The site is created and maintained by James Trostle, Mary Ellen Rodgers, and Christine Maltese.

World Health Organization. Emerging diseases.

http://www.who.int/health_topics/emerging_diseases/en/

This page provides links to descriptions of activities, reports, news and events, as well as contacts and cooperating partners in the various WHO programs and offices working on this topic. Also shown are links to related web sites and topics.

World Health Organization. Communicable Disease Surveillance and Response

http://www.who.int/csr/en/

The above site describes information on how the Department of Communicable Disease Surveillance and Response (CSR) is working towards global health security; epidemic alert and response. One can also discover CSR's main areas of work and their activities at the global, regional and country level. One can also review a comprehensive list of disease outbreaks, investigate other information resources and learn how CSR is organized. Highlights of the most recent disease outbreaks, CSR's latest activities, and useful links are listed on CSR's home page.

Prevention (Including Preparation for Deliberate Epidemics)

Addressing emerging infectious disease threats: a prevention strategy for the United States, 1994.

http://www.cdc.gov/ncidod/publications/eid_plan/default.htm

This plan was developed by the Centers for Disease Control and Prevention (CDC) in partnership with local and state public health officials in 1994 along with federal agencies, medical and public health professional associations, infectious disease experts from academia and clinical practice, and international and public service organizations. The plan addresses the urgent need to improve the nation's ability to identify infectious disease threats and respond to them effectively. It will serve as a guide for CDC's work in collaboration with its partners in safeguarding the nation from the threat of emerging infectious diseases. The CDC's prevention strategy is outlined and great emphasis is put on the critical role of partnerships in addressing emerging infectious disease threats.

Centers for Disease Control and Prevention: Preventing Emerging Infectious Diseases: A Strategy for the 21st Century.

http://www.cdc.gov/mmwr/preview/mmwrhtml/00054779.htm

CDC scientists have designed a plan to respond to the emergence and resurgence of microbial threats in the new millennium — *Preventing Emerging Infectious Diseases: A Strategy for the 21st Century.* This Web site contains the full version of CDC's plan and its supporting materials, which describe the steps that CDC and its public health partners must take to ensure that people are better protected from infectious diseases. Visitors to the site will find the strategic plan, booklets, cases, a slide set that can be used to discuss the plan, and can listen to audio. The nine target areas are antimicrobial resistance; foodborne and waterborne diseases; diseases transmitted through blood transfusions or blood products; chronic diseases caused by infectious agents; vaccine development and use; diseases of people with impaired host defenses; diseases of pregnant women and newborns; and diseases of travelers, immigrants, and refugees.

INSPEAR - International Network for the Study and Prevention of Emerging Antimicrobial Resistance

http://www.cdc.gov/ncidod/hip/surveill/inspear.HTM

INSPEAR is a response to the global emergence of drug-resistant organisms and the resulting need for international surveillance programs and the strengthening of the microbiologic and epidemiologic capacities of hospitals worldwide. The main purposes of INSPEAR are to:

1. serve as an early warning system for emerging antimicrobial resistance,
2. rapid distribution of information about the acquired resistance, and
3. provide microbiologic and epidemiologic assistance to INSPEAR members.

Of special interest to readers are the sections on outbreak management, antimicrobial resistance, bloodborne pathogens, guidelines and recommendations, surveillance and training.

The United States Army Medical Research Institute for infectious diseases - USAMRIID

http://www.usamriid.army.mil/index.htm

As the Department of Defense's lead laboratory for medical aspects of biological warfare defense, USAMRIID conducts research to develop vaccines, drugs and diagnostics for laboratory and field use. In addition to developing medical measures, USAMRIID is also engaged in formulating strategies, information, procedures, and training programs for medical defense against biological threats. Available information consists of USAMRIID programs, education, publications and related links.

World Health Organization. Preparedness for Deliberate Epidemics

http://www.who.int/csr/delibepidemics/en/

Site describes the available Information resources including publications, FAQs, program reports and links to other sites. It also covers information on national and international preparedness and global alert and response.This page lists the pathogens most likely to be used: anthrax, botulism, plague, smallpox and tularemia. The page also describes preventive measures to be taken for specific diseases.

World Health Organization. Strategies for Fighting Emerging/Reemerging Diseases: preparedness and response plans.

http://www.who.int/csr/delibepidemics/en/Preparednessproject.pdf

It defines WHO's response to deliberately caused epidemics. It also gives some background, WHO's role and key partners.

World Health Organization. Global Outbreak Alert & Response Network

http://www.who.int/csr/outbreaknetwork/en/

The subheading for this page is "containing outbreaks in the field." The Global Outbreak Alert and Response Network is "a technical collaboration of existing institutions and networks who pool human and technical resources for the rapid identification, confirmation and response to outbreaks of international importance." Site describes the WHO's partners, its guiding principles for international outbreak alert and response and its planned response to intentional use of biological agents.

International Travel

Centers for Disease Control and Prevention. Travelers' Health
http://www.cdc.gov/travel/

This page provides valuable information for international travelers. CDC has made a selection of guidelines, outbreak information, vaccination recommendations, and safe food and water suggestions. Those traveling to specific destinations can click on the Destinations link or on the specific region from a drop down box and get specific information about the infectious disease that can be caught while in that locale. Instructions for parents traveling with children are also available on this site. Special needs travelers such as those with disabilities, HIV or who are pregnant or breast-feeding are also provided with suggestions and information. Visitors can also check up on specific diseases that affect travelers and read content from the CDC's *Health Information for International Travel 2003-2204* (more commonly known as the "Yellow Book."

World Health Organization.International travel and health
http://www.who.int/ith/

Health advice for travelers. Full description under Books.

Related Links

FASTATS: Infectious Disease. National Center for Health Statistics
http://www.cdc.gov/nchs/fastats/infectis.htm

This site provides high quality morbidity data for specified infectious diseases with links to additional sources of data and information.

Notifiable Diseases On-Line. Health Canada
http://dsol-smed.phac-aspc.gc.ca/dsol-smed/ndis/index_e.html

The site is intended for dissemination of surveillance data collected on Canadian notifiable diseases. Site includes information on emerging/re-emerging diseases.

National Notifiable Disease Surveillance System (NEDDS).Centers for Disease Control and Prevention (U.S.). Epidemiology Program Office. Division of Public Health Surveillance and Informatics
http://www.cdc.gov/epo/dphsi/nndsshis.htm

Data collected through the NEDDS on selected notifiable infectious diseases are published weekly in the MMWR (http://www.cdc.gov/mmwr/) and at year-end in the annual Summary of Notifiable Diseases, United States (http://www.cdc.gov/

epo/dphsi/annsum/index.htm). Case definitions for infectious conditions under public health surveillance are available for reading.

Surveillance Resources, National Center for Infectious Diseases

http://www.cdc.gov/ncidod/osr/index.htm

This site offers information about disease surveillance systems operating within or collaborating with the National Center for Infectious Diseases (NCID). The site also maintains links to disease reports, data, and publications that these systems publish.

Books and Reports

Brower, Jennifer and Peter Chalk. The Global Threat of New and Reemerging Infectious Diseases: Reconciling U.S. National Security and Public Health Policy, $20.00 (paperback, 163 pp.) ISBN: 0-8330-3293-3, MR-1602-RC, © 2003

http://www.rand.org/publications/MR/MR1602/

This book covers the topic of reemerging infectious diseases in six chapters that can be purchased or downloaded for free in PDF format. The book "offers a more comprehensive analysis of the security implications of the spread of infectious diseases than has been done to date" and examines the effects that the HIV/AIDS epidemic had in South Africa. It also examines the threat posed by specific diseases in the United States and "offers specific actions that can be taken to improve the country's ability to meet this emerging challenge."

Forum on Microbial Threats:

Learning from SARS: Preparing for the Next Disease Outbreak -- Workshop Summary / Stacey Knobler, Adel Mahmoud, Stanley Lemon, Alison Mack, Laura Sivitz, and Katherine Oberholtzer, Editors, Forum on Microbial Threats. Washington, DC : National Academies Press, 2004.,

http://books.nap.edu/catalog/10915.html

Global defence against the infectious disease threat: Progress Report 2002 (WHO) [PDF]

http://www.who.int/infectious-disease-news/cds2002/index.html

While all chapters from this publication are worth examining, three are of immediate interest: Chapter 1: Global health security; Chapter 3: Defence behind the front lines (covers geographic information systems and mapping, vector control and social mobilization and training); and Chapter 4: Emerging and epidemic-prone diseases.

Institute of Medicine (U.S.).Board on Global Health. The Emergence of zoonotic diseases: understanding the impact on animal and human health : workshop summary / Tom Burroughs, Stacey Knobler, and Joshua Lederber, editors. -- Washington, D.C.: National Academies Press, 2002.

http://books.nap.edu/books/0309083273/html/index.html

The book describes the importance of zoonotic diseases, describes the factors of emergence, their diagnoses and control, surveillance and management of zoonotic disease outbreaks and gives a summary and assessment. A good glossary and acronym list is included.

Institute of Medicine.Emerging Infections: Microbial Threats to Health in the United States, 1992.

http://www.nap.edu/books/0309047412/html/

Institute of Medicine (U.S.). Informing the Future: Critical Issues in Health. Second edition, Washington, DC : National Academies Press, 2003.

http://www.nap.edu/catalog/10853.html

International Travel and Health (World Health Organization)

http://www.who.int/ith/index.html

This online book examines the risks to which travelers are exposed and the precautions needed to protect their health. The book offers guidance on the full range of health risks likely to be encountered at specific destinations and associated with different types of travel – from business, humanitarian and leisure travel to backpacking and adventure tours.

Microbial Threats to Health: Emergence, Detection, and Response. Mark S. Smolinski, Margaret A. Hamburg, and Joshua Lederberg, Editors, Committee onEmerging Microbial Threats to Health in the 21st Century. Washington, DC : National Academies Press, 2003. (398 pages).

http://books.nap.edu/catalog/10636.html

National Institute of Allergy and Infectious Diseases. Microbes in Sickness and in Health

http://www.niaid.nih.gov/publications/microbes.htm

The Resistance Phenomenon in Microbes and Infectious Disease Vectors: Implications for Human Health and Strategies for Containment -- Workshop Summary. Board on Global Health , Institute of Medicine (IOM (http://www.iom.edu)) Stacey L. Knobler ... [et al.], editors -- Washington, DC: National Academy Press, 2003.

http://www.nap.edu/books/0309088542/html/

e-Journals and Newsletters

Action Against Infection. (WHO)

http://www.who.int/health-topics/idindex.htm

Listing most of the major infectious diseases and their accompanying fact sheets, surveillance and response information, WHO publications and research and development content.

BMC Infectious Diseases published by BioMed Central

http://www.biomedcentral.com/1471-2334/

Disease Outbreak News (WHO)

http://www.who.int/csr/don/en/

Brief news items often in the form of press releases on recent outbreaks. Outbreak information first appears by date and then is archived by disease (http://www.who.int/csr/don/archive/disease/en/), year (http://www.who.int/csr/don/archive/year/en/) and country (http://www.who.int/csr/don/archive/country/en/).

EID Weekly Updates: Emerging and Reemerging Infectious Diseases, Region of the Americas

http://www.paho.org/English/AD/DPC/CD/eid-eer-ew.htm

EID Weekly Updates started on 10 July 2003.The purpose of these updates is to disseminate information on the latest outbreaks of emerging and reemerging infectious diseases (EID) in the Region of the Americas. The updates also include information on communicable disease prevention and control measures; new research developments; new technical materials on EID on the PAHO web; and information on courses, seminars and conferences related to disease prevention and control.

Emerging Infectious Diseases (CDC)

http://www.cdc.gov/ncidod/EID/index.htm

Emerging Infectious Diseases is published monthly by the National Center for Infectious Diseases, Centers for Disease Control and Prevention (CDC). It is a critically important, peer reviewed journal that publishes articles that investigate new, known, and reemerging infections. Additional articles explore the understanding of factors involved in disease emergence, prevention, and elimination. A primary goal is the quick dissemination of reliable information on emerging infectious diseases.

Eurosurveillance Weekly and Eurosurveillance Monthly

http://www.eurosurveillance.org/index-02.asp

Good source of European peer-reviewed informationon communicable disease surveillance and control. Topics deal with a variety of infectious disease topics. Recent articles covered HIV in eastern Europe and central Asia, European surveillance of travel associated Legionnaires' disease, HIV infections, avian to human transmission of influenza A (H7N7) during an outbreak in the Netherlands in 2003 as well as other case reports. Links are provided to a special SARS page of resources and to titles from other current national bulletins.

Weekly Epidemiological Record (WHO)

http://www.who.int/wer/en/

Timely epidemiologic information on cases and outbreaks of infectious diseases that is distributed every Friday in a bilingual English/French edition. The electronic edition is accessible free of charge.

ProMed Mail

http://www.promedmail.org/pls/askus/f?p=2400:1000

The global electronic reporting system for outbreaks of emerging infectious diseases& toxins, open to all sources.
ProMED-mail, the Program for Monitoring Emerging Diseases, is a program of the International Society for Infectious Diseases. Archives are kept for readers to search for previous messages sent to the list. Of special interest is the selection of maps of outbreaks. Currently available are maps that examine West Nile Virus, Anthrax, Influenza, Avian Influenza and animal diseases.

World Health Organization. Resources Forum. Email Discussion Groups & Lists

http://www.who.int/tdr/kh/bittdre.htm

Listing subscription information and brief description about a variety of email discussion groups related to infectious diseases.

Webcasts

http://www.phppo.cdc.gov/phtn/sars/

This Webcast and course provides information on Severe Acute Respiratory Syndrome (SARS) reported in patients from Asia, North America, and Europe. This global broadcast discusses the latest findings regarding the SARS outbreak and prevention of transmission in healthcare settings. The faculty includes representatives from WHO, CDC, and several affected countries who have reported their experiences with SARS.

(Reproduced from www.nwcphp.org/nph/s2004/ with permission from authorLaura Larsson, Department of Health Services, Department in the University of Washington School of Public Health and Community Medicine).

Indian Portal: (http://www.indiaenvironmentportal.org.in/taxonomy/term/9257).

REFERENCES

1. The World Health Report (2004) http://www.who.int/whr/2004/annex/topic/en/annex_2_en.pdf.
2. Molyneux P, Hotez J, Fenwick A. 2005. Rapid-impact interventions: how a policy of integrated control for Africa's neglected tropical diseases could benefit the poor. *PLoS Medicine* 2(No 11 e336).
3. World Health Organization (2007) Global plan to combat neglected tropical diseases 2008-2015. Available from:whqlibdoc.who.int/hq/2007/WHO_CDS_NTD_2007.3_eng.pdf
4. World Health Organization (2009) Neglected tropical diseases: hidden successes, emerging opportunities.
5. Centers for Disease Control. (2004) Schistosomiasis fact sheet. Available from: www.cdc.gov/
6. World Health Organization. Lymphatic Filariasis: The disease and its epidemiology Available from: www.who.int/lymphatic_filariasis/epidemiology/en/]
7. www.who.int/leishmaniasis/burden/en/

8. World Health Organization. Chagas disease information, Available from: www.who.int/tdr/diseases/chagas/diseaseinfo.htm

9. Reddy M, Gill SS, Kalkar SR, Wu W, Anderson PJ, Rochon PA (2007) Oral drug therapy for multiple neglected tropical diseases: a systematic review. *JAMA* 298(16):1911-24

10. Simarro PP, Jannin J, Cattand P. 2008. Eliminating human African trypanosomiasis: where do we stand and what comes next? PLoS Med 5(2): e55.

11. Heymann DL, ed. Control of communicable diseases manual with addendeum to influenza data: updated April 18, 2006. 18 ed. Washington, DC: American Public Health Association, 2004.]

12. World Health Organization. Dengue and dengue haemorrhagic fever: fact sheet no. 117. Available from: www.who.int/mediacentre/factsheets/fs117/en/]

13. Ebert, D (2008). "Host–parasite coevolution: Insights from the Daphnia–parasite model system". Curr. Opin. in Microbiol. 11: 290–301.

14. Woolhouse, MEJ; Webster, JP; Domingo, E; Charlesworth, B: Levin, BR (2002). "Biological and biomedical implications of the coevolution of pathogens and their hosts". Nature Genetics32: 569-7.

15. Jokela, Jukka; Liveley, Curtis M.; Dydahl, Mark F.; Fox, Jennifer A. Genetic variation in sexual and clonal lineages of a freshwater snail. Biol. Jr. Linn. Soc. 79: 165-181.

16. Bérénos, C; Schmid-Hempel P; Wegner, KM "Evolution of host resistance and trade-offs between virulence and transmission potential in an obligately killing parasite". Journal of Evolutionary Biology10: 2049-56.

17. Fischer O, Schmid-Hempel P.(2005):Selection by parasites may increase host recombination frequency. Biology Letters 22,1(2):193-5.

18a. Watts S (1997) Epidemics and History: Disease, Power and Imperialism. New Heaven, Conn.: Yale University Press, 1997

18b. Langmuir AD, Worthen TD, Solomon J, Ray CG, Petersen E., (1985) N Engl J Med.313(16):1027-30.

19. Morens DM, Littman RJ. (1994) Thucydides syndrome reconsidered: new thoughts on the "Plague of Athens". Am J Epidemiol. 140:621-8

20. Fauci AS. World Health Day, April 7, (1997) Available at http://ww3.niaid.nih.gov/news/newsreleases/1997/world.htm

21.a nowden F. (2008) Emerging and reemerging diseases: a historical perspective. Immunol Rev. 225:9-26.

21.b Fauci, A. S (2001) Infectious diseases: considerations for the 21st century. Clin. Infect Dis. 32, 675–685.

22.a Krause, R. M. The Restless Tide: The Persistent Challenge of the Microbial World National Foundation for Infectious Diseases, Washington DC, 1981

22.b Coordinating Committee of the Global HIV/AIDS Vaccine Enterprise (2005) The Global HIV/AIDS Vaccine Enterprise: Scientific Strategic Plan. PLoS Med 2: e25. doi:10.1371/journal.pmed.0020025

23. Subbarao, K. & Katz, J. Avian influenza viruses infecting humans. Cell Mol. Life Sci. 571, 1770–1784(2000).

24. Gao R, Cao B, Hu Y, Feng Z, Wang D, Hu W, Chen J, et al. (2013). Human Infection with a Novel Avian-Origin Influenza A (H7N9) Virus. N Engl J Med. 368: 1888-1897.

25. Cyranoski D. Health labs focus on mystery pneumonia (2003) Nature 422, 247

26. WHO report: http://www.who.int/csr/don/2003_03_16/en/index.html.

27. Zhi Y, Wilson JM, Shen H (2005) SARSvaccine: progress and challenge.Cell Mol Immunol. Apr; 2(2):101-105

28. Sambri V, Capobianchi M, Charrel R, et al. (2013) West Nile virus in Europe: emergence, epidemiology, diagnosis, treatment, and prevention.ClinMicrobiol Infect. Mar 14.doi: 10.1111/1469-0691.12211

29. Rappole JH, Derrickson SR, Hubálek Z (2000) Migratory birds and spread of West Nile virus in the Western Hemisphere.Emerg Infect Dis. 6:319-28.

30. Takasaki T.West Nile fever/ Encephalitis.Uirusu. 2007 Dec;57(2):199-205.

31. Khan SA, Dutta P, Khan AM, Chowdhury P, Borah J, Doloi P, et al. (2011) West Nile virus infection, Assam, India [letter]. Emerg Infect Dis .2011. http://dx.doi.org/10.3201/eid1705.100479

32. Centers for Disease Control and Prevention (2005). Outbreak of Marburg virus hemorrhagic fever–Angola, October 1, 2004–March 29, MMWR Morb Mortal Wkly Rep.; 54:308–9.

33. Nussinovitch, M., B. Volovitz, and I. Varsano. (1995) Complications of mumps requiring hospitalization in children. Eur. J. Pediatr. 154:732–734

34. Carbone, K. M., and S. Rubin. 2007. Mumps virus, p. 1527–1550.In D. M. Knipe, P. M. Howley, et al. (ed.), Fields virology, 5th ed. Lippincott-Raven, Philadelphia, PA.

35. Kanra, G., P. Isik, A. Kara, A. B. Cengiz, G. Sec¸meer, and M. Ceyhan. 2004. Complementary findings in clinical and epidemiologic features of mumps and mumps meningoencephalitis in children without mumps vaccination.Pediatr. Int. 46:663–668.

36. Koskiniemi, M., M. Donner, and O. Pettay. 1983. Clinical appearance andoutcome in mumps encephalitis in children. Acta Paediatr. Scand. 72:603–609.

37. Vazeille M, Moutailler S, Coudrier D, Rousseaux C, Khun H, et al. (2007) Two Chikungunya isolates from the outbreak of La Reunion (Indian Ocean) exhibit different patterns of infection in the mosquito, Aedes albopictus. PLoS ONE 2: e1168.

38. Queyriaux B, Simon F, Grandadam M, Michel R, Tolou H, Boutin, J.P (2008). Clinical burden of chikungunya virus infection. Lancet Infect Dis 8: 2–3.

39. Tsetsarkin KA, Vanlandingham DL, McGee CE, Higgs S (2007) A single mutation in chikungunya virus affects vector specificity and epidemic potential. PLoS Pathog 3: e201. doi:10.1371/journal.ppat.0030201

40. Kumar NP, Joseph R, Kamaraj T, Jambulingam P (2008) A226V mutation in virus during the 2007 chikungunya outbreak in Kerala, India. J Gen Virol 89: 1945–1948

41. Cherian SS, Walimbe AM, Jadhav SM, Gandhe SS, Hundekar SL, et al. (2009) Evolutionary rates and timescale comparison of Chikungunya viruses inferred from the whole genome/E1 gene with special reference to the 2005–07 outbreak in the Indian subcontinent. Infect Genet Evol 9: 16–23.

42. Jain M, Rai S, Chakravarti A (2008) Chikungunya: a review. Trop Doct 38: 70–72.

43. Arankalle VA, Shrivastava S, Cherian S, Gunjikar RS, Walimbe AM, et al. (2007) Genetic divergence of Chikungunya viruses in India (1963–2006) with special reference to the 2005–2006 explosive epidemic. J Gen Virol 88: 1967–1976.

44. Sam IC, AbuBakar S (2006) Chikungunya virus infection. Med J Malaysia 61: 264–269.

45. Thiboutot MM, Kannan S, Kawalekar OU, Shedlock DJ, Khan AS, Sarangan G, Srikanth P, Weiner DB, Muthumani K (2010) Chikungunya: a potentially emerging epidemic? PLoS Negl Trop Dis. 4:e623.

46. Harendra S. Chahar, PreetiBharaj, Lalit Dar, Randeep Guleria, Sushil K. Kabra, and ShobhaBroor. (2009) Co-infections with Chikungunya Virus and Dengue Virus in Delhi, India. Emerging Infectious Diseases •www.cdc.gov/eid • Vol. 15, No. 7

47. Laddy DJ, Yan J, Khan AS, Andersen H, Cohn A, et al. (2009) Electroporation of synthetic DNA antigens offers protection in nonhuman primates challenged with highly pathogenic avian influenza virus. J Virol 83: 4624–4630.

48. Kutzler MA, Weiner DB (2008) DNA vaccines: ready for prime time? Nat Rev Gent. 10:776-88.

49. Krauss H, Weber A, Appel M, et al. (2003). Zoonoses: Infectious Diseases Transmissible from Animals to Humans (3rd ed.). Washington, D.C.: ASM Press. ISBN 1-55581-236-8.

50. Anderson, R. M., Swinton, J. & Garnett, G. P. Potential impact of low-efficacy HIV-1 vaccines in populations with high rates of infection (1995) Proc. R. Soc. Lond. 261, 147–161.

51. John TJ, Samuel R.Herd immunity and herd effect: new insights and definitions. Eur J Epidemiol. 2000;16:601-6.

52. R. M. Anderson, R. M. May, Infectious Diseases of Humans: Dynamics and Control. Oxford Univ. Press, New York, 1991.

53. Andrew W. Park, Janet M. Daly,Nicola S. Lewis, Derek J. Smith, James L. N. Wood, Bryan T. Grenfell (2009). Quantifying the Impact of Immune Escape on Transmission Dynamics of Influenza. Science, 326: 726-8.

54. Thompson, W. W. et al. (2003) Mortality associated with influenza and respiratory syncytial virus in the United States. J. Am. Med. Assoc. 289, 179–186

55. Jernigan, D. B. et al. Investigation of bioterrorism-related anthrax, United States, 2001: epidemiologic findings. (2002) Emerg. Infect. Dis. 8, 1019–1028

56. Turnbull, W., and P. Abhayaratne. (2003) 2002 WMD terrorism chronology: incidents involving sub-national actors and chemical, biological, radiological, and nuclear materials. Monterey Institute of International Studies: Center for Nonproliferation Studies, Washington, D.C.

57. Darling, R. G., Catlett, C. L., Huebner, K. D. & Jarrett, D. G. (2002).Threats in bioterrorism: I. CDC category A agents. Emerg. Med. Clin. North Am. 20, 273–309.

58. Wagner, Michael M.; Espino, Jeremy; et al. (2004), “The role of clinical information systems in public health surveillance”, Healthcare Information Management Systems (3 ed.), New York: Springer-Verlag, pp. 513–539

3

2

Skin Infections

Chapter Outline

2.1 Normal skin flora, mode of colonization and general Epidemiology
2.2 Bacterial infections- Pathogenesis, Epidemiology, Treatment and prevention
2.3 Viral infections- Pathogenesis, Epidemiology, Treatment and Prevention
2.4 Fungal Infections- Pathogenesis, Epidemiology, Treatment and Prevention

2.1 Normal Skin Flora

An average human adult has about 2m2 of skin surface that can vary greatly in chemical composition and moisture content. Most skin microorganisms are associated directly or indirectly with the sweat glands of several kinds. Hair follicles provide an attractive habitat for microorganisms. Each hair follicle is associated with a sebaceous gland, which secretes a lubricant fluid. A variety of aerobic and anaerobic bacteria and fungi inhabit these regions, mostly within the area just below the surface of the skin. The secretions of the skin glands are rich in microbial nutrients like urea, amino acids, salts, lactic acid, and lipids which are present in considerable amounts. The pH of the human secretions is almost always acidic, the usual range being between pH 4 to 6.

The microorganisms of the normal flora of the skin are either transient or resident populations of microorganisms. Normal resident flora of the skin includes the following species.

Bacteria: Aerobic Gram positive cocci: Staphylococci, Micrococci, Gram-positive rods and diphtheroids.

Anaerobes: *Propionibacterium* species: in the deeper part of the hair follicles, aerobic cocci are present on the surface of the hair follicles and sebaceous glands. In the pubertal age, *P. acnes*, *P. granulosum* reside on the face and upper trunk region. These microorganisms participate in the pathogenesis of acne.

Fungal Infections: Superficial fungal infections are found on the top layers of the skin and mucous membranes, the hair, and the nails. Examples of fungal infections encountered on the skin and other external surfaces include athlete's foot, jock itch, ringworm and candida. These can affect the growth of hair and nails as well. Deep fungal infections invade deeper layers of the skin and attack hair follicles and subsequently spread to the blood or internal organs.

2.1.1 Pathological Manifestations

The manifestation of the disease starts with an inflammation that helps in destroying harmful microorganisms and parasites. It also inactivates and dilutes toxins, rejects foreign bodies and seals off the damaged tissue that cannot be rejected. Inflammation results in regeneration or repair of the damaged tissue. Acute inflammatory response is normally followed by fairly rapid repair.

Inflammatory changes often vary in intensity and form. Regenerated skin does not retain identical histological features as that before the damage.

Repair is the healing of wounds resulting in distortion of the original tissues, or it replaces the damaged tissue with elements like scar etc. The duration between the initiation of inflammatory changes and its subsequent repair depends on:

- Virulence pattern of the organism
- Condition of the skin, where infection may be more severe in devitalized and lacerated skin.
- The body immune system.
- Nature of the causative factor.
- Extent of the damage.

Different species of microflora resides normally on the skin surface without causing harmful effects under the optimum conditions of the skin. Some of this resident flora may change their behavior and become pathogenic during the process of wound and repair.

1. *Mode of Colonization* – The skin becomes colonized by normal flora since birth. Colonization of the skin by normal skin flora is related to different factors:

Age: In infants and children, Micrococci are more prominent than in adult's skin. In adults Propionibacterium are more due to the increased skin lipids. In old age streptococci and enterococci become residents of skin especially in moist areas.

Weather: The weather may cause an increase in temperature and humidity, which increases the density of the skin microflora. The skin is a dry, acidic environment that is conducive to the growth of most microorganisms. Moist areas around sweat glands often get colonized by Gram positive bacteria. Organisms that cannot survive on the skin generally succumb from either skin's low moisture content or low pH.

Personal hygiene: Personal hygiene influences the resident microflora, and unclean individuals usually have higher microbial population densities on their skin.

2. *Epidemiology of Skin Infections* – The precise proportion of skin diseases seen in general practice in a community setting (Fig. 2.1) will vary with age, structure of population served, the amount of industry in the area and socioeconomic factors.

Demographic studies may reveal a trend e.g. for unknown reasons, atopic exzema has become more common over the last 20 years.

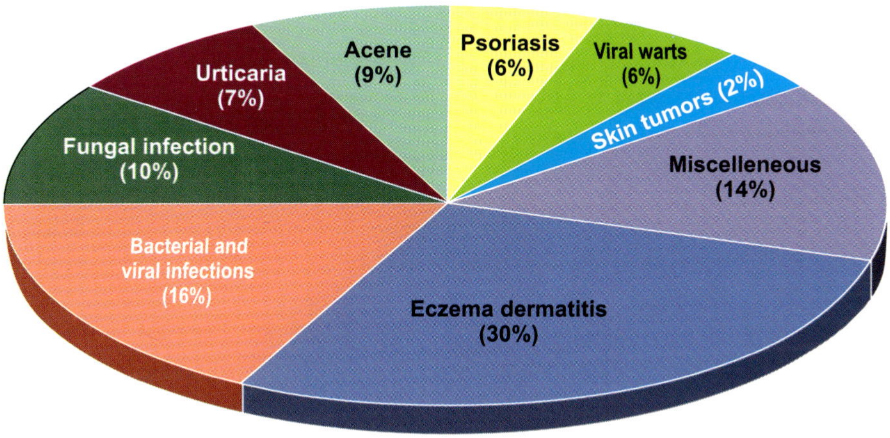

Fig. 2.1: Proportion of skin diseases seen in general practice (% of total)—Adapted from Dermatology (IIIrd edition) by David J. Gawkrodger, Churchill Livingtone, an Imprint of Elsevier Limited.

3. ***Prevalence of Skin diseases in India*** – Skin disorders affect 20–30% of the general population at any one time. Socio demographic factors play pivotal roles in determining the pattern of skin disease. Studies from India reveal wide variations in prevalence of skin disease among children. The prevalence of skin diseases amongst children in reports from various parts of India range from 8.7% to 35% in school based surveys (Table 2.1) and a prevalence of 15% was considered for sample size estimation.

Table 2.1: Skin disease prevalence studies among children in India

Researcher	Study population	Year	Sample	Prevalence	Findings
Dogra S[1]	School children 6–14 years, Chandigarh, India	2003	12,586	38.8%	Skin infestations 11.4%
Rao SG[2]	School children 6–12 years, Mangalore, India	1999	1,161	76.65%	Comminicable dermatoses 19%, Nutritinal deficiency 6.71 %
Valia RA[3]	School children, Above 4 years, Varanasi, India	1991	12,481	54%	Pediculosis capitis 35%, Pityriasis alba 12%
Sharma NK[4]	School children, India	1986	3,697	14.3%	Communicable dermatoses and nutritional deficiencies

Pediatric dermatoses should be addressed differently from adult dermatoses as there are important differences in clinical presentation, treatment and prognosis of the disease. The factors influencing dermatoses in children are socio-economic status, climatic exposure, dietary habits and external environment as compared to adults. Dermatoses leading to cutaneous infections are common in children during school going years. Those who develop cutaneous diseases due to intrinsic genetic abnormalities also have onset in the pediatric age-group.

A retrospective study from a large pediatric hospital in Delhi[5], a referral centre of north India compared the pattern of dermatoses in different age groups within the pediatric population. Overall most common dermatoses in all age groups were bacterial infections[6], seborrheic dermatitis (10.49%), scabies (10.16%), pityriasis alba (5.85%), miliaria (5.46%), atopic dermatitis (5.27%), fungal infection (4.65%), urticaria/angioedema (4.46%), viral infections (3.68%) and papular urticaria. Prevalence of transmissible skin disease in a study population of Calicut, Kerala, India is represented in Fig. 2.2a and 2.2b[7]. Skin disease constitutes a public health problem in this population and is not given due attention. Thus appropriate interventions are needed.

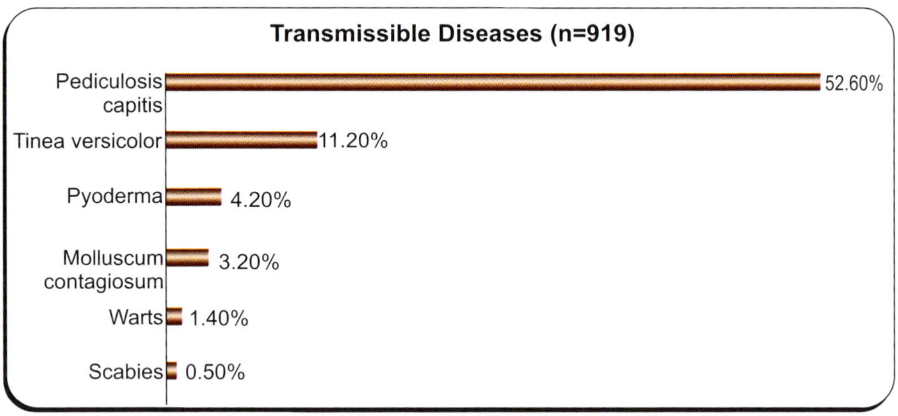

Fig. 2.2A: Prevalence of transmissible skin disease in the study population

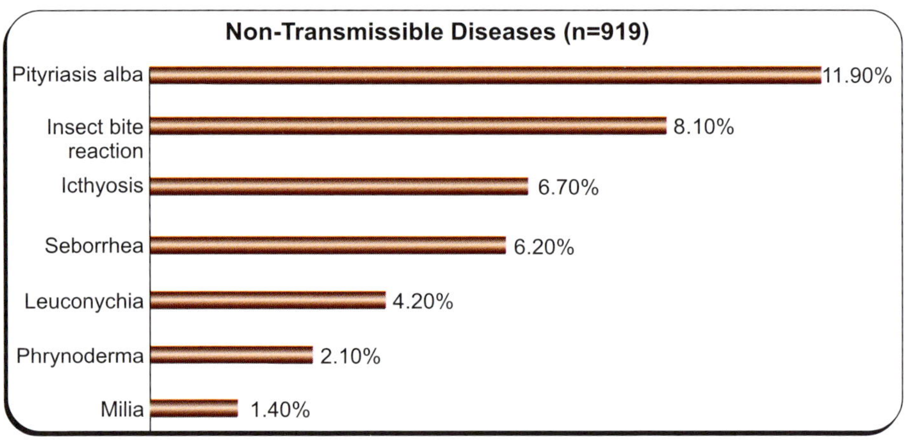

Fig. 2.2B: Prevalence of non transmissible skin disease in the study population (adapted from Dibu et. al. www.imakmj.com[7])

2.2 Bacterial Infections—Pathogenesis, Epidemiology, Treatment and Prevention

Only a few species of bacteria commonly invade the intact skin directly. Many bacterial pathogens can invade the skin through injuries or via the blood stream. Hair follicle infections exemplify direct invasion.

2.2.1 Infections of the Hair Follicles-Cause Folliculitis, Furuncles and Carbuncles

Pathogenesis

Most furuncles, carbuncles as well as many cases of folliculitis, are caused by *Staphylococcus aureus*, a facultative anaerobic, Gram-positive coccus. The infection induces an inflammatory response with swelling and redness, followed by congregation of polymorphonuclear leukocytes. If the infection continues, the follicle becomes a plug of inflammatory cells and necrotic tissue. Once the organism enters the blood stream, the infection can be disseminated to other parts of the body, such as heart, bones or brain.

Folliculitis

They represent different outcomes of hair follicle infections. Folliculitis is an infection that is localized to the hair follicle. A folliculitis looks like confined to the hair follicle. Factors leading to the development of a folliculitis include: Trauma, chronic friction, occlusive clothing, occlusive chemicals, excessive sweating and exposure to water. Hot Tub folliculitis is a special case of folliculitis that is not caused by *S. aureus*. This self-limited infection is caused by *Pseudomonas aeruginosa* contracted from hot tubs, whirlpools, and pools that are not adequately chlorinated.

Furuncle

The furuncle infection involves the sebaceous gland and happens to be more extensive than a folliculitis. A furuncle may appear on different parts of the body like neck, face, armpits and buttocks. At the onset, a small, tender, red nodule appears and it becomes painful at a later stage by developing pus. The pus often drains out spontaneously and the furuncle may resolve on its own. Some of the factors like obesity, blood disorders, consumption of steroids and diabetes may contribute in the development of furuncle.

Carbuncle

A carbuncle can simply be defined as multiple furuncles grouped together. A carbuncle usually involves the deeper layers of the skin - the subcutaneous fat. It looks like a large, red nodule that is hot and may have visible layers of pus just beneath the surface of the skin. Fever is often present, along with other signs of a serious infection.

Epidemiology

Staphylococcus aureus inhabits the nostrils of virtually everyone at some or other time. The organisms are mainly disseminated to other parts of the body and to the environment by the hands. Therefore, hand washing is an important control measure. Besides nasal chamber, moist areas of skin are also frequently colonized. Patients with surgical wounds or chronic infections are at risk to get infected with *S. aureus.* The bacteria survive well in the environment, which favours their transmission from one host to another.

Treatment of Folliculitis, Furuncles, and Carbuncles

Mild cases of folliculitis and small furuncles may heal on their own with good hygiene and wound care. More extensive furuncles and all carbuncles need to be treated with antibiotics such as dicloxacillin or cephalexin. When pus appears, in addition to antibiotics, a procedure called incision and drainage (I&D) should be performed to drain the pus. This procedure allows the lesion to heal from the inside out. Antibiotic treatment at times is complicated because of the fact that 90% of *S. aureus* strains produce the penicillin-destroying enzyme penicillinase (beta-lactamase), so penicillin cannot routinely be used in treatment.

2.2.2 Atopic Dermatitis (AD)

Atopic dermatitis (Fig. 2.3) is a chronic eczematous disease initiating at an early infantile hood and childhood. This disease may follow a remitting or flaring course that may continue throughout life. It develops as a result of a complex interrelationship of environmental, immunologic, genetic and pharmacologic factors. The disease may get triggered due to various factors like infection, psychological stress, seasonal changes, irritants or allergens. The rapid rise in prevalence of AD is thought to be primarily related to changes in our environment. It is characterized by the itch-scratch cycle: Affected persons have the sensation of itch, followed by scratching and the subsequently rash develops. This disease is often referred as a classic triad including eczema, asthma, and allergies. A wide range of environmental factors, such as contact allergens, stress, food, skin flora, and humidity, play roles in the development and severity of atopic dermatitis

Pathogenesis

The skin of patients with AD is characterized by higher levels of colonization by aerobes. Significantly higher colony forming units per surface unit could be isolated from clinically normal skin of patients with AD than from corresponding skin sites in normal controls. Acute exudative lesion yielded a mean of 10^7 colony forming units (CFU)/cm^2. Thus, the atopic skin must provide a favorable environment for the colonization and multiplication of aerobic bacteria. Coagulase-positive staphylococci (*Staphylococcus aureus*), which are usually not found on normal skin, accounted for the majority of CFU isolated from AD skin. *Staphylococcus epidermidis* was the predominant organism isolated from the clinically uninvolved skin of AD patients, whereas *Staphylococcus epidermidis* was second to S. aureus found in lesional skin.

Epidemiology

Relatively few studies have focused on potential risk factors, but the strongest risk factor is a parental history of atopy or eczema. Currently, the epidemiological data available on atopic dermatitis is scanty. However, it is now estimated that atopic dermatitis affects 10% to 20% of children worldwide, especially those less than five years old[8]. Maternal atopy is considered a greater risk for atopic disorders in offspring than paternal atopy. AD and atopy in general seem to be diseases of the advantaged classes. The prevalence of AD is inversely related to the number of siblings[9]. The larger the family size, the less the likelihood of having AD. It occurs more among the migrants. Its increasing incidence strongly points towards the environment as a determining factor for many clinical representation of atopic dermatitis.

Effect of Genetics

Numerous studies have confirmed the role of genetic factors in the development of atopic dermatitis. The disease has been referred as a genetic disorder based on the interactions of unknown number of genes and environmental factors. The inheritance pattern seems to be complex and does not follow the Mendelian pattern. Parental atopy (parents with allergies) carries the highest risk factor. Atopic dermatitis is often accompanied by the symptoms of asthma and hay fever. A multifactorial cause for atopic dermatitis suggests more investigation required to understand the genetic and environmental factors influencing disease development. Identification of the responsible genes would be of enormous help in diagnosis and treatment of the disease.

Treatment

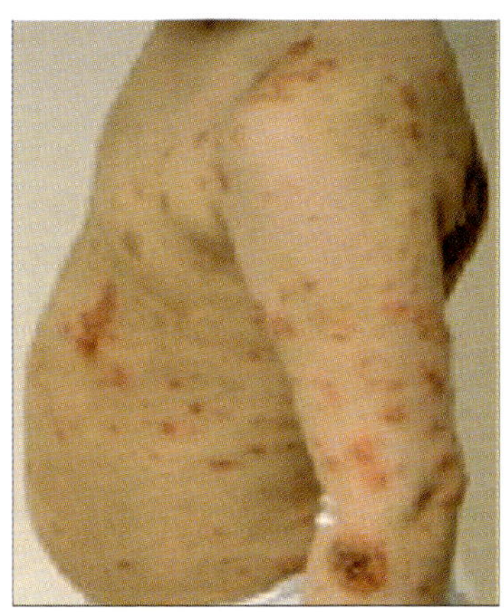

Fig. 2.3: Patient suffering from Atopic dermatitis (AD)

Topical corticosteroids have been the main drug of choice in atopic dermatitis for the last 50 years. These drugs are available at different level of potency. Application of high potency drugs produces more adverse side effects. So drugs with lower potency for longer duration are often preferred. Absorption of corticosteroids often depends on the skin thickness. Generally, the less potent corticosteroids are used on areas with thinner skins, such as, eyes, genitalia and face. Areas with thicker skin layers like palms and soles may need more potent corticosteroids. In children having a low body volume to skin surface area ratio, the rate of corticosteroid absorption is usually higher, therefore potent corticosteroids are avoided in children for possible side effects. Low potency corticosteroids are generally used at the beginning of treatment and for maintenance therapy whereas, higher potency corticosteroids are used when the disease is in a severe stage.

2.2.3 Scalded Skin Syndrome

Scalded skin syndrome is a skin infection characterized by damage to the skin with marked shedding (Fig. 2.4). Scalded skin syndrome usually affects infants and children under the age of five. Scalded skin syndrome is also called Ritter disease or Staphylococcal scalded skin syndrome (SSS). During the infection, the staphylococci produce a poison that causes the skin damage. The damaged skin appears as blisters as if the skin were scalded, hence the name scalded skin syndrome is given.

Symptoms and Pathogenesis

SSS is caused by an epidermolytic toxin i.e. destroying the material that binds the layers of the skin cells together. The toxin exfoliation produced by certain strains of *Staphylococcus aureus*. The infection begins with general redness of the skin (erythema), which spreads to cover most of the body. Other symptoms such as malaise, irritability and fever are also present. The nose, mouth and genitalia may be painful for 24 to 48h before the typical features of the disease become apparent. The skin becomes wrinkled, and large blisters filled with clear fluid develop followed by fever. The skin becomes tender to touch and looks like sandpaper.

Staphylococci growing at the site of infection, usually in the skin, produce toxin that is absorbed and carried by the blood stream to large areas of the skin. Skin slips off with gentle pressure, leaving wet red areas (Nikolsky's sign). SSSS is caused by group II coagulase-positive staphylococci and usually phages 71, 55, 3c and 3A were found to be predominant. Over a period of two or three days the skin exfoliates in large sheets, exposing a red moist surface. The organism secretes elaborate exfoliatin also known as epidermolysin, a toxin that splits the upper part of the epidermis just beneath the granular cell layer. Two types of epidermolytic toxin exists: one is coded by a plasmid gene, while the other is chromosomal.

Epidemiology

Staphylococcal scalded skin syndrome (SSSS) is more prevalent in children below 6 yrs of age. It rarely occurs in older patients unless they have renal failure or are immunocompromised. Epidemics may occur in nurseries, presumably transmitted by the hands of the care taker who are in contact with an infected infant or who are nasal carriers of *Staphylococcus aureus*. Sporadic cases may also occur. The male to female ratio of this disease was found to be 5:1.

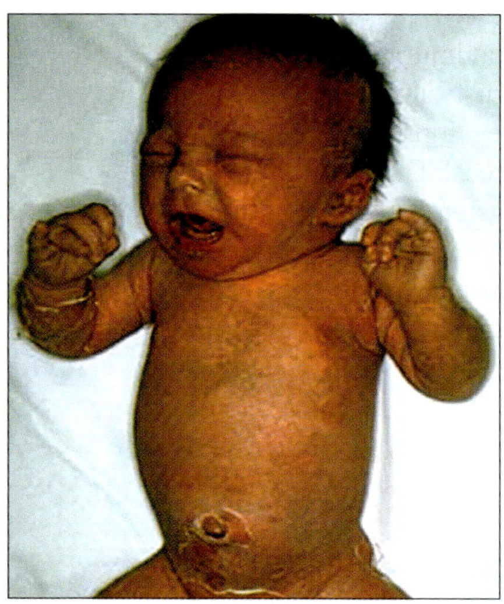

Fig. 2.4: Patient suffering from Staphylococcal Scalded skin syndrome (SSS)

Treatment

For preventing the loss of fluid through open skin during the disease, it is important to introduce intravenous fluid to prevent dehydration for managing the disease. Intravenous antibiotics for staphylococcus are given to help fight the infection. Moist compresses to the skin may improve comfort. One can apply an emollient to keep the skin moist. With prompt diagnosis and therapy, death rarely occurs. The stratum corneum is quickly replaced, and healing usually occurs as the infection subsides within 5 to 7 days after start of treatment. Recommended doses of antibiotics are Nafcillin12.5 to 25 mg/kg IV q 6 h for newborns > 2 kg and 25 to 50 mg/kg for older children is given until improvement is noted, followed by oral cloxacillin 12.5 mg/kg q 6 h (for infants and children weighing = 20 kg) and 250 to 500 mg q 6 h (for older children).

2.2.4 Pyoderma

A skin infection characterized by superficial, pus-filled blisters is called pyoderma. Pyoderma is a group name for pyococcal dermatoses which are generally purulent. The disease is more common during summer and monsoon seasons in tropical countries. The two important pyogenic organisms are the *Staphylococcus aureus* and the *Streptococcus pyogenes*. The former is more important in skin infections than the latter. Follicular infections are mainly due to staphylococci; while erysipelas and cellulitis are caused by streptococci. Besides these, other organisms which occasionally cause pyodermas are bacillus, proteus, pseudomonas and coliform bacilli.

Symptoms and Pathogenesis

The most common type of pyoderma is generally known as Impetigo. It is a superficial skin infection, involving patches of epidermis just beneath the dead, scaly outer layer. The organisms normally enter the skin through minor breaks and injuries. Initially blisters are performed and subsequently they develop yellow crust due to drying of plasma that is released out of the skin. The patient often develops fever and pain due to enlargement of lymph nodes near the involved areas. This is often an indication of bacterial products entering in the lymphatic system.

The infection is caused either by *Staphylococcus aureus* or by *Streptococcus pyogenes*. These gram-postive , chain forming cocci are β-hemolytic and are referred as "β-hemolytic Group A Streptococci" as their cell walls contain a polysachharide called group a carbohydrate. The former organism produce catalase enzyme whereas the latter is catalase negative. The organisms secrete an array of enzymes to degrade host tissue that include 1. proteases to break proteins 2. nucleases to break nucleic acids; 3. Hyaluronidase to break hyaluronic acid that is an important component of host tissue. A cell-wall component known as the M-protein, is very important in enabling *S. pyogenes* to cause pathogenesis because they interfere with the phagocytosis. Acute glomerulonephritis may develop as a serious complication of *S. pyogenes* pyoderma.

Epidemiology

The disease spreads by direct contact with carriers or patients with impetigo. The incubation period is 1–3 days. Dried streptococci in the air are not infectious to intact skin. Scratching may spread the lesions. A molecular epidemiology study conducted on aboriginal communities of northern Australia reported that the majority of the group A streptococcal isolates from the Northern Territory are not typeable by M typing and developed a novel genotyping method called Vir typing. A preliminary study using this method discriminated all the M-nontypeable (MNT) isolates. Vir typing is based on restriction fragment length polymorphisms of the 4- to 7-kb Vir regulon of group A streptococci, which contains a number of genes, including emm (the gene for M protein). A total of 407 isolates of group A streptococci obtained from four Aboriginal communities over a 4-year period were typed by this genotyping method. Forty-two distinct genotypes were found among the isolates, including 22 among the MNT isolates. The correlation between Vir type and M type was good. This genotyping method allows the characterization of all group A streptococcal isolates and proposed that Vir typing be used in conjunction with M typing for epidemiological surveillance in geographical regions where the majority of isolates are MNT[10]

Treatment

Commonly the disease is treated with an application of the antiseptic gentian violet. However, currently topical or oral antibiotics are usually prescribed. Treatment may involve washing with soap and water and letting the impetigo dry in the air. Mild cases may be treated with bactericidal ointment, such as fusidic acid, mupirocin, chloramphenicol or neosporin. These antibiotics are commonly available in some countries over-the-counter. However to treat severe cases, oral antibiotics are prescribed. They are dicloxacillin, flucloxacillin or erythromycin. Alternatively amoxicillin combined with clavulanate potassium, cephalosporins (1st generation) and many others may also be used as an antibiotic treatment. Hydrogen peroxide is often used as an alternative to topical antibiotics in the treatment of Impetigo. Hydrogen peroxide has been shown to be as effective as antibiotics in the treatment of Impetigo Contagiosa.

2.2.5 Lyme Disease

Borrelioses are an important group of spirochetal infection. Lyme disease or Lyme borreliosis is often referred as a tick-born disease caused by *Borrelia burgdorferi*. Infection was first described in children suffering from arthritis belonging to Lyme, Connecticut, but was later recognized along the northeastern coast of the United States. Borrelioses are found in the temperate zones Europe, North America and Asia. Ticks of the genus Ixodes (*Ixodes ricinus*) are infected while biting deer and mice[11]. Three species of *Borrelia burgdorferi* have been identified as human pathogens: *B. burgdorferi* sensu stricto, *B. garinii* and *B. afzelii* .Three days to four weeks after inoculation, a papule develops. A ring-like rash expands outward from the site of the bite and lasts from a month to a year (Fig. 2.5). The disease has a protean manifestation affecting many systems that include skin, joints, muscle, central nervous system, heart and eye.

Symptoms and Pathogenesis

In case the patients suffering from Lyme disease are left untreated, the disease progress in stages from mild symptoms to serious, long-term disabilities. There are three stages of Lyme disease: early localized, early disseminated and late persistent.

Stage 1: Early Localized Infection (1 to 4 weeks)

People with Lyme disease have a rash at the site of the tick bite called erythema migrans which is the hallmark of Lyme disease. The rash begins as a red spot or bump and is usually circular and it gets larger over time to a median diameter of 15 cm. Infected people with Lyme disease develop a rash within 1 to 4 weeks. About half of these cases develop satellite lesions which are smaller in size.

Stage 2: Early Disseminated Infection (1 to 4 months)

If Lyme disease is not detected and treated while early symptoms are present, the infection may affect the different regions of the body like skin, joints, nervous system, and heart within weeks to months after the initial infection.

Symptoms at this stage may include:

- Tiredness.
- Additional skin rashes in several places on the body and the infection spreads.
- Pain, weakness, or numbness in the arms or legs.
- Paralysis of the facial nerves.
- Recurring headaches or fainting.
- Poor memory and reduced ability to concentrate.
- Conjunctivitis may develop or sometimes may damage deep tissue in the eyes.
- Palpitations or, in rare cases may lead to serious heart problems.

Stage 3: Late Persistent Infections (Late Lyme disease)

If Lyme disease is not promptly or effectively treated, damage to the joints, nerves, and brain may develop after several months of infection. Symptoms at this stage may include:

- Swelling and inflammation in the knee joints.
- Numbness and tingling in the hands, feet, or back.
- Severe fatigue.
- Partial facial nerve paralysis occurs within the first few months after the tick bite.
- Neurologic changes including problems with memory, mood, or sleep and sometimes the patient finds problem in speaking.

Chronic Lyme arthritis which causes recurring episodes of swelling, redness, and fluid buildup in one or more joints that last up to 6 months at a time. In these patients, early symptoms of infection like rash may not develop and instead they develop heart, nervous system, and joint symptoms as first signs of Lyme disease.

Epidemiology

Lyme borreliosis (LB) is the most common tick-borne disease in Europe and North America. LB occurs in all age groups, with equal prevalence in men and women. Ticks, the disease vector, thrive in moist, forested and grassy areas, and their population peak during the warmer summer months. Persons of all ages are thought to be equally susceptible to the spirochete bacterium, but certain groups of individuals are at higher risk. Individuals who work in wooded areas or are involved in outdoors-recreational activities have an increased risk of infection. The highest reported rates of Lyme disease occur in children aged greater than 15 and adults aged between 30-59 years. Over 62,000 cases of Lyme Disease were reported to the CDC from 1993-1997, making Lyme Disease accountable for more than 95% of vector-borne illness cases in the United States. Ticks attach and take blood meals from their hosts only three times in their lives, at each separate stage of their life, with each blood meal lasting several days. During a blood meal, spirochete bacterium is transferred to the host from the tick's saliva. A tick bite does not ensure transmission, especially if the tick is removed within twenty-four hours. In addition, only few ticks often carry the infectious spirochete. In Connecticut, for example, only one percent of the tick population is infected.

Treatment

Various steps can be adopted to prevent Lyme disease by: 1. Using insect repellent, 2. Removing ticks promptly, 3. Practicing proper landscaping, and integrated pest management. The ticks *Ixodes* ricinus that transmit Lyme disease can occasionally transmit other tick-borne diseases as well. Early lyme disease responds readily to several oral agents such as doxycycline, amoxicillin, or cefuroxime that are usually prescribed for 2-3 weeks. There are few published, controlled trials that compare different regimens for late Lyme disease. Intravenous therapy, usually ceftriaxone or penicillin, is used for 2-3 weeks.

- *Erythema migrans.* In EM, oral antibiotic therapy with doxycycline shortens the duration of the rash and prevents the development of late sequelae. Amoxicillin is also effective and preferred for children under 9 years of age and in pregnant or lactating women.
- *Musculoskeletal disease.* Treatment for one month with oral doxycycline or amoxicillin is usually effective. For refractory cases, intravenous therapy with ceftriaxone or penicillin G, and arthroscopic synovectomy may lead to clinical improvement. Analgesics such as acetaminophen or nonsteroidal anti-inflammatory agents should be used in patients with symptomatic arthritis.
- *Neurologic disease.* Patients with facial nerve palsy alone can be treated with oral doxycycline or amoxicillin. Intravenous penicillin G, ceftriaxone, or cefotaxime is effective for meningitis, cranial or peripheral neuropathies, encephalitis, or other late neurologic complications.

There is no surgical treatment for Lyme disease. In severe cases of chronic Lyme arthritis, removal of the lining of the affected joints (synovectomy) may be needed if the arthritis has not responded to antibiotics.This remedy may perhaps be specially powerful for those who live in high-risk areas. Nonsteroidal anti-inflammatory drugs (NSAIDs), for instance aspirin or ibuprofen, are typically helpful for signs or symptoms of arthritis that may occur with late Lyme disease.

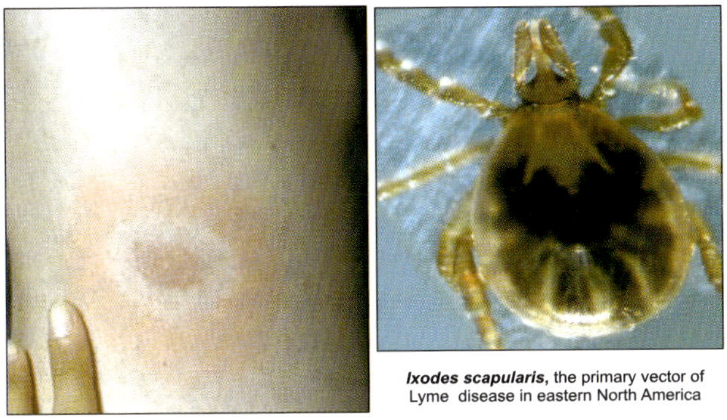

Ixodes scapularis, the primary vector of Lyme disease in eastern North America

Fig. 2.5: Common bulls eye rash pattern associated with Lyme disease

2.3 Viral Infections–Pathogenesis, Epidemiology, Treatment and Prevention

Viral infections can cause a variety of rashes. In children they are characterized by distinctive skin rashes caused by viruses carried to the skin by the blood from sites of infection in the upper respiratory tract. They are normally diagnosed by inspection of rashes and other clinical findings. Many individuals develop a variety of infections in their lifetime, but quickly overcome them. However, some individuals develop chronic or persistent infections. In the majority of cases, persistent infections are caused by viruses and not bacteria. The common viruses that can cause chronic infection include measles, hepatitis, various prion infections that affect the brain (e.g. mad cow), herpes, infectious mononucleosis and Cytomegalovirus (CMV).

2.3.1 Human Papilloma Virus (HPV)

This is a DNA virus from the papillomavirus family that is capable of infecting humans. Like all papillomaviruses, HPVs establish productive infections only in the stratified epithelium of the skin or mucous membranes. While the majority of the nearly 200 known types of HPV cause no symptoms in most people, some types can cause warts (verrucae), while few of them in some cases may lead to cancers of the cervix, vulva, vagina, and anus in women or cancers of the anus and penis in men. Warts can be differentiated depending on the surface infected and its relative moisture, and the patterns of their growth: common warts, plantar warts, juvenile warts and genital warts. Over 30 types of HPV get transmitted from one person to another mainly through sexual contact. The only way to prevent getting an HPV infection is to avoid direct contact with the virus, which is transmitted by skin-to-skin contact.

Epidemiology

Infection with cutaneous HPVs is very common. Some HPV types, such as HPV-5, may establish infections that persist for the lifetime in an individual without ever manifesting any clinical symptoms[12]. These HPV types are often referred as human commensals. Other cutaneous HPVs, such as HPV types 1 or 2, may cause common warts in some infected individuals. Skin warts are most common in childhood and they typically appear and regress spontaneously over a time period of few weeks to months. About 10% of adults also suffer from recurring skin warts. All HPVs are believed to be capable of establishing long-term "latent" infections in small numbers of stem cells present in the skin. These latent infections may never be fully eradicated. Immune system of the body prevent from the appearance of symptoms such as warts. They most often affect and cause symptoms and persist longer in immunocompromised hosts. Anti-HPV-directed, cell-mediated immunologic response plays a role in spontaneous regression of plane warts and that preservation of nonspecific immunity is prerequisite for spontaneous regression of plane warts.

Pathogenesis

Some HPV infections can cause warts (verrucae), which are noncancerous skin growths. Infection with these types of HPV causes a rapid growth of cells on the outer layer of the skin (Fig. 2.6).

Various types of warts include

Common warts: These are caused by "cutaneous" HPV types. Common sites of these warts are surfaces like hands and feet but can also occur in other areas such as the elbows or knees Fig. 2.6B. Common warts show a characteristic cauliflower-like appearance and are typically slightly raised above the surrounding skin. Cutaneous HPV types can also cause genital warts but are not associated with the development of cancer. These are particularly common during childhood when immune system is not fully developed and at the same time transmission rate of the viral infection through skin-to-skin contact is high. Another group who exhibit a high infection rate are meat, poultry and fish handlers[13,14].

Plantar warts: are found on the soles of the feet (Fig. 2.6D). Plantar warts grow inward, generally causing pain while walking. HPV 1 is the most common etiology of plantar warts, although HPV 2 and other viruses may also cause them[15].

Subungual or periungual warts: form under the fingernail (subungual), around the fingernail, or on the cuticle (periungual) Fig. 2.6A. They may be more difficult to treat than warts formed in other location.

Flat warts: Flat warts mostly appear on the arms, face or forehead (Fig. 2.6E). Manifestation of flat warts are common in children and teens as observed in case of common warts. In immunocometent individuals, flat warts are not associated with the development of cancer. Flat warts are commonly caused by HPV 3, 10 and occasionally by HPV 2. These are frequently presented as a bunch of subtle papules 2-4 mm in diameter that are elevated above the surface by less than a millimeter.

Genital warts: These are classified as contagious warts and are quite aggressive compared to flat, and plantar warts and are more likely to spread from person to person (Fig. 2.6C). Human papillomavirus types 6 and 11 are the most prevalent viral types found in condyloma acuminata (genital warts)

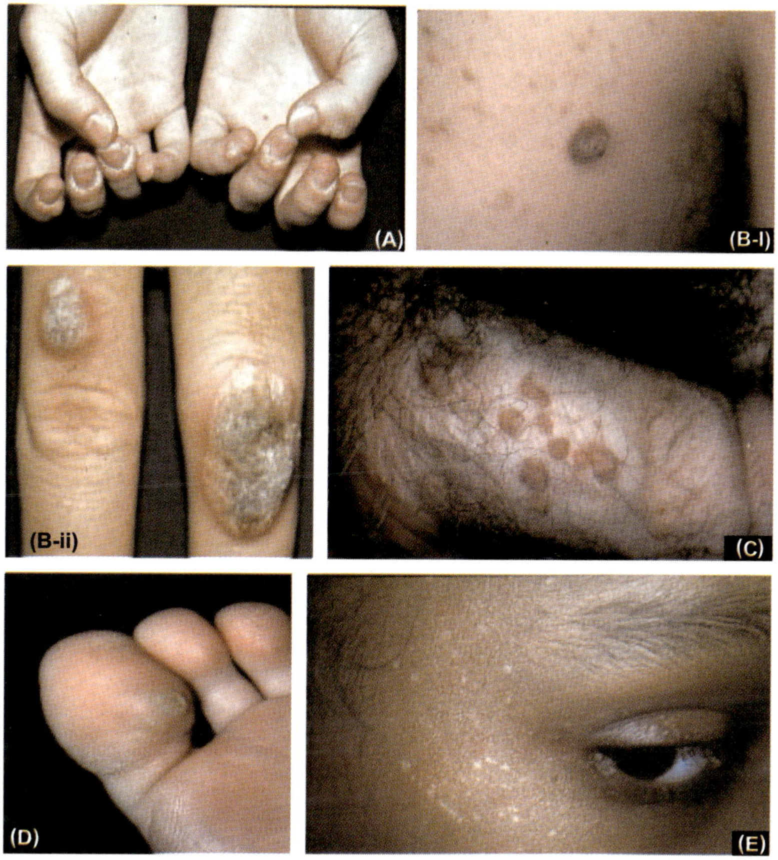

Fig. 2.6: (A) Periungal warts, (B) Common warts (i) type A, (ii) type B, (I and II) (C) Genital warts, (D) Plantar warts, (E) Flat warts

Treatment

Treatment for warts can be divided into ablative and medical approaches. In the ablative approach, the affected tissue is destroyed using either physical or chemical methods. The tissue is excised using electro desiccation, laser or liquid nitrogen. Sometimes vesicants like cantharidin, alone or in combination with podophyllin or other agents are found to be quite effective. These methods are often painless and therefore more suitable for children. Aggressive cryotherapy appears to be effective but with side effects like pain, blistering and scarring.

Topical medical therapy usually includes once or twice daily application of cytotoxic or antiviral agents. 5-fluorouracil, retinoic acid, podophyllin or podofilox are often used for this purpose. Keratolytic agents such as salicylic acid/lactic acid combinations are used commonly, often in combination with a cytotoxic agent for a period of 4-6 weeks. Duration of treatment varies depending on the wart size and intensity of pain.

Topical immunotherapy, which involves the controlled induction of an allergic contact reaction in the wart itself, can be done with powerful haptens such as diphencyprone (DPC), squaric acid dibutyl ester (SADBE) or dinitrochlorobenzene (DNCB)[16]. Topical immunotherapy is very effective form of therapy to treat warts as it does not leave any scar and can be applied by the patient easily. However, the doses and any allergic reactions arising out of the treatment need to be discussed thoroughly with the attending physician and necessary precautions should be taken to minimize side effects.

2.3.2 Human Herpesviruses (HHV)

HHV family of virus causing disease in humans consists of eight types of virus. HHV-1 causes cold sores/genital herpes, HHV-2 is also known as the herpes simplex virus (HSV) that causes cold sores/genital herpes and viral encephalitis, HHV-3 causes chickenpox and shingles, HHV-4 (Epstein-Barr virus = EBV) causes infectious mononucleosis, Burkitt's lymphoma and nasopharyngeal carcinoma, HHV-5 causes mononucleosis, retinitis, etc., HHV-6 causes "sixth disease" (roseola infantum, exanthem subitum), HHV-7 is closely related to HHV-6 and causes roughly the same symptoms, HHV-8 causes Kaposi's sarcoma.

Since the viruses use the host machinery for their survival and growth, therefore eradication of viruses is more difficult. The best approach to deal with viral infections is to develop effective vaccines in order to provide resistance to infection. Drugs are also developed that treat the symptoms of viral infections.

2.3.2.1 Herpes Simplex Virus

The symptoms due to Herpes simplex virus infection (Fig. 2.7b) are frequent episodes of small, painful, fluid-filled blisters on the skin, mouth, lips (cold sores), eyes, or genitals. The disease is quite contagious and usually spread by direct contact with sores. With the first infection, the patient often develops fever and a feeling of illness. The virus sometimes infects other parts of the body including the eyes and brain. Besides the symptoms, blood tests or biopsy from the sore area are required to be carried out for proper diagnosis. Complete eradication of the virus is difficult by using drugs however the antiviral drugs can offer some relief by reduce the symptoms.

There are two types of herpes simplex virus (HSV): HSV-1 and HSV-2. HSV-1 is the usual cause of cold sores on the lips (*Herpes labialis*) (Fig. 2.7b) and sores on the cornea of the eye, *Herpes simplex keratitis*. HSV-2 is the usual cause of genital herpes. This distinction is not always considered to be absolute as genital infections are sometimes caused by HSV-1 as well. Infection also can manifest in other parts of the body such as the brain causing a serious illness or even in the gastrointestinal tract. Widespread infections have been recorded in newborns or in immunocompromised individuals.

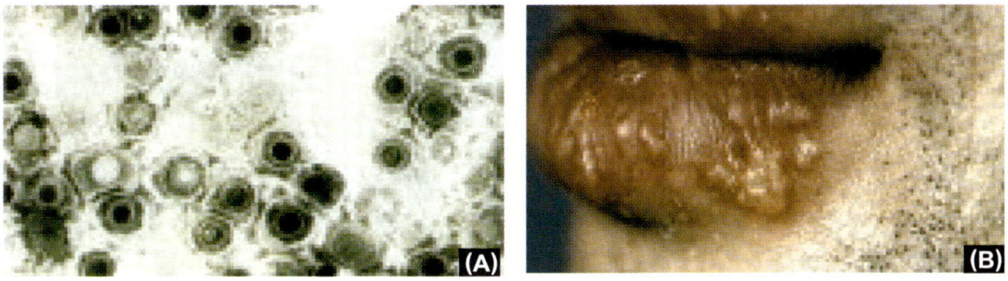

Fig. 2.7(A): HSV-1 Virus, and **(B)** Herpes simplex virus causing cold sores

HSV is very contagious and can be spread by direct contact with sores and sometimes by contact with the oral and genital areas of HSV infected people having frequent episodes of sores. HSV infections originate with an eruption of tiny blisters. The first eruption is called primary herpes. Once the blisters subside, the virus remains in a dormant state inside the nerve cells (ganglia) near the spinal cord that supply the nerve fibers to the infected area. At a periodic interval, the virus gets reactivated and travels from nerve fibers back to the same area of the skin causing eruption of blisters again. The virus may also chose to stay on the skin or in the mucous membranes without showing any symptoms of blisters.

Oral Infection: The first oral infection with HSV usually causes multiple painful sores inside the mouth (herpetic gingivostomatitis). The infected person develops various symptoms like fever, headache, and body aches. The mouth sores normally last for 10 to 14 days and often the patient finds difficulty in eating and drinking. In some individuals, oral infections lead to swollen gums. Herpetic gingivostomatitis is more frequently seen in children. Unlike the first oral infection, during second episode only a single sore develops. Since the sore triggers by cold or fever, commonly known as cold sore or fever blister. Various other triggers have been identified like sunburn on the lips, certain foods, anxiety, certain dental procedures, and any condition that reduces the body's resistance to infection. In case of cold sore it is advised not to go for dental visits until the sore heals.

HSV-1 may also infect the cornea of the eye. This infection causes a painful sore with tearing pain. The infected person becomes very sensitive to light and often experiences blurred vision. If the patient remains untreated, in due course of time, the cornea can become cloudy causing a significant loss of vision and in terminal cases may require corneal transplantation.

Treatment

Antiviral Drugs: None of the antiviral drugs currently available can either eradicate HSV infection completely or prevent chronic infection of nerves arising out of first oral or genital infections. However, during recurrences, antiviral drugs, such as acyclovir ZOVIRAX, valacyclovir (VALTREX), or famciclovir (FAMVIR) help in relieving discomfort to some extent and help in resolving symptoms little sooner. Treatment is more effective at the initial stage of the infection just before the blisters start appearing. Those who are more prone to frequent attacks, number of episodes can be reduced by continuous therapy with antiviral drugs. Antiviral drugs are available by prescription only.

2.3.2.2 HHV-6

Human herpes virus 6 (HHV-6) was discovered in patients with lymphoproliferative diseases in 1986[17]. Although primary infection with variant B HHV-6 causes exanthem subitum[18], the clinical features of variant A HHV-6 infection remain unclear. Most infants aged 6 months to 1 year acquire a primary infection of variant B HHV-6 in Japan on the basis of seroepidemiological analysis[19]. The typical clinical course of exanthem subitum begins with an abrupt onset of fever and with the appearance of skin eruptions coinciding with the fever resolution on the third or fourth day. The disease is more common in pediatric population. Histological examination of skin tissue is difficult to carry out in this set of population therefore, exact pathogenesis of skin eruptions are therefore not very clearly defined. Like other herpes viruses, HHV-6 establishes a latent infection after the primary infection and then reactivates upon immunosuppression. HHV-6 has recently been recognized as an opportunistic pathogen in organ transplant recipients.

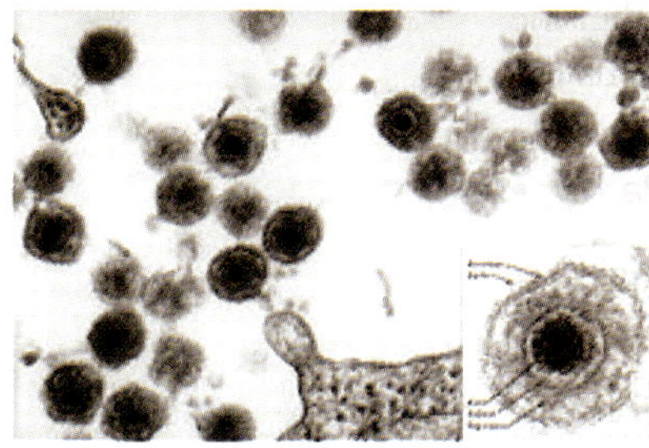

Fig. 2.8: Electron micrograph of HHV-6

Pathogenesis

The newly isolated virus, human herpesvirus 6 (HHV-6), causes exanthem subitum (or roseola infantum) as the primary infection (Fig. 2.8). Liver dysfunction and neurological disorders are observed during exanthem subitum. Seroepidemiological studies have shown that almost all children are infected with HHV-6 about 6-24 months after birth. This virus also causes other diseases including lymphadenitis and infectious mononucleosis-like disease. HHV-6 has also been reported in multiple sclerosis patients[20]. HHV-6 has cellular tropism for CD4+ lymphocytes, in which it replicates in vivo. Transmission of this virus may occur by air. Mother to child transmission is also common at an early age. Antibodies against HHV-6 virus have been detected in the saliva of a healthy individual. Latent infection follows the primary infection and the virus can often be reactivated under conditions of immunosuppression such as AIDS or organ transplantations. The site of its latent infection is not yet known, but could be monocytes/macrophages. The pathogenic mechanisms of HHV-6 at the molecular, cellular, and tissue level remain largely obscure.

Treatment

For treating HHV-6 encephalitis infection, many doctors treat with intravenous ganciclovir (Cytovene) or foscarnet (Foscavir). Ganciclovir is also available in an oral form (Valcyte). However these drugs are not approved by FDA. There is an urgent need for antiviral and immune modulating therapies for active HHV-6 infection. Thus, the efficacy of the anti-herpetic drugs currently used is based upon compilation of case studies.

2.3.3 Varicella (Chicken pox)

Chicken pox is a highly contagious illness caused by primary infection with varicella zoster virus (VZV)[21]. The causative virus is a member of the Herpes virus family. Like other members of this family VZV also possess the characteristic feature of a latent state infection following initial infection and may reactivate at a later stage. Varicella-zoster virus is known by many names, including: chickenpox virus, varicella virus, zoster virus, and human herpes virus type 3 (HHV-3). It is an enveloped, medium-sized (150-200 nm) DNA virus showing close similarity with other members of herpes viruses. It can be cultivated in tissue cell cultures, and the virus can be diagnosed and identified using specific antiserum.

Pathogenesis

Initial symptoms of the disease are skin rashes appearing on the body and head like small vesicles followed by strong itchy sensations (Fig. 2.9). These rashes turn into raw pockmarks and often heal without leaving any scar. Chicken pox virus transmits easily through coughs or sneezes of infected individuals or through direct contact with secretions from the rash. Primary VZV infection results in chickenpox (varicella). In rare occasion, some complications are noticed including encephalitis or pneumonia. Even when clinical symptoms of chickenpox have resolved, VZV remains dormant in the nervous system of the infected person (virus latency), in the trigeminal and dorsal root ganglia[22]. In about 10-20% of cases, VZV reactivates later in life producing a disease known as herpes zoster or shingles. Interestingly, an exposure to VZV in a healthy child initiates the production of host immunoglobulin G (IgG), immunoglobulin M (IgM), and immunoglobulin A (IgA) antibodies. IgG antibodies persist for life and confer immunity. It is observed that the cell-mediated immune responses are also important in limiting the duration of primary varicella infection. After primary infection, VZV often spreads from mucosal and epidermal lesions to local sensory nerves.

Epidemiology

Historically, the disease is predominantly affecting preschool and school-aged children. In adults the pock marks are darker and the scars more prominent than in children (Fig. 2.9). Primary varicella is considered to be an endemic disease. Cases of varicella are reported throughout the year but incidence rates are high during winter and early spring. Varicella is one of the classic diseases of childhood, with the highest prevalence

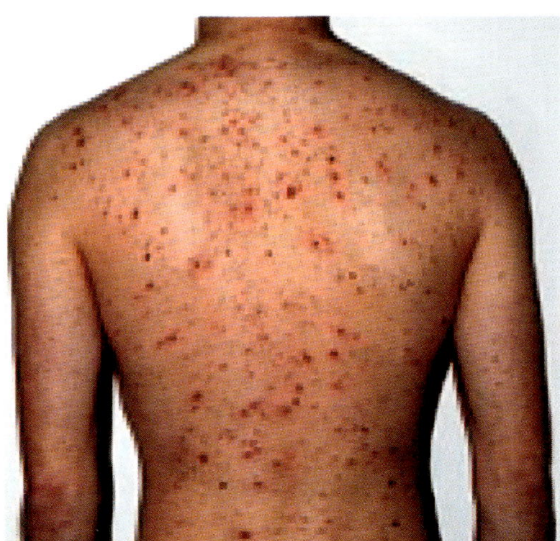

Fig. 2.9: The back of a 30-year-old male, taken on day 5 of the rash due to Varicella (chicken pox)

in the age group of 4–10 years. Varicella is highly communicable disease with infection rate of 90% through close contacts. Most people become infected before adulthood but 10% of young adults remain susceptible. Varicella infection in pregnant women can lead to viral transmission to the child via the placenta and infection of the fetus. If infection occurs during the first 28 weeks of gestation, this can lead to fetal varicella syndrome referred as *congenital varicella syndrome*[23]. Effects on the fetus can range in severity from underdeveloped toes and fingers to severe anal and bladder malformation.

Other complications that can arise due to the disease are as follows[24]:

- *Damage to brain:* encephalitis, microcephaly, hydrocephaly, aplasia of brain
- *Damage to the eye:* optic stalk, optic cup, and lens vesicles, microphthalmia, cataracts, chorioretinitis, optic atrophy
- Other neurological disorders include damage to cervical and lumbosacral spinal cord, motor/sensory deficits, often show absence of deep tendon reflexes, anisocoria/Horner's syndrome
- *Damage to other organs include:* hypoplasia of upper/lower extremities, anal and bladder sphincter dysfunction
- Skin disorders include skin lesions, hypopigmentation

Infection occurring during late gestation period or immediately following birth is referred to as "*neonatal varicella*"[25]. Maternal infection may result into premature delivery. The risk of the disease in the baby increases following exposure to infection in the period 7 days prior to delivery and up to 7 days following the birth. The baby may also be exposed to the virus through other infectious siblings or other contacts, but this is of less concern if the mother is immune.

Prevention and Treatment

Hygiene Measures: Since the disease is highly contagious, the best measure to prevent transmission is to keep the infected individual in an isolation ward. The virus gets transmitted through exposure to respiratory droplets, or direct contact with lesions. The virus remains highly infectious within a period lasting from three days prior to the onset of the rash to four days after the onset of the rash. Therefore, avoiding physical contact or close proximity with the affected individual during that period will prevent contagion. The chicken pox virus (VZV) is susceptible to disinfectants like notably chlorine bleach *i.e.,* sodium hypochlorite. Like other enveloped viruses, VZV is sensitive to desiccation, heat and detergents. Therefore these viruses can be eradicated quickly.

Vaccine: A varicella vaccine was first developed by Michiaki Takahashi in 1974 derived from the Oka strain. The vaccine is not able to give lifelong protection and further vaccination is necessary five years after the initial immunization[26].

Treatment

Number of drugs and therapeutic agents used to treat the infection include acyclovir for the chicken pox; famciclovir, valaciclovir for the shingles; zoster-immune globulin (ZIG), and vidarabine. VZV immune globulin are also used for treatment.

2.3.4 Mumps Virus

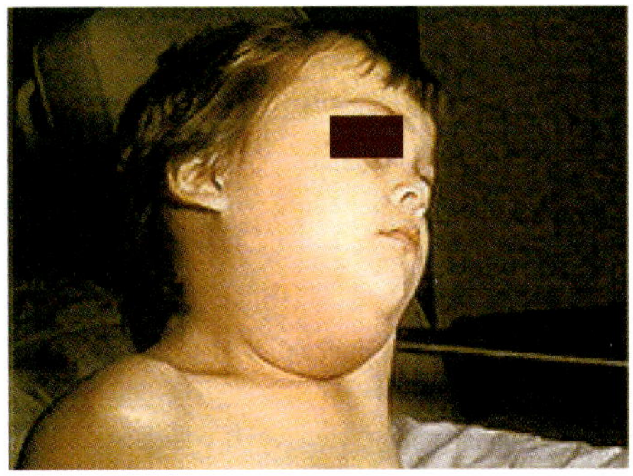

Fig. 2.10: Child with mumps

Mumps virus (MuV) is the etiological agent responsible for an acute viral infection which presents clinically with parotitis. (Fig. 2.10). The patient develops symptoms consisting of a low-grade fever, headache, malaise, anorexia, rash, and abdominal pain[27]. MuV is enveloped, nonsegmented, negative-sense, single stranded RNA virus and is classified as a member of the family *Paramyxoviridae*, subfamily *Paramyxovirinae* and genus *Rubulavirus*. RNA virus is covered by a glycoprotein envelope. There are two glycoproteins on the RNA envelope: one regulates hemagglutinin-neuraminidase activity, the other is responsible for attachment to the lipid membrane of a host cell. *Rubalavirus* can be detected in host culture of saliva, urine, and cerebrospinal fluid. Polymerase chain reaction (PCR) assay can specifically detect mumps RNA, and gives a rapid confirmation for the diagnosis. In a case of mumps infection, elevated level of mumps-specific immunoglobulin G (IgG) antibodies are observed in the serum.

Epidemiology

The symptoms of the disease remains benign and normally not life threatening however, in ten percent of cases, the patients need to be hospitalized following MuV infection. Certain complications that may arise out of the disease include pancreatitis, orchitis, oophoritis, mastitis, and meningoencephalitis. Meningoencephalitis can result in deafness and other severe neurological sequelae[28]. Invasion of the central nervous system (CNS) by MuV appears in 50 percent of patients with clinical mumps as evidenced by cerebrospinal fluid (CSF) pleocytosis; however, symptomatic CNS infection like aseptic meningitis is much less frequent, occurring in 10 percent of cases and encephalitis occurs in 1 percent of mumps cases. Interestingly, there is a gender bias for neurological manifestations, with males disproportionately affected (3:1 male/female ratio)[29].

Based on the sequence of the SH gene coding for putatively membrane-associated small hydrophobic protein (SH), WHO has recognized 12 distinct genotypes of MuV[30]. MuV shows a distinct geographic distribution where more than one genotype may circulate simultaneously in a region. WHO recommended monitoring of MuV that are circulating in various populations. The magnitude and epidemiology of mumps infection in India are poorly understood. In a study conducted in Punjab, India for the first time, revealed the status of mumps virus genotypes circulating in the Indian subcontinent as subtype G2 of genotype G[31].

Pathogenesis

Mumps is transmitted by spread of droplet or by direct contact. The primary site of viral replication is on the epithelium of the upper respiratory or the GI tract or eye. Upon entry, the virus quickly spreads to the local lymphoid tissue resulting in a primary viraemia. Then the virus spreads to distant sites in the body. The major target being parotid gland, the disease may also involve other organs like the CNS, testis, pancreas and ovary. A few days after the onset of illness, virus can again be isolated from the blood, indicating that virus multiplication in target organs leads to a secondary viraemia. Parotitis is the most frequent presentation, occurring in 95% of those with clinical symptoms. Often meningitis may appear one week before parotitis. Infectious virus is excreted in the urine during the period of two weeks following the onset of clinical illness. It is still not known whether virus actually multiplies in renal tissues or whether the virus presented is of haematogenous origin. Life-long immunity is the rule after natural infection, but reinfections can occur and 1 – 2 percent of cases.

Clinical Features

The incubation period is 16 - 18 days but may vary from 14 - 25 days. Parotid swelling develops in 95% of those with clinical illness. Cases of subclinical infections are also reported in a small proportion of patients. Typical symptoms consisting of headache, malaise, myalgia and low grade fever occurs 1 - 2 days before the onset of parotid enlargement. Patients with classic mumps develop enlargement of one parotid gland first followed by enlargement of the contralateral gland after 1-5 days. The patient complains of pain and tenderness in the area of the gland. The sub mandibular and sublingual glands also get involved occasionally. The parotid swelling generally starts to subside after 4 to 7 days. Virus shedding into the saliva begins a couple of days before the onset of parotitis and ends 7 to 8 days later.

Complications

All the other manifestations of mumps can be regarded as systemic complications of mumps rather than as true complications. They could be of following kinds:

1. **Meningitis:** Nearly 10% of patients suffering from mumps develop aseptic meningitis. Some of these patients also develop abnormalities in CSF. The symptoms may develop in these patients even before parotid swelling appear, a typical feature of mumps. Aseptic meningitis can manifest any time, one week before parotid swelling to 3 weeks after it. The CSF report suggests lymphocytosis level below 500 lymphocytes/mm^3, with elevated level of protein. Virus can be isolated from the CSF during the first 2 to 3 days after onset. Later, specific antibodies can be demonstrated in the CSF. Symptoms of meningitis subside 3 to 10 days after onset and recovery is usually complete. In some cases the aseptic meningitis may develop without apparent symptoms of parotiditis.
2. **Encephalitis:** Incidence of encephalitis in mumps patients remains low, around 1 in 6000 cases may develop this secondary complication where certain lesions are found in the brain and spinal cord. The disease probably develops as a result of allergic inflammatory reactions or due to direct viral invasion in the nervous tissue. Clinical symptoms that could be detected are

convulsions, focal neurological signs, movement disorder and changes in sensory perception. Sometimes polio-like paralytic symptoms may develop and in rare cases it may be fatal.

3. **Hearing Loss:** Mumps used to be a leading cause of hearing loss in affected individuals before vaccines were developed. This was common both in children and adults. In most cases, the hearing loss is transient but permanent damages may also occur. Hearing problems appear due to direct damage to the cochlea. This damage does not appear as a result of meningitis. The incidence of hearing loss is estimated to be 1 per 15,000 cases.
4. **Orchitis and oophoritis:** Orchitis is represented as an inflammation of one or both testicles. It causes pain and tenderness of the testicles. This can affect fertility. Orchitis and oophoritis are more likely to occur after puberty. Oohoritis is also refereed as ovaritis causing inflammation of ovaries. Incidence rate is 20 – 30 percent in patients suffering from mumps
5. **Pancreatitis:** The exact incidence of pancreatitis has not been determined but assumed to be as high as 5%.
6. **Arthralgia:** Arthralgia affecting a large joint may develop 2 weeks after parotitis. They are more frequent in young male adults.
7. **Myocarditis:** The disease develops in 10-15 percent of patients and can be diagnosed by ECG examination. Heart failure and deaths have been reported to be rare.
8. **Transient Renal Dysfunction:** This is a frequent complication of clinical mumps. Cases of symptomatic nephritis following mumps are unusual.
9. **Insulin Dependent Diabetes:** Few epidemiological evidences suggest that mumps may be a triggering mechanism for IDDM. It is thought that immunological mechanisms are involved and certain HLA-D haplotypes are particularly susceptible.
10. **Abortion:** When a pregnant woman contracts mumps during pregnancy, there is increased risk for abortion. This is thought to be due to hormonal imbalances caused by virus infection.
11. **Thyroiditis:** Some unconfirmed evidences show that mumps virus may contribute as a causal factor for the development of subacute thyroditis.

Management and Prevention

Mumps without associated major complications is a self-limited disease, therefore can be managed on an outpatient basis with supportive care and good follow-up. Patients diagnosed with mumps should be isolated at least for five to six days from the onset of symptoms. For good management of the disease, the patients are advised to take rest and consume large quantity of oral fluid for adequate hydration. The patient should refrain from acidic foods and liquids as they may cause swallowing difficulty, as well as gastric irritation. Physicians often prescribe analgesics (acetaminophen, ibuprofen) for severe headaches or discomfort due to parotitis. Aspirin is not used due to a hypothetical link with Reye's syndrome. Topical application of warm or cold packs to the swollen parotid or other affected areas may soothe the area. Warm salt water gargles are also recommended.

Mumps is the commonest cause of meningitis and encephalitis in many countries. An inactivated vaccine was first used in the 1940s successfully. This has now been replaced by a live attenuated vaccine. The protection rate is over 95 percent and no adverse reactions are associated with the vaccine. For prevention, the MMR (measles, mumps, rubella) vaccine are now routinely given to children globally. There is no effective post-exposure treatment. The WHO (World Health Organization) recommends the use of mumps vaccines in all countries with well-functioning childhood vaccination programs. The efficacy of the vaccine depends on the strain of the vaccine, but is usually around 80 percent[32]. The Jeryl Lynn strain is most commonly used in developed countries but has been shown to have reduced efficacy during epidemic situations. The Leningrad-Zagreb strain commonly used in developing countries appears to have superior efficacy in epidemic situations.

2.4 Fungal Infections

Certain species of molds can invade the skin, hair and nails causing various infections named as Athelete's foot, Jock itch and ringworm. These skin-invading fungi belong mainly to the genera *Trichophyton, Microsporum* and *Epidermophyton* and they are collectively referred as dermatophytes. Dermatophytes can grow only in a special type of tissue containing keratin and are generally unable to grow in living tissues. They multiply in keratin-containing structures such as the nails, hair and upper layer of epidermis. Most of their unpleasant effects result from secondary infection or itchy allergic reactions due to the presence of fungi and causes thickening of nails and breaking of hair.

2.4.1 Athlete's Foot (*Tinea Pedis*)

Athlete's foot is more common in men and tends to show up in the web spaces between the 4th and 5th toes. If not treated, athlete's foot can also spread to the bottom and top of the feet. There are four patterns of foot fungal infection (Fig. 2.11):

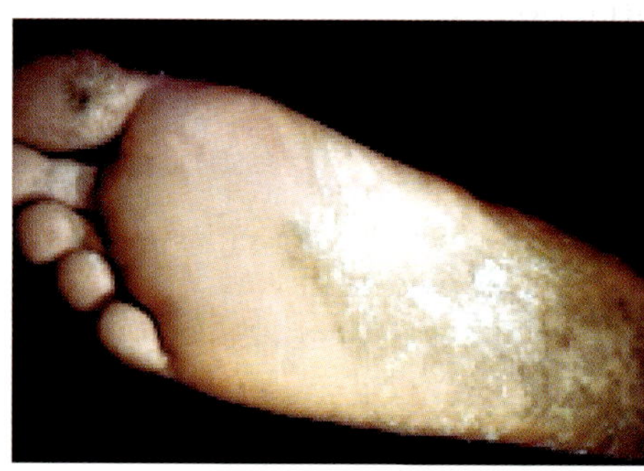

Fig. 2.11: Infection with Tinea pedis (photograph, courtesy of CDC)

A. Interdigital (between the outer toes)

Affected skin gets softened and macerated. It further leads to scaling and some inflammation can be noticed especially in the web space between the outer toes. This often spread to surrounding skin surfaces.

B. Ulcerative

Ulcers and erosions of the tissue can develop between the toes. These symptoms develop as a result of secondary bacterial infection.

C. Moccasin type

Infection develops on the soles of the feet. Very fine scaly eruptions with redness of the skin are seen. The infection may extend to the side of the heel and at the front of the foot as well.

D. Vesicular type

Vesicular type of infection appears due to an immunological response to the fungus which is referred to as an eczema like reaction. Small blisters appear usually on the instep of the foot. It may also appear on the hands.

2.4.2 Jock Itch (*Tinea Cruris*)

Jock itch is commonly seen in boys and men whose body physiology allows for a warm, moist environment which is ideal for fungal growth:

- The affected individual develops itchy sensation in the creases of the groin without affecting the scrotum. The infection may spread to the inner and upper regions of the thigh, buttocks and to the lower abdomen.
- The disease is more prevalent in warm climates.
- The disease appears as a red-brown scaly rash, sometimes with small blisters or pustules at the edge.
- Due to continuous rubbing and itching of the affected are, the skin shows a thickened appearance, a process known as lichenification.

2.4.3 Ringworm (*Tinea Corporis*)

Ringworm causes red scaly patches on the skin that are typically ring-shaped. Symptoms are normally seen on the upper portion of the body or the hands and feet. One should refrain using topical cortisone creams for treatment of ring worms. This treatment may worsen the condition of the patient.

Following persons are at risk for contracting with ringworm infection:

- those who have a suppressed immune system
- those who are involved in contact sports or recreational activities
- children contracting from infected parents.
- those who are genetically-predisposed to this type of infection.

2.4.4 Tinea Versicolor

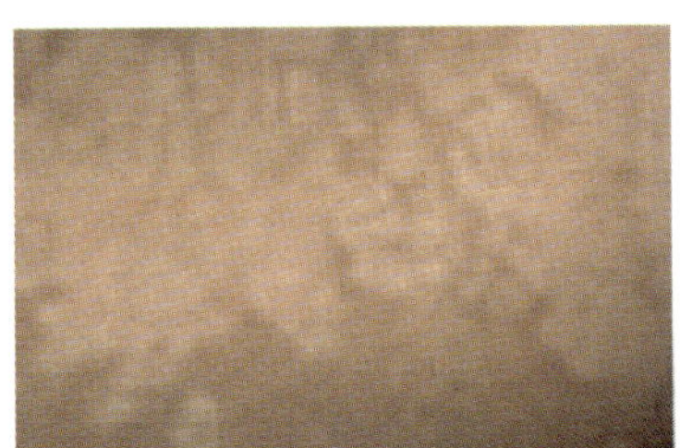

Fig. 2.12: Infection of *Tenia versicolor* on skin

This is a yeast type of skin fungus found on normal skin. If the skin is oily, warm and moist enough, it starts growing into small "colonies" on the surface of the skin. Colonies of the yeast normally grows very fast and spread. The organism leaks out acidic bleach that changes the skin color of the affected area. The patches are reddish brown in color that can be clearly seen on very pale skin but they normally do not tan (Fig. 2.12). Because of lack of any tanning, they look like white spots on darker or tanned skin. These patches may appear on different areas of the body like the neck, upper chest, upper arms and back. The rash can be brown, red or white hence the name "versicolor". The patches have a tendency to join together in the center. The skin gets dry with a scaly appearance. Typically the white patches remains for months after the rest of the rash has disappeared.

Treatment

Topical treatments including anti-dandruff shampoos containing selenium sulfide (Selsun), ketoconazole shampoo (Nizoral), terbinafine spray (Lamisil) and antifungal creams and oral medications (ketoconazole [Nizoral], itraconazole [Sporanox], fluconazole [Diflucan]) are available to treat Tinea versicolor. Topical treatments usually are used one or two times each day. Treatment can last days or weeks, depending on how the rash responds.

REFERENCES

1. Dogra S, Kumar B. Epidemiology of skin diseases in school children: a study from northern India. Pediatr Dermatol. 2003;20:470-3.
2. Rao SG, Kumar P, Kuruvilla M. Prevalence of various dermatoses in school children. Indian J Dermatol Venereol Leprol 1999;65:126-7
3. Valia RG, Valia AR, editors. IADVL Textbook and Atlas of Dermatology, Vol I & II, Mumbai: Bhalani Publishing House; 1994.
4. Sharma N K, Garg B K and Goel M : Pattern of skin diseases in urban school children, Ind J Dermatol Venereol Leprol, 1986;52:330-1.
5. Jain N, Khandpur S. Pediatric dermatoses in India. Indian J Dermatol Venereol Leprol 2010;76:451-4
6. Dhar S. Kanwar AJ. Epidemiology and clinical pattern of atopic dermatitis in a North Indian pediatric population. Pediatr Dermatol 1998;15:347-51.
7. Libu G.K. et al, Prevalence and socio-demographic determinants of skin disease among lower primary school children in Calicut, Kerala, www.imakmj.com
8. Larsen FS, Hanifin JM. Epidemiology of atopic dermatitis. (2002) Immunol Allergy Clinics NA;22:1-25.
9. Strachan DP. Hay fever, hygiene, and household size. BMJ 1989;299:1259-60.
10. D L Gardiner and K S Sriprakash 1996, Molecular epidemiology of impetiginous group A streptococcal infections in aboriginal communities of northern Australia. J Clin Microbiol. 34(6): 1448–1452.
11. Steere AC, Broderic IE, Malawista SE, et al.(1978) Erythema Chronicum Migrans and Lyme Arthritis: Epidemiologic evidence for a tick vector. Ann J Epidemiol ;108: 312-5.
12. Antonsson A, Forslund O, Ekberg H, Sterner G, Hansson BG (2000). The ubiquity and impressive genomic diversity of human skin papillomaviruses suggest a commensalic nature of these viruses. J. Virol. 74 (24): 11636–41
13. Kilkenny M, Marks R (1996) The descriptive epidemiology of warts in the community. Australas J Dermatol 37(2):80-86.
14. Rudlinger R, Bunney M, Grob R, et al.(1989). Warts in fish handlers. Br J Dermatol 120(3):375-381.
15. Corley E, Pueyo S, Goc B, et al. (1988) Papillomaviruses in human skin warts and their incidence in an Argentine population. Diagn Microbiol Infect Dis 10(2):93-101.

16. Naylor M, Neldner K, Yarbrough G, et al. (1988) Contact immunotherapy of resistant warts. J Amer Acad Dermatol 19(4):679-683,
17. Salahuddin SZ, Ablashi DV, Markham PD, Josephs SF, Strurzenegger S, Kaplan M, Halligan G, Biberfeld P, Wong-Staal F,Kramarsky B, Gallo RC. (1986) Isolation of a new virus, HBLV, in patients with lympho proliferative disorders. Science 234:596–601.
18. Yamanishi K, Okuno T, Shiraki K, Takahashi M, Kondo T, Asano Y, Kurata T (1988) Identification of human herpesvirus-6 as a causal agent for exanthem subitum. Lancet 1:1065–1067.
19. Yoshikawa T, Suga S, Asano Y, Yazaki T, Kodama H, Ozaki T. 1989. Distribution of antibodies to a causative agent of exanthemsubitum (human herpesvirus-6) in healthy individuals. Pediatrics 84:675–677.
20. Alvarez-Lafuente R, Martín-Estefanía C, de las Heras V, et al. (2002). Prevalence of herpesvirus DNA in MS patients and healthy blood donors. Acta Neurol. Scand. 105 (2): 95–9
21. Wood MJ (2000). "History of Varicella Zoster Virus". Herpes 7 (3): 60–65.
22. Steiner I, Kennedy PG, Pachner AR (2007) The neurotropic herpes viruses: herpes simplex and varicella-zoster. Lancet Neurol 6 (11): 1015–28
23. Boussault P, Boralevi F, Labbe L, Sarlangue J, Taïeb A, Leaute-Labreze C (2007). Chronic varicella-zoster skin infection complicating the congenital varicella syndrome. Pediatr Dermatol 24 (4): 429–32
24. Matsuo T, Koyama M, Matsuo N (1990). Acute retinal necrosis as a novel complication of chickenpox in adults. Br J Ophthalmol 74 (7): 443–4
25. Sauerbrei, Andreas; Wutzler, Peter (2001). Neonatal Varicella. Journal of Perinatology. 21 (8): 545–549.
26. Chaves SS, Gargiullo P, Zhang JX, et al. (2007). Loss of vaccine-induced immunity to varicella over time. N Engl J Med 356 (11): 1121–9.
27. Carbone, K. M., and S. Rubin. 2007. Mumps virus, p. 1527–1550. In D. M. Knipe, P. M. Howley, et al. (ed.), Fields virology, 5th ed. Lippincott-Raven,Philadelphia, PA.
28. Nussinovitch, M., B. Volovitz, and I. Varsano. 1995. Complications of mumps requiring hospitalization in children. Eur. J. Pediatr. 154:732–734.
29. Kanra, G., P. Isik, A. Kara, A. B. Cengiz, G. Sec¸meer, and M. Ceyhan. (2004) Complementary findings in clinical and epidemiologic features of mumps and mumps meningoencephalitis in children without mumps vaccination. Pediatr. Int. 46:663–668.
30. Dayan GH, Quinlisk MP, Parker AA, Barskey AE, Harris ML, Schwartz JM, Hunt K, Finley CG, Leschinsky DP, O'Keefe AL, Clayton J, Kightlinger LK, Dietle EG, Berg J, Kenyon CL, Goldstein ST, Stokley SK, Redd SB, Rota PA, Rota J, Bi D, Roush SW, Bridges CB, Santibanez TA, Parashar U, Bellini WJ, Seward JF (2008) Recent resurgence of mumps in the United States. New Engl. J. Med. 358:1580–1589
31. Mishra, B.; Pujhari, S. K.; Dhiman, V.; Mahalakshmi, P.; Bharadwaj, A.; Pokhrel, S.; Sharma, D.; Sharma, M.; Bhatia, D.; Ratho, R. K. (2013) Genotyping and subtyping of mumps virus isolates from the Indian subcontinent. Archives of Virology. doi:10.1007/s00705-013-1717-4.
32. Schlegel M, Osterwalder JJ, Galeazzi RL, Vernazza PL (1999). Comparative efficacy of three mumps vaccines during disease outbreak in Eastern Switzerland: cohort study. BMJ 319 : 352

3

Respiratory Tract Infections

Chapter Outline

3.1 Upper respiratory tract Infections

3.2 Lower Respiratory tract infections

3.1 Upper Respiratory Tract Infections

The upper respiratory tract includes the sinuses, nasal passages, pharynx, and larynx, which serve as gateways to the trachea, bronchi, and pulmonary alveolar spaces. Rhinitis, pharyngitis, sinusitis, epiglottitis, laryngitis, and tracheitis are specific manifestations of URIs. It is considered as one of the most prevalent and acute illness observed in any outpatient department. URIs ranges from the common cold, typically a mild, self-limited, catarrhal syndrome of the nasopharynx to life-threatening illnesses such as epiglottitis. Viruses are the most dominating organisms infecting upper respiratory tract. Bacterial primary infection or superinfection are also encountered requiring targeted therapy.

3.1.1 Normal Upper Respiratory Tract Flora and mode of Colonization

Normal bacterial flora colonizing various surfaces of the upper respiratory tract including nasal and oral passages, nasopharynx, oropharynx, and trachea are regular inhabitants and rarely cause disease. In order to maintain a healthy state of the body, two important functions are performed by the normal flora inhabiting the upper respiratory tract. These organisms compete with pathogenic organisms for potential attachment sites, and they often produce substances that are bactericidal and prevent infection by pathogens. Common causes of various respiratory diseases are listed in Table 3.1.

Table 3.1: Common causes of various respiratory diseases by location

Disease Location	Disease	Group of Pathogen	Comments
Upper respiratory tract			
Nasal passages	Common cold	Viruses	Most common cause rhinovirus
Nasal sinuses	Rhinosinusitis	Viruses Bacteria	Viruses are most common cause of rhinosinusitis
Pharynx	Pharyngitis	*Streptococcus pyogenes and Corynebacterium diphtheria and Viruses*	Viruses cause 90% of these infections
Respiratory airways			
Epiglottis	Epiglottitis	Bacteria	Usually *Haemophilus influenzae* type b
Trachea and bronchi	Bronchitis, tracheobronchitis, croup, laryngitis	Viruses	Usually caused by viruses
Bronchioles	Bronchiolitis	Viruses	Most common cause is respiratory syncytial virus
Lower respiratory tract			
Alveoli and alveolar sacs	Pneumonia	Bacteria	Most common cause in adults is *Streptococcus pneumonia*

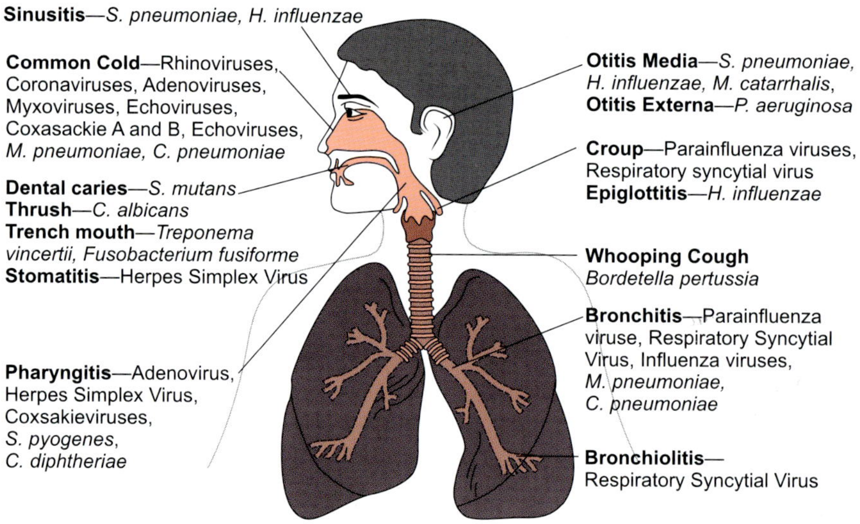

Fig. 3.1: Location of various respiratory diseases

Mode of Colonization

The respiratory tract happens to be the most potential site for infection by pathogens because of the following reasons: A. This site is in frequent contact with the physical environment directly and therefore get exposed to microorganisms present in the air. B. The human respiratory tract is exposed to many potential pathogens via the smoke, soot, and dust that is inhaled from the air. It has been calculated that the average individual ingests about 8 microorganisms per minute or 10,000 per day.

Tiny hair-like projections known as cilia are present all over in the cells lining the nasal chamber, tear ducts, sinuses, mastoid air cells and middle ear. These cilia are present at their exposed free border. These cilia beat synchronously at a rate of about 1000 times a minute and thereby continually propel a film of mucous. The mucous is secreted by the goblet cells and spans the whole of the respiratory system. Through the mucous present in the nose and throat, the microorganisms get entrapped and get swallowed. The microorganisms and viruses are then exposed to the action of acidic gastric juices that help in destroying the microorganisms. Viral infection as well as consumption of alcohol, tobacco and narcotics all together impair ciliary movement and thereby promote infection. Parts of the upper respiratory tract are normally considered sterile though other parts are colonized by numerous species of bacteria (Fig. 3.1).

Predominating bacteria inhabiting the nose are staphylococci. Two main species of staphylococci are *Staphylococcus aureus* and *S. epidermidis*. In addition to the staphylococci, aerobic corynebacteria ("diphtheroids") have been successfully cultured from the nasal surfaces.

Streptococcus pneumoniae, *Neisseria meningitidis*, and *Haemophilus influenzae* are identified as subdominating bacteria found in the region of nasopharynx. However, most of these strains are not encapsulated or virulent. An exception was observed with nonencapsulated form of *H. influenzae* which may contribute in the pathogenesis of otitis media.

Abundance of *Staphylococcus* aureus and *Staphylococcus epidermis* are observed as the normal flora in the region of oropharynx. Other important groups of native microorganisms observed in this niche are the alpha-hemolytic streptococci or viridans streptococci. Members of this group include *S. mitis, S. mutans, S. milleri,* and *S. salivarius*. Interestingly, these members work as antagonist and prevent against invasion by pathogenic streptococci. In addition to that, cultures from this region usually show large numbers of diphtheroids, *Moraxella* (formerly Branhamella) *catarrhalis*, and small Gram-negative cocci related to Neisseria species. Therefore, two important functions these normal microflora play in maintaining the healthy state of the host by

1. Blocking the potential attachment sites for the pathogenic microorganisms
2. Secreting bactericidal compounds in the form of toxins or acids thereby prevent the entry of pathogenic microbes

Asymptomatic colonization of *S. pneumoniae, H. influenzae, M. catarrhalis,* and *Staphylococcus aureus* typically occur at young age in the nasopharynx area.

Antibiotics and vaccines modify the flora by removing organisms that are part of the commensal flora[1]. These organisms exist as a complex and maintain a symbiotic relationship among each other. Any manipulation of one microorganism may trigger effects on other components of the flora because multiple species simultaneously colonize the same niche. The survival of an organism may depend on its ability to compete with co inhabitants of the niche. In order to maintain the homoeostasis, the host factors also contribute and influence on the outcome of this competition.

Mode of Transmission

The transmission of pathogenic organisms causing URIs occurs by aerosol, droplet, or direct hand-to-hand contact with infected secretions. This may lead to subsequent passage to the noses or eyes[2]. Thus, the rate of transmission is higher in densely populated areas. Direct invasion of the respiratory epithelium may result into symptoms depending on the corresponding affected areas.

Sinusitis and acute bronchitis are the two common disease that occur as a result of common cold. Sinonasal allergies may lead to temporary anatomic abnormalities such as a deviated nasal septum, sinus ostial blockade. This happens as a result of either mucosal edema or immunodeficiency disorders such as hypo gammaglobulinemia and human immunodeficiency virus infection. Even cocaine abuse may predispose a person to the development of acute sinusitis[3].

Twentieth century has witnessed influenza caused predominantly by the influenza A virus where few cases were reported caused by the influenza B virus. The mode of transmission of this epidemic was believed to be through infected school children to their families. The rate of mutation is very high in this virus. Thereby, annual influenza epidemics result from the transmission of a mutated influenza virus for which the affected population did not develop immunity. This phenomenon is known as antigenic drift. Pandemics, on the other hand, occur when a totally new influenza virus is transmitted to humans from other species, most commonly swine and birds. This is defined as antigenic shift. Those who are older than 65 years and suffered from influenza sometime in the early period of life are at higher risk of hospitalization or death than healthy people.

Once a specific pathogen gets an entry in the respiratory tract, it must colonize the surfaces of the respiratory tract before it can cause obvious disease. Most microorganisms cause disease by adopting some or other mechanism of pathogenesis.

1. Bacterial adherence factors: F and M proteins of *Streptococcus pyogenes*, Hemagglutinins of *B. pertussis*.
2. Extracellular toxins: diphtheria toxin; pertussis toxin.
3. Growth in host tissue: *Chlamydia sp*.
4. Evasion of host defense mechanism: capsules of *Streptococcus pyogenes*, *S. pneumoniae* and *H. influenzae*
5. Inhibiting the process of phagocytosis

3.1.2 Bacterial Infections- Pathogenesis, Epidemiology, Treatment and Prevention

Respiratory tract infections are common in humans because of either malfunction of a complex array of defense mechanisms or because of the virulence of the specific microorganism. Bacterial infections in the mucosal layer of the respiratory tract represent a dynamic interaction where both host and bacterial factors contribute. Two chronic respiratory tract infections like cystic fibrosis or *bronchiectasis* result due to abnormal host defense associated with these diseases. The bacteria that cause bronchial infections possess a wide array of potential virulence factors that contribute to their pathogenicity. The first line of defense against these virulence factors is offered by the mucociliary system. In case the bacteria gets established in the host it multiplies, spreads within the bronchial lumen. Persistence of bacteria within the bronchial lumen, and consequent damage to the epithelium, stimulates a chronic inflammatory response. This also leads to impairment of mucociliary clearance and damages lung tissue.

A broad range of microorganisms including commensal bacteria and few potential pathogens may colonize the nasopharynx. The organisms such as *S. pneumoniae, H. influenzae* predominantly nontypeable strains, and *M. catarrhalis* are the predominant colonizers. Mostly they remain harmless without causing any clinical symptoms. However, during the condition of dysbiosis, some of the potential pathogenic microorganisms invade adjacent sites and subsequently invade the bloodstream to cause the disease.

3.1.2.1 Streptococcal Pharyngitis: Pathogenesis, Epidemiology and Treatment

Pathogenesis This disease is commonly known as strep throat. It is an upper respiratory tract infection caused by group AB haemolytic streptococci -*Streptococcus pyogenes*. The pathogen is resistant to phagocytosis. This organism can be detected in the throat of many people who are asymptomatic carriers. This is characterized by local inflammation and fever with tonsillitis.

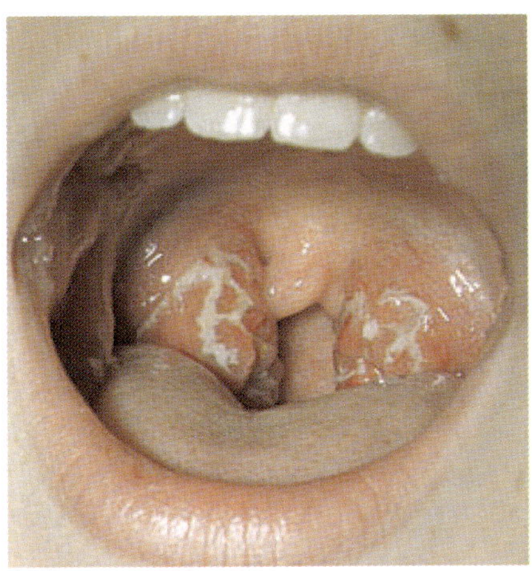

Fig. 3.2: Positive case of streptococcal pharyngitis with typical tonsillar exudate

The typical symptoms of streptococcal pharyngitis are a sore throat, fever of greater than 38°C (100°F), tonsillar exudates consisting of pus on the tonsils and may lead to enlarged cervical lymph nodes (Fig. 3.2).

Other symptoms include:

- Headache
- Nausea and vomiting
- Abdominal pain
- Muscular pain
- Scarlatiniform rash

The symptoms for strep throat develop within one to three days after a person gets infected with the pathogen. Strep throat is caused by group A beta-hemolytic streptococcus (GAS)[4]. A very similar disease known as pharyngitis is caused by non–group A beta-hemolytic streptococci and fusobacterium[5]. Invasive GAS disease is a severe and sometimes life-threatening infection in which the bacteria invade various parts of the body, such as the blood, deep muscle and fat tissue or the lungs. Two of the most severe, but least common, forms of invasive GAS disease are called "necrotizing fasciitis" (infection of muscle and fat tissue) and "streptococcal toxic shock syndrome" (STSS, a rapidly progressing infection causing low blood pressure/shock and injury to organs such as the kidneys, liver and lungs). Necrotizing fasciitis (NF), a life-threatening rare infection of the soft tissues, is a medical and surgical emergency. It is characterized by subtle, rapid onset of spreading inflammation. This follows necrosis starting from the fascia, muscles, and subcutaneous fat, with subsequent necrosis of the overlying skin.

Transmission

The disease normally gets transmitted by direct or close contact with an infected person and thus highly populated areas like military and schools increase the rate of transmission. Nose and throat discharge of affected individual or infected skin lesions are the potential source of transmission of the disease. Dried bacteria present in the dust are not infectious however the pathogen can survive as long as two weeks on moist surfaces like toothbrushes or any other accessories used for cleaning mouth. Rarely, contaminated food can result in outbreaks. Around 12 percent of children are carriers of GAS in the pharynx and may remain asymptomatic[6]. A throat culture is the gold standard for the diagnosis of streptococcal pharyngitis with a sensitivity of 90–95%[7]. A rapid strep test (also called rapid antigen detection testing or RADT) is also used for the detection purpose. While the rapid strep test is quicker, it has a lower sensitivity (70%) and statistically equal specificity (98%) as throat culture. In doubtful cases, a combination of test like a positive throat culture or RADT in association with symptoms establishes a positive diagnosis.

Epidemiology

Mortality rates are still high and have shown no tendency to decrease in the last 100 years[8]. Approximately 20 percent of patients with necrotizing fasciitis and 60 percent with STSS die. Only about 10-15 percent of patients with other forms of invasive group A streptococcal disease die.

Treatment

Upon detection, the disease must be treated with adequate antibiotics. In rare cases the strep bacteria can spread to other parts of the body causing ear or sinus infections or an abscess behind or around the tonsils (peritonsillar abscess). It is important to take the full course of antibiotics as prescribed and not to stop the medication when symptoms resolve. Premature termination of antibiotics may lead to adverse consequences or relapse of the infection.

Streptococcus responds fairly well to penicillin and the cephalosporin antibiotics. Oral penicillin V (Pen-Vee-K, Veetids) is the preferred oral form of penicillin for strep throat. The usual dose is 250 milligrams three times a day or 500 milligrams twice a day. A full 10 day course must be completed even though patients usually feel better only after two to three days.

Injectable penicillin G (Bicillin) is also very effective and may be used in individuals where the disease is more severe. It is important that appropriate dose of the drug should be available in the body of the infected person for up to 21 days to treat the infection. Other penicillin derivatives such as amoxicillin (Amoxil), amoxicillin-clavulanate (Augmentin), cloxacillin (Cloxapen, Tegopen), and dicloxacillin (Dynapen) are all adequate treatments for strep. These second generation antibiotics are more effective than penicillin because of better absorption and greater potency. Cephalosporin antibiotics are also a very effective in treating group *A streptococcus*.

Complications

Streptococcal pharyngitis can cause serious complications like

1. ***Scarlet fever*** – Some strains of *S. pyogenes* produce a toxin that is absorbed and carried by the blood stream throughout the body, causing rash and whitish coating of the tongue. It is characterized by[9]:
   - Sore throat
   - Fever
   - Bright red tongue with a "strawberry" appearance
   - Characteristic rash appears as
     - fine, red, and rough-textured; it blanches upon pressure.
     - appears 12–48 hours after the fever.
     - starts on the chest, armpits, and behind the ears.
     - worse in the skin folds.
     - rashes may run together in the armpits and groin

   The rash begins to fade three to four days after onset and desquamation (peeling) begins. This phase begins within a week time, with flakes peeling from the face, palms and around the fingers.

   Clinical diagnosis of scarlet fever is carried by blood test. The blood test shows marked leukocytosis with neutrophilia and increased eosinophils, high erythrocyte sedimentation rate (ESR) and C-reactive protein (CRP), and elevation of antistreptolysin O titer. The complications of scarlet fever include septic complications due to spread of streptococcus in blood and may also generate aberrant immune response.

2. ***Glomerulonephritis*** – Streptococcal throat infections may lead to acute Glomerulonephritis, a disease of kidney as a result of antibody reactions with streptococcal products. It typically sets in 10–14 days after a skin or pharyngeal infection with this bacterium. This condition appears during convalescence in patients who remain untreated due to *S. pyogenes* infections and characterized by the abrupt appearance of fever and high blood pressure and detection of blood and protein in the urine sample.

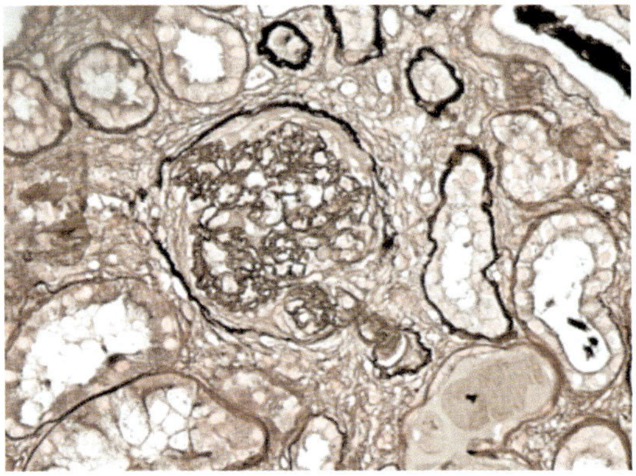

Fig. 3.3: Photomicrograph of renal biopsy from a Glomerulonephritis patient showing crescent formation and tuft narrowing. Periodic acid silver methenamine stain.

Source: http://en.wikipedia.org/wiki/Glomerulonephritis

Damage to the kidney is caused by immune complexes that are deposited in the glomeruli responsible to provoke an inflammatory reaction (Fig. 3.3).

3. ***Rheumatic fever*** – Rheumatic fever primarily affects children between ages 5 and 17 years. It usually occurs about three weeks after the onset of a streptococcal sore throat and is characterized by inflammation of various tissues, including the joints, heart, skin and occasionally affects brain. It is a systemic disease affecting the peri-arteriolar connective tissue and develops in individuals remaining untreated due to Group A Beta hemolytic streptococcal pharyngeal infection. This results due to antibody cross-reactivity. This cross-reactivity is a Type II hypersensitivity reaction and is often termed as molecular mimicry. During a Streptococcus infection, mature antigen presenting cells such as B cells present the bacterial antigen to CD4-T cells which differentiate into helper T_2 cells. Helper T_2 cells subsequently activate the B cells to become plasma cells and induce the production of antibodies against the cell wall of Streptococcus. However the antibodies may also react against the myocardium and joints producing the symptoms of rheumatic fever[10]. The rate of development of rheumatic fever in individuals with untreated strep infection is estimated to be 3 percent. The incidence of recurrence with a subsequent untreated infection is substantially greater and may reach as high as 50 percent[11]. Incidence rate is much lower in individuals who have received antibiotic treatment. Persons who have suffered a case of rheumatic fever have a tendency to develop flare-ups with repeated strep infections.

   The recurrence of rheumatic fever is relatively common in the absence of maintenance of low dose antibiotics, especially during the first three to five years after the first episode. Heart complications may also develop. Heart complications may be long-term and severe, particularly if valves are involved.

3.1.2.2 Diphtheria

Diphtheria is referred as a highly contagious and potentially life-threatening bacterial disease caused by *Corynebacterium diphtheriae*. There are two types of diphtheria: respiratory and cutaneous. The respiratory diphtheria involves the nose, throat and tonsils, and cutaneous diphtheria involves the skin. This disease primarily affects the mucous membranes of the respiratory tract leading to respiratory diphtheria. Occasionally it also affects the skin causing cutaneous diphtheria where the affected areas involve lining tissues in the ear, eye, and the genital areas.

Pathogenesis

Symptoms: This communicable disease spread by respiratory droplets from the throat through coughing and sneezing. The disease normally breaks out 2 to 5 days after infection. It usually affects the tonsils, pharynx, larynx and occasionally the skin. Symptoms range from a moderately sore throat to toxic life-threatening diphtheria of the larynx or of the lower and upper respiratory tracts. Diphtheria is often complicated by diphtheric myocarditis causing toxic damage to heart muscles or neuritis causing toxic damage to peripheral nerves. The disease usually begins with a mild sore throat and slight fever, accompanied by great deal of fatigue and malaise. Swelling of the neck ensues as the disease progresses. A whitish gray membrane forms on the tonsils and the throat, or in the nasal cavity. This thick membrane, referred as pseudomembrane, often forms over the site(*s*) of infection as a result of the combined effects of bacterial growth, toxin production, necrosis of underlying tissue, and the host immune response. This disease may develop serious consequences like heart and kidney failure and paralysis.

Skin infections with diphtheria are still common in tropical countries and are even more contagious than respiratory diphtheria. Skin wounds are characterized by a scaling rash, sores or by blisters that may appear anywhere on the body. Skin wounds may be painful, swollen and reddened. The skin infection is treated by thorough cleansing with soap and water and by administering appropriate antibiotics.

Cause: Corynebacterium diphtheriae, an aerobic gram-positive bacillus. Toxin-production occurs only when the bacillus is itself infected that is lysogenized by a specific virus referred as bacteriophage carrying the genetic information for the toxin gene (tox gene). Only toxigenic strains can cause severe disease. Three biotypes have been identified - gravis, intermedius, and mitis. The most severe disease is associated with the gravis biotype, but any strain may produce toxin. All isolates of *C. diphtheriae* should be tested in the laboratory for toxigenicity.

Epidemiology

The disease remains contagious for two to three weeks. As per the available data, the disease is fatal in 5%-10% of patients even after treatment. If remains untreated, the disease claims even more lives. Untreated patients are infectious for 2 to 3 weeks. In children under 5 years and adults over 40 years, the fatality rate may be as high as 20%[12]. Few outbreaks have been reported worldwide from developed nations such as Germany and Canada. After the breakup of the former Soviet Union in the late 1980s, vaccination rates in its constituent countries fell so low that there was an explosion of diphtheria cases. By 1998, according to Red Cross estimates, there were as many as 200,000 cases in the Commonwealth of Independent States, with 5,000 deaths[13] .This increase in incidence of diphtheria has been cited in the Guinness Book of World Records as "most resurgent disease".

Indian Scenario

Universal Immunization Program in India, offer 2 booster doses at 18 months and between 54 and 72 months of age. After 3 doses of primary vaccine, protective levels of antitoxin develop in 94% to 100% of children[14]. However, without booster doses, over time toxoid-induced antibody drops below protective levels[15].

In 2005, India contributed 5,826 (71%) of the 8,229 diphtheria cases reported globally. Of the total cases from India, 4,161 (71%) were from the state of Andhra Pradesh[16]. The administrative coverage of primary vaccination among children 12–23 months of age (a performance indicator for Universal Immunization Programme [UIP]) ranged from 98% to 100% in the city from 1995 through 2006. A study conducted on Hyderabad city of Andhra Pradesh, India revealed an increase in annual incidence from 11/100,000 to 23/100,000 from 2003 through 2006 ($\chi 2$ trend 152; $p = 0.00001$)[17] . Median age of case-patients was 17 years (range 9 months–80 years). Attack rates were lowest among infants, increased with age, and reached a maximum among children 10–14 years of age. Vaccine was efficacious among those who received >4 doses. The proportion of the population receiving boosters was low, especially among Muslims. Reappearance of positive confirmed cases of diphtheria were recorded in a tertiary care hospital in north India[18]. Poor immunisation coverage, population migrations, and overcrowded urban slums were implicated as contributory factors.

Action of Diphtheria Toxin

Susceptible persons may acquire toxigenic diphtheria bacilli in the nasopharynx. The mode of action of the toxin is now well defined. The organism produces a toxin that inhibits cellular protein synthesis. The toxin further destroys the local tissue and inhibits membrane formation. The toxin produced at the site of the membrane is absorbed into the bloodstream and then distributed to the tissues of the body. The toxin is responsible for the major complications of myocarditis and neuritis and can also cause low platelet counts (thrombocytopenia) and protein in the urine (proteinuria). Clinical disease associated with non-toxin-producing strains is generally milder.

Mechanism of action.

The diphtheria toxin is a small protein of 58 kD and composed of two parts, A and B. Fragment B gains access to the host cell by binding to the plasma membrane at specific sites called the EGF-like domain of Heparin-binding EGF-like growth factor (HB-EGF). The HB-EGF then help in engulfing the fragments through receptor-mediated endocytosis. The low pH of the endosome (sac which transports the HB-EGF and bound fragments A and B) induces the pore formation through fragment B and further catalyses the release of a catalytic fragment A into the cytosol or intracellular fluid. The toxin catalyses the ADP-ribosylation of the elongation factor eEF-2 that inactivates it. The elongation factor eEF-2 is considered as an essential protein in the process of protein synthesis. The inactivated elongation factor inhibits the translation of the protein synthesis. Once the toxin enters the host cell, it gets hydrolyzed by a trypsin-like protease to release the toxic fragment. The toxin then transfers an ADP-ribose from NAD+ to a diphthamide residue usually referred as a histidine amino acid found within the EF-2 protein. EF-2 is needed for the moving of tRNA from the A-site to the P-site of the ribosome during translation. The ADP-ribosylation is a reversible process. The process can be reversed by giving high doses of nicotinamide (or vitamin B3), one of the reaction's products. Diphtheria toxin is produced by *C. diphtheriae* only when it is infected with a bacteriophage. The bacteriophage integrates a gene into the bacteria that causes the toxin to be produced[19].

Treatment and Prevention

Treatment consists of immediate administration of diphtheria antitoxin and antibiotics. Antibiotic treatment usually renders patients non-infectious within 24 hours. The bacteria are sensitive to antibiotics such as erythromycin and penicillin but such treatment stops only the transmission of the disease, however, it remains inactive for the toxin already absorbed by the body. Unless immunized, children and adults may repeatedly be infected with the disease.

Action of the vaccine: Since the disease results primarily from toxin absorption rather than microbial invasion, its control can be accomplished most effectively by immunization with toxoid. Toxoid, prepared by formalin treatment of diphtheria toxin, is used for immunization against diphtheria. Toxoid administration results in the production of antibodies that specifically neutralize the diphtheria toxin. Childhood vaccination of DPT consists of combination of Diphtheria, tetanus toxoid and pertussis vaccine. Immunization program for infants and children consists of five doses of DTP vaccinations generally given at ages 2, 4, and 6 months, with the fourth dose being administered between 15-18 months, and the fifth dose at ages 4-6 years. At age 11-12 years, children should receive a single Tdap vaccination if they have completed the recommended childhood vaccination schedule. Since the immunity loses over time, subsequent booster immunization is required every 10 years thereafter to maintain protective antibody levels. Travelers visiting the endemic areas where diphtheria is endemic must take medical advice and accordingly take necessary vaccinations if required.

PUNCH, OR THE LONDON CHARIVARI.-JULY 3, 1858.

FATHER THAMES INTRODUCING HIS OFFSPRING TO THE FAIR CITY OF LONDON

A Punch Magazine Cartoon from 1858 Shows Father Thames with 'His Offspring', Diphtheria, Scrofula and Cholera

With the increasing industrialization of the nineteenth century in Western Europe, disease began to spread more quickly, in part because of increased foreign travel and trade, and also because of newly realized sanitation problems. Numerous diseases and illnesses were on the rise, some surfacing in England for the first time. Cholera, for instance, was a disease endemic to tropical areas, but as Europeans began to travel to these places, they also began bringing disease back to their native countries.(Harrison 87) Among these diseases were the three noted in the drawing: Cholera, scrofula, and diphtheria, as depicted by the children in the picture. It was initially believed by some that Cholera was being spread via waterways, such as the River Thames. In the middle of the nineteenth century, a medic named John Snow Medicine theorized cholera was waterborne, and pointed to the Thames as one of the many culprits. (Gilbert 107 Scrofula, also known as tuberculous adenitis, which is an "inflammation of the lymph nodes due to the bacillus of tuberculosis," was not generally fatal, but could disfigure if left untreated. (Bloch 11) at the time of the picture, it was believed this disease could be cured with a 'royal touch,' which is ironic since contact with an infected person is generally how the disease is contracted. It was also later discovered it could be spread through contaminated milk, just as diphtheria. Diphtheria was rampant in England in the mid-nineteenth century, and was also contracted via human contact, as well as through contaminated milk, although it was believed, at the time, to be spread through contaminated water, such as the Thames River. (Hardy 91)

3.1.2.3 Otitis Media (OM) and Sinusitis

Pathogenesis

Cause: Important organisms responsible for causing purulent otitis media are the *Pneumococcus*, nontypeable strains of *Haemophilus influenzae*, and *Moraxella catarrhalis*. Another organism that was more prevalent in earlier days was group A streptococci leading to aggressive form of the disease compared to other pathogens. Other organisms that can cause acute otitis media are coagulase-negative staphylococci and anaerobic bacteria. Gram-negative bacilli and *Staphylococcus aureus* can cause acute otitis media in neonates. Among children, *Streptococcus pneumoniae* is the most common etiologic agent responsible for OM. It may also cause invasive bacterial infections in children of all age groups. *S pneumoniae* is a gram-positive diplococcus with 90 identified serotypes (classified based on the polysaccharide antigen), the frequency of which varies between age groups and geography. Most prevalent serotypes causing high episodes of Pneumococcal infections have been identified by culture methods. Serotypes 4, 6B, 9V, 14, 18C, 19F, and 23F are responsible for most invasive pneumococcal disease in America. These serotypes can be isolated from the ear aspirates of the patients. The polyvalent pneumococcal vaccine can confer immunity to approximately 85% of those serotypes responsible for OM.

H influenzae is the second most frequently isolated bacterium causing OM. This organism is responsible for approximately 20% of episodes in preschool children. The frequency may increase in susceptible children and even in adults. The bacterium is a small, pleomorphic, gram-negative coccobacillus. The types of bacteria isolates having polysaccharide capsule coating are classified into 6 distinct types(a-f), while nonencapsulated types are termed as nontypeable strains. The nontypeable strains are responsible for the majority of OM episodes.

Symptoms: The symptoms of this infection in children appear in the region of middle ear as they approach two years of age. Usually Otitis media is preceded by infection of nose and throat. It is assumed that the infection spreads upward through the Eustachian tube, moving cell-to-cell along the epithelium. This may result in changes of airway pressure that force secretion generated out of the infection to move upward in the Eustachian tube.

The infection damages the ciliated cells of the airway tube, resulting in inflammation and swelling. This helps in establishment of an acute inflammatory reaction that can be characterized by typical vasodilatation, exudation, leukocyte invasion, phagocytosis, and local immunological responses within the middle ear cleft. Thus the clinical pattern of acute Otitis media is established.

Epidemiology

Globally around seventy percent of children experience one or more episodes of acute otitis media (AOM) within two years of their age. A recent study from Pittsburgh prospectively followed urban and rural children for the first 2 years of life[20]. The study indicates that children experience incidence rates of middle ear effusion episodes of approximately 48%, 79%, and 91% at age 6 months, 1 year, and 2 years, respectively. The peak incidence of OM is in children aged 3-18 months. Some infants may experience their first attack shortly after birth and they become sensitive to OM and therefore they may get recurrent attack of OM. In the same study it was also observed that the incidence was highest among poor urban children. (http://emedicine.medscape.com/article/859316-overview). The incidence rate also depends on differences in racial, socioeconomic, and climatic factors. The incidence is slightly higher in boys than in girls. A study has also revealed that racial differences exist in the incidence of OM. Native Americans have very high rates of acute and chronic ear infection, while African Americans appear to have a slightly lower rate than white children living in the same communities.

Treatment

Amoxicillin remains the antibiotic of choice for the initial treatment of AOM. The dosage needs to be increased in patients at risk for drug-resistant *S. pneumoniae*. In cases refractory to high-dose amoxicillin, TMP/SMX are prescribed. Ceftriaxone is normally reserved as third-line treatment in case the previous antibiotics do not respond. With the increase in drug-resistant *S. pneumoniae*, it is very important to use the antibiotics more judiciously especially in established cases. Alternative treatment approaches are also explored for the prevention of OM. Removal of modifiable risk factors should be first-line therapy for prevention of OM [21].

3.2 Viral Infections of Upper Respiratory Tract- Pathogenesis, Epidemiology, Treatment and Prevention

Viral infections of upper resipratory tract involves common cold, eye infections, sinuses and throat infections.

3.2.1 Common Cold

It is caused primarily by rhinoviruses and coronaviruses. Sore throat, runny nose and cough are the common symptoms of common cold. There is currently no known treatment that shortens the duration; however, symptoms usually resolve spontaneously in 7 to 10 days, with some symptoms possibly lasting for up to three weeks[22].

Pathogenesis

The common cold is the most frequent infectious disease in human with on average two to four infections a year in adults and up to 6–12 episodes in children. This includes colds, influenza, and other infections with similar symptoms lead to influenza-like illness. In case of patients with underlying diseases such as asthma and COPD, rhinovirus can precipitate asthma attacks and exacerbate wheezing. Since rhinovirus exists as multiple genotypes, young children and adults may come down with multiple infections per year.

Causal agents: The common cold is a viral infection of the upper respiratory tract. The most commonly implicated virus is a rhinovirus (30–50%), a type of picornavirus with 99 known serotypes[23]. Others include: coronavirus (10–15%), influenza (5–15%), human parainfluenza viruses, human respiratory syncytial virus, adenoviruses, enteroviruses, and metapneumovirus. More than 200 serologically different viral types may cause cold. Coronaviruses are particularly responsible for common colds among adults. So far around 30 coronaviruses have been identified, however only 3 or 4 types cause infections in humans. The viruses are difficult to cultivate in laboratory conditions therefore, they are not well understood. Due to the many different types of viruses and their high mutation rates, it is impossible to gain complete immunity to the common cold.

Mode of transmission: The common cold virus is transmitted mainly from contact with saliva or nasal secretions of an infected person. The transmission may occur by direct contact or either directly, when a healthy person breathes in the virus-laden aerosol generated when an infected person coughs or sneezes. It may also spread by touching a contaminated surface and then touching the nose or eyes. It is difficult to identify the virus type through symptoms however influenza can be distinguished by its sudden onset, fever, and cough.

The major entry point for the virus is normally the nose, but can also be the eyes. Virus introduced into the eye is promptly transmitted to the nasal passage via the nasolacrimal duct. From there, it is transported to the back of the nose and the adenoid area. The virus then attaches to a receptor, ICAM-1, located on the surface of cells of the lining of the nasopharynx. Inter-Cellular Adhesion Molecule

1, ICAM1) also known as CD54 (Cluster of Differentiation 54) is a protein that in humans is encoded by the ICAM1 gene. ICAM-1 serves as the binding site for entry of the major group of human rhinovirus (HRV) into various cell types (Fig. 3.4).

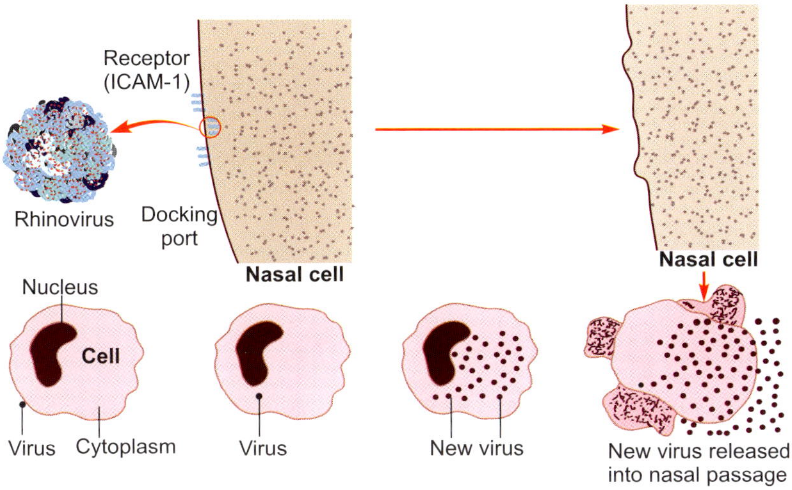

Fig. 3.4: Mode of attachment of Rhinovirus with the nasal cell

The receptor fits into a docking port on the surface of the virus. Large numbers of virus receptor are present on cells of the adenoid. After attachment to the receptor, the virus gets an entry into the cell, where it starts an infection[24]. Rhinovirus normally does not damage the nasal epithelium. After the Rhinoviruses settle on the respiratory epithelium cells, they replicate and infect adjacent cells. This leads to the stimulation of nervous reflexes resulting in an increase in nasal secretions, sneezing, and swelling of the mucosa and nasal erectile tissue. This swelling causes the airway to become partially or completely obstructed. The production of cytokines gets triggered by the macrophages. This further causes the systemic effects. Among other factors that contribute in the manifestation of local symptoms are the mediators known as bradykinin. This leads to the development of local symptoms such as sore throat and nasal irritation[25].

The common cold is self-limiting, and the host's immune system effectively deals with the infection. Within a few days, as a result of the body's humoral immune response, specific antibodies are generated that can prevent the virus from infecting cells. Additionally, as part of the cell-mediated immune response, leukocytes destroy the virus through phagocytosis. Destruction of the infected cells prevents further replication of the virus. When an individual is healthy and immunocompetent, the common cold may resolve in seven days on average.

Epidemiology

Humans are the only source of cold viruses. The disease is contracted when airborne droplets containing virus particles are inhaled, or when virus containing secretions are rubbed into eyes or nose by contaminated hands. Available reports suggest that the duration of illness may extend more by three to four days in a person with a history of smoking[26]. Getting fewer than seven hours of sleep per night has been associated with a risk three times higher of developing an infection when exposed to a rhinovirus, compared to those who sleep more than eight hours per night[27]. The population staying outside the tropical zone usually suffers from common cold during winter months. This is believed to be partly due to behavioral change such as increased time spent indoors, which puts infected people in close proximity to other people, rather than to exposure to cold temperature[28]. Low humidity again increases viral transmission rates. It has been suggested that dry air causes evaporation of water, thus allowing small viral droplets to disperse farther and thus stay in the air for a longer duration.

Prevention and Treatment

Avoiding contacts with the infected persons is the best prevention since the majority of infections are acquired by inhaling virus-laden air that an infected person has coughed, sneezed, or breathed out. Clinical evidences also suggest that exposure to cold temperatures may compromise the immune system. Exposure to cold temperatures should be avoided thus be avoided especially by vulnerable populations during winter months[29]. First-generation oral antihistamines, such as clemastine fumarate, chlorpheniramine, and brompheniramine have been shown to decrease sneezing during experimental rhinovirus colds and oral ibuprofen reduced the number of sneezes during natural viral URIs[30]. Efforts to develop a vaccine against the common cold have been unsuccessful. Common colds are produced by a large variety of rapidly mutating viruses therefore creation of effective vaccine against rhinovirus has not been successful so far. There are no specific approved antiviral drugs for the common cold.

3.3 Lower Respiratory Tract Infections

3.3.1 Normal Lower Respiratory Tract Flora and Mode of Colonization

Compared to the upper respiratory tract, the lower respiratory tract is well protected from colonization by microorganisms. The lower respiratory tract consists of:

- trachea or wind pipe
- two bronchial tubes one to each lung
- bronchioles
- lungs

The lungs are surrounded by two membranes: one adheres to the lung and the other to the walls of the chest and diaphragm. These membranes are named as "pleura". For the normal function of the lungs, these membranes slide against each other as the lung expands and contracts. The space between the two membranes is quite vulnerable for infection. As the infection grows it may compress the lungs. The lower respiratory tract is lined with ciliated cells and a film of mucus known as 'mucociliary escalator'. This lining helps to trap and remove the microorganisms or any other foreign bodies. Extraneous factors like tobacco smoke and chilling may hamper the ciliary action of the lower tract. Prolonged exposure to such irritants such as tobacco smoke leads to loss of cilia and the person becomes more vulnerable for infection.

The cough reflex has both sensory known as afferent and motor termed as efferent components. These cough receptors also known as pulmonary irritant receptors are present in the epithelium of the respiratory tract. These are sensitive to both mechanical and chemical stimuli. Stimulation of the cough receptors by dust or other foreign particles produces a sensation that leads to cough. This is necessary to remove the foreign material from the respiratory tract before it reaches the lungs. However the cough reflex can get subsided by external factors like alcohol and narcotics. The lung tissues contain large number of phagocytic macrophages which can migrate readily into the alveoli and airways as and when required to engulf and destroy the infectious agents. These protective mechanisms are very efficient against bacteria and therefore bacterial lung infections are rare except in immune compromised persons.

3.3.2 Bacterial infections-Pathogenesis, Epidemiology, Treatment and Prevention

There are a number of acute and chronic infections that can affect the lower respiratory tract.

3.3.2.1 Bacterial Pneumonia

The disease sets in very quickly as soon as a person gets the infection. Typical symptoms of pneumonia involve shaking chills with high fever. It usually leads to the production of yellow or brown sputum when coughing. A patient normally developes chest pain as the infection progresses. The pain aggravates with coughing and the patient also feels shortness of breath. The condition worsens in case of any chronic lung conditions such as asthma or emphysema. Older people may suffer due to a change in their mental abilities as a sign of pneumonia. Infectious pneumonia results due to inhalation of pathogenic microorganisms through respiration that subsequently escape the defense mechanisms of the upper respiratory tract and finally gets lodged in the lower respiratory tract

A. Streptococcus Pneumonia

Around 60 percent of bacterial pneumonia is caused by *S. pneumoniae*. This infection may often require hospitalization of the patients. These bacteria are gram positive diplococcus and possess a thick capsule. This capsule layer is implicated for the virulence of this organism. There are over 80 different serotypes of *S. pneumonia* which differ in capsular antigens.

Pathogenesis

This organism remains the most common cause of community-acquired pneumonia worldwide. *Streptococcus pneumoniae* is well adapted organism in human and therefore happens to be frequent inhabitant of the upper airways in healthy hosts. The process of colonization by the organism results in a competition where the pathogen attempts to engage the host, proliferate, and invade the lower airways (Fig. 3.5). The host in turn continuously deploys an array of innate and acquired cellular and humoral defenses to prevent pneumococci from invading the tissue barriers. Genetic variations have evolved in *S. pneumoniae* due to differences in the structure of the capsular polysachharides and thereby compete with each other to infect humans. The genetic variation prevents individuals from developing an effective immunological memory against all strains. Genetic variation in *S. pneumoniae* has evolved as a result of selection by the immune response of its human hosts. Therefore designing appropriate vaccine has been a challenge.A multifaceted approach with many vaccine antigens, antibiotic combinations, and immunoadjuvant therapies are needed to control this pathogen.

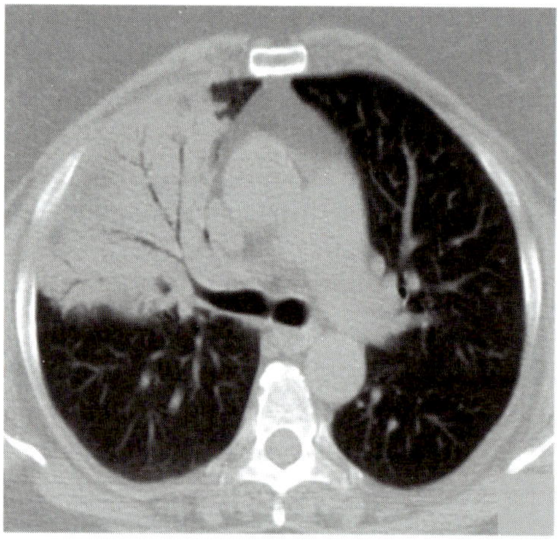

Fig. 3.5: *Pneumococcal pneumonia* produces consolidation in the right upper lobe with multiple air bronchograms (black branching structures) present since the spaces surrounding the air-filled bronchi normally contain air but now are filled with inflammatory exudate. There is no cavitation, the disease is in the lower lobe and it contains air bronchograms, all characteristics of pneumonia caused by *Streptococcus pneumoniae*.

Historical Background

Streptococcus pneumoniae has been one of the most extensively studied microorganisms since its first isolation in 1881. It was the object of many investigations that led to important scientific discoveries[31]. In 1928, Griffith observed that when heat-killed encapsulated pneumococci and live strains constitutively lacking any capsule were concomitantly injected into mice, the nonencapsulated could be converted into encapsulated pneumococci with the same capsular type as the heat killed strain. Years later, the nature of this "transforming principle," or carrier of genetic information, was shown to be DNA[32]. Other important discoveries resulting from investigations on pneumococci were the therapeutic efficacy of penicillin, the role of the bacterial capsule in resistance to phagocytosis, the ability of polysaccharides (PS) to induce antibodies (Ab), the first demonstration of antigen-specific tolerance or immunological unresponsiveness, the discovery of regulatory thymus-derived T lymphocytes, and the putative use of PS antigens as vaccines[33,34].

S. pneumoniae is carried in the upper respiratory tract by many healthy individuals. It has been suggested that attachment of pneumococci is mediated by a disaccharide receptor on fibronectin, present on human pharyngeal epithelial cells. Adherence of pneumococci to tracheal epithelial cells can be enhanced by prior influenza virus infection.

Epidemiology and Serotype Distribution

Globally Invasive Pneumococcal Disease is responsible for more than one million child deaths every year according to the WHO. The nasopharynx region is the natural reservoir of *S pneumonia*. The disease gets transmitted from one individual to other through respiratory droplets. Children are more prone to get colonized with one or more strains of *S pneumoniae* and become a carrier during early years of life. In most cases carriage remain asymptomatic. When the disease becomes symptomatic in a minority of persons, the bacteria normally spread locally from the rhinopharynx into the sinuses and middle ear cavity or to the lungs and results in systemic infections including bacteremia and meningitis. Infection of the blood stream and subsequent infection of secondary sites is referred to as invasive pneumococcal disease. The incidence of pneumococcal disease is highest in infants under 2 years of age and in people over 60 years of age[35]. After *Haemophilus influenzae* type b, Pneumococci are the second most frequent cause of bacterial meningitis and otitis media in children. The recent introduction of conjugate vaccines for *H. influenzae* type b, pneumococcal meningitis is gaining importance to prevent meningitis. *S. pneumonia* is the most important etiological agent of community acquired pneumonia in adults and is the second most common cause of bacterial meningitis behind *Neisseria meningitidis*. Invasive Pneumococcal Disease (IPD) including meningitis, pneumonia and septicemia contributes significantly to morbidity and mortality in all age groups worldwide. In spite of the availability of antibiotics, the mortality of pneumococcal disease remains high. The distribution of serotypes causing IPD varies with age and geographic region. However, the mortality of pneumococcal bacteremia in the last four decades has remained stable between 25 and 29%[36]. The rates of pneumococcal infections are reported to be highest among young children and elderly. On the basis of differences in capsular PS structure, pneumococci can be divided into more than 80 serotypes. Out of this, the most important pediatric serotypes (6A, 14, 19F, and 23F) were responsible for almost 60% of all infections. In adults, however, serotypes 3, 19F, and 6A accounted for only 31% of the isolations[37].

Indian Scenario

Study on child health epidemiology in India has not been carried out systematically. Other key issues include the importance of malnutrition as an underlying cause of child deaths associated with infectious diseases. The effects of multiple concurrent illnesses, and recognition that pneumonia and diarrhoea remain the diseases that are most often associated with child deaths[38] (Fig.6). In India 25% of all child deaths are due to pneumonia. In a case study in Southern India, 50% of infants were reported to be colonized by *S pneumoniae* by 2 months of age and 80% were carriers by the age of 6 months[39]. In India, the Invasive Bacterial Infection Surveillance (IBIS) network and South Asian Pneumococcal Alliance (SAPNA) have been involved in collection of important data regarding serotype distribution and antimicrobial resistance of pneumococcal infections for more than 12 years. The IBIS study was planned to generate evidence that would be useful to policy-makers for decisions regarding use of pneumococcal vaccines in India.

The IBIS network included 6 tertiary care hospitals in Vellore, New Delhi, Mumbai, Lucknow, Trivandrum and Chennai. Children and adults presenting to the hospitals with clinical symptoms of pneumonia, meningitis or septicemia were enrolled in the study. The participants were recruited from both in-patient and outpatient settings. Blood, CSF or other fluids from normally sterile sites were collected for identification of *S. pneumoniae* by cultural methods. Serotypes and anti microbial resistance pattern of the isolates were determined at the reference laboratory at Christian Medical College, Vellore.

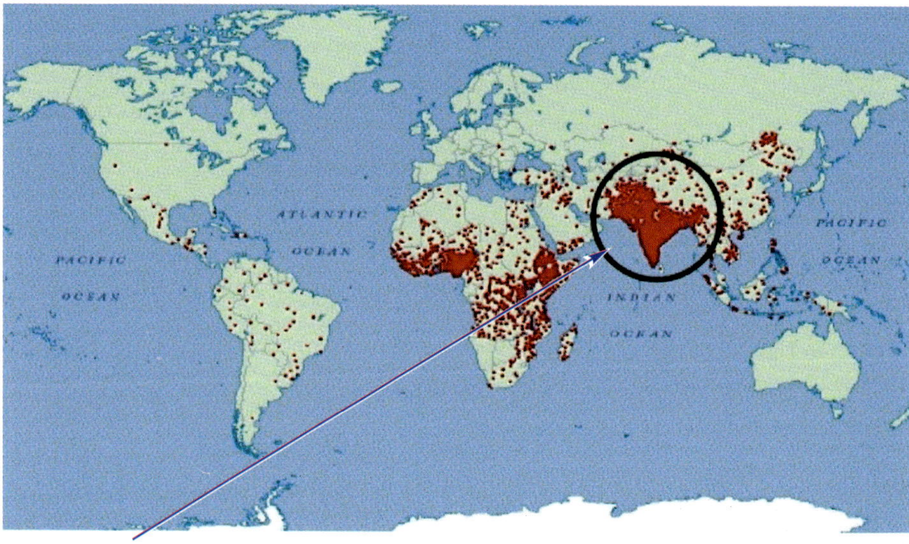

Fig. 3.6: Global mortality rate in different countries due to Pneumococcal infection[38]

Treatment and Prevention

A 23-valent pneumococcal polysaccharide vaccine is available for use in adults. This vaccine contains pneumococcal serotypes 1, 2, 3, 4, 5, 6B, 7F, 8, 9N, 9V, 10A, 11A, 12F, 14, 15B, 17F, 18C, 19F, 19A, 20, 22F, 23F and 33F. More than 80% of the healthy adults develop antibodies against these serotypes within 2 to 3 weeks after vaccination. The overall effectiveness of the vaccine is 60–70% against adult invasive disease. The adult polysaccharide vaccine is not immunogenic in infants.

Prevenar, a vaccine for Invasive Pneumococcal Disease (IPD), was recently launched for the first time in India. Prevenar is the first and only pneumococcal conjugate vaccine that protects a child against pneumococcal diseases like meningitis, bacterial pnuemonia, septicaemia and bacteraemia. Prevenar vaccine was officially launched by Wyeth Limited, one of the world's largest research-driven healthcare products companies, in Mumbai. Prevenar is a seven-valent vaccine. It is administered in three doses to children below the age of two years; 6-10-14 weeks after birth and a booster dose at 12-15 months.

B. Pneumonia Caused by Klebsiella Species

A number of species of Gram-negative rods can cause pneumonia, especially if hist defenses are impaired. These pneumonias are less common and more serious than pneumococcal pneumonia. *Klebsiella pneumonia* is an important example.

Pathogenesis and Epidemiology

The causal organism *Klebsiella pneumonia* is an encapsulated, non-motile enterobacterium. *K. pneumoniae* has long been recognized as a possible cause of community-acquired pneumonia. The pathogenesis is very similar to pneumococcal pneumonia, except that klebsiellae commonly cause permanent damage to the lung and leads to more complications leading to a mortality rate of 50%. It produces a highly lethal pneumonia in those people who have been hospitalized, typically after two days of hospitalization and is therefore commonly known as hospital-acquired pneumonia (HAP) or nosocomial pneumonia. The most common symptom of this type of pneumonia is cough with sputum secretion. The patient may also suffer from chills, chest pain, high fever, shortness of breath and flu-like symptoms. Diagnosis of *Klebsiella pneumonia* is done on the basis of patient's symptoms, physical examination with further investigations like chest X-ray, blood test and sputum cultures.

The virulence of klebsiellae is due partly to the antiphagocytic property of their capsules and endotoxin probably plays a role in the tissue destruction. Host factors significantly affect the outcome once a *K. pneumoniae* infection has developed, as do bacterial factors, such as presence of receptors, degree of piliation, and capsular components such as endotoxin. The capacity to adhere to a mucosal surface may be a determining factor in the ability of a microorganism to overcome the defenses of the host and cause infection. The presence of pili or fimbriae that can attach to the host cell has been described in *K. pneumoniae*. The presence of a capsule is known to be essential for the virulence of *K. pneumonia*[40]. At least two pathogenic mechanisms appear to be involved: 1) protection of the bacteria from phagocytosis and 2) direct modification of the immune response. In *K. pneumoniae,* the large polysaccharide capsule and slime that surround the cell envelope protect the cell from phagocytosis and are, therefore, largely responsible for its survival and capacity to cause disease. A characteristic feature of *K. pneumoniae,* is that they produce endotoxin. Endotoxin is a lipopolysaccharide (LPS)-protein-phospholipid complex which has numerous and diverse biologic activities. The role of endotoxin in virulence is unclear. Recent studies have shown that strains of *K. pneumoniae* differ in the quantity of endotoxin produced and that these differences correlate with the amount of low molecular weight lipid A produced. Differences in endotoxin production or in lipid A or LPS structure may be associated with differences in virulence because even minor differences in lipid A structure can profoundly alter endotoxin activity. Endotoxin can alter the host's cellular immune system and produce pathophysiologic changes as well as interfere with the immune response to other antigens, such as bacterial challenge[41]. Marked inter- species differences in endotoxin production correlate with virulence. Infections by K1 and K2 capsular serotype, the mucoid phenotype (Fig. 3.7), and aerobactin production were important determinants of virulence. The mucoid phenotype was seen in 94% of isolates in patients with community-acquired pneumonia and in 100% of isolates that caused the invasive syndrome in Taiwan and South Africa, compared with only 2% of isolates elsewhere[42].

Fig. 3.7: Mucoid phenotype of *Klebsiella pneumoniae*. When colonies were touched with a loop and the loop lifted vertically from the surface of the agar plate, mucoid isolates adhered to the loop as it was lifted from the plate.[42]

Epidemiology

In the past decade, geographic differences have been recognized in the spectrum of disease caused by *Klebsiella pneumoniae*. These differences include a preponderance of severe invasive disease in Taiwan and other parts of Asia[43]. *K. pneumoniae* can cause infections most common in the very young and very old age groups and those whose immune system is compromised either through disease (alcoholism, diabetes) or therapy. In studies from the 1920s to the 1960s, *K. pneumoniae* was considered an important cause of community-acquired pneumonia; however, in the last decade *K. pneumoniae* accounted for <1% of cases of pneumonia requiring hospitalization in North America. A striking clinical finding concerning a new manifestation of community acquired *K. pneumoniae* infection has been documented.

An unusual invasive presentation of *K. pneumoniae* infection in the form of primary bacteremic liver abscess has been described by numerous investigators in Asia; > 900 patients with Klebsiella liver abscess have been reported from Taiwan in the last 10 years[44]. In addition, case reports and small series from Korea, Singapore, Japan, India, and Thailand have been published. The Taiwanese patients with *K. pneumoniae* liver abscess have no history of hepatobiliary disease. Seventy percent of such patients have diabetes mellitus; 11% to 12% of the reported patients with Klebsiella liver abscess have other septic metastatic lesions, including pulmonary emboli or abscess, brain abscess, pyogenic meningitis, endophthalmitis, prostatic abscess, osteomyelitis, septic arthritis, or psoas abscess.

Treatment and Prevention

Klebsiella pneumonia is difficult to treat and repetitive occurrence of the disease is seen, which the patient continues to suffer for a longer duration. Treatment for *Klebsiella pneumonia* involves using antibiotics such as aminoglycosides, bactericidals such as cephalosporins, amikacin and Tobramycin. Klebsiella infections often respond slowly to therapy. Treatment depends upon the patient's health condition, medical history and severity of the infection. It is always recommended to take precautions in order to prevent from such infections.

C. Mycoplasma Pneumonia

The disease is commonly mild and difficult to diagnose.This species lacks a peptidoglycan cell wall and instead, it has a cell membrane that incorporates sterol compounds, similar to eukaryotic cells. It obtains these sterols from the host serum, allowing it to retain a simple structure. The outer most envelop is three layered plasma membrane called trilaminar unit membrane. Since these organisms lack typical cell wall, therefore these organisms are resistant to penicillins and other beta-lactam antibiotics. These antibiotics act by disrupting the bacterial cell wall.

Pathogenesis

Transmission of *Mycoplasma pneumoniae* occurs through respiratory droplets. Once attached to the mucosa of a host organism, *M. pneumoniae* extracts nutrients from the host for its growth and multiplication. Attachment sites include the upper and lower respiratory tract, causing pharyngitis, bronchitis and pneumonia. The infection caused by this bacterium is designated as atypical pneumonia. There is a lack of sputum production and often show extra-pulmonary symptoms. Chronic mycoplasma infections have been implicated in the pathogenesis of rheumatoid arthritis and other rheumatological diseases. The organisms attach to specific receptors on the respiratory epithelium, interfere with ciliary action and damage the ciliated cells. An inflammatory response characterized by infiltration of lymphocytes and macrophages causes the walls of the bronchial tubes and alveoli to thicken.

Epidemiology

It is the leading cause of pneumonia in college students and military recruits. These infections mimic viral respiratory syndromes. Patients are normally treated at OPD. Frequency of epidemic is usually seen in an interval of 4-7 years. The incidence rate is highest among school children and second highest among children below 5 years of age. Persons who have already suffered from *M. pneumoniae* pneumonia once, rates of subsequent infection with this organism are low. The body also develops immunity with increase in age. Once an individual is contracted with the disease it may last for several months. Patients with humoral immunodeficiency often develop severe infections due to *M. pneumoniae,* with involvement of the joints. The disease persists in these individuals for a longer period of time. *M. pneumoniae* has been isolated from bronchial washings from children with AIDS. These children can recover from mycoplasmal infection with appropriate antibiotic treatment[45]. *M. pneumoniae* infection spread slowly but effectively between neighbors through children playmates. Transmission appears to occur primarily during acute disease.

Treatment and Prevention

No practical preventive measures exist for mycoplasma pneumonia. In the treatment of mycoplasmal pneumonia, antimicrobials against *M pneumoniae* are bacteriostatic, not bactericidal. Tetracycline and erythromycin compounds are very effective, and the second-generation tetracyclines (doxycycline) and macrolides are the drugs of choice. Penicillins and cephalosporins are ineffective because the organism lacks a cell wall.

3.3.2.2 Pertussis Caused by Bordetella Pertussis

Whooping cough still remains as one of the ten most common causes of death from infectious disease worldwide. Despite a high vaccine uptake, rise in frequency of this disease have been observed in several countries.

Pathogenesis

B. pertussis is a pathogen that causes a respiratory disease and restricted to only in human. The organism fails to survive outside its host. In order to maintain its survival within the host, it has efficiently designed its transmission potential from one host to another. Pertussis is an acute, communicable infection of the respiratory tract caused by the gram-negative bacterium, *Bordetella pertussis*; another species of Bordetella, *B. parapertussis*, causes a mild pertussis-like illness. Descriptions of "whooping cough" dates back to 1500, although the first epidemiologic account of the disease was available only in1640[46]. The illness was named in 1670 by Thomas Sydenham; as "violent cough." Pertussis has generally been considered as an infection of children. The disease is associated with considerable morbidity and mortality. The infection rate is also increasing in adults. "Bordet- Gengou medium" (or "BG medium"), is now widely used for cultivating this organism (Fig 3.8)[47].

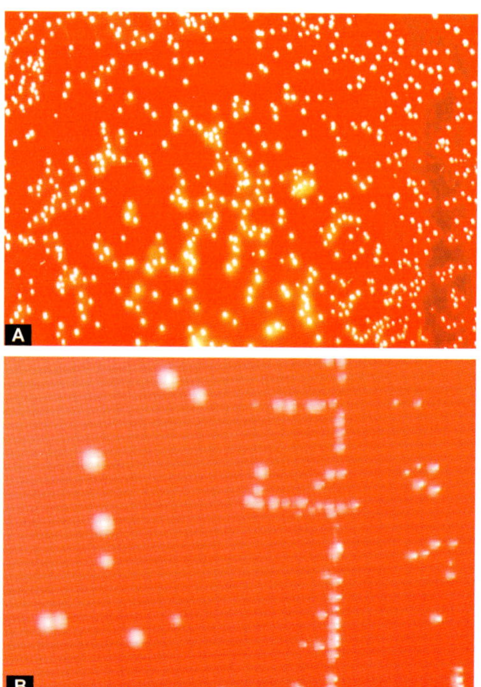

Fig. 3.8: *Bordetella pertussis* on Bordet-Gengou medium. **(A)** Freshly collected isolate, **(B)** Subcultured isolate. (Reproduced with permission from Right Links of Oxford journals)[47]

The manifestation of the disease pertussis occurs in two stages. The first stage involves the process of colonization. During this stage is the disease manifests in the upper respiratory tract. Common symptoms at this stage are restricted to fever, malaise and coughing, which increases in intensity over a period of ten days. During this stage the organism can be recovered in large numbers from pharyngeal cultures. The severity and the duration of the disease can be controlled at this stage by antimicrobial treatment.

Transmission occurs by direct contact, or via respiratory aerosol droplets, or fomites. The Bacteria initially adhere to ciliated epithelial cells in the nasopharyngeal region. This interaction with epithelial cells is mediated by a series of protein adhesins including filamentous haemaglutinin, pertactin and fimbriae. The expression of pertussis toxin which is unique to the organism helps in the adhesion process. The symptoms at the initial catarrhal phase of infection are very similar to those of the common cold. Due to rapid multiplication of the bacteria, large numbers of bacteria can be recovered from the pharynx. Then the bacteria proliferate to spread into the respiratory tract, where the secretion of toxins causes ciliostasis. This process further facilitates the entry of bacteria to tracheal/ bronchial ciliated cells. This toxin is a tracheal cytotoxin which is a derived from disaccharide-tetrapeptide of peptidoglycan layer of the bacteria. Unlike most other toxins, tracheal cytotoxin is expressed constitutively and this is a normal product of the breakdown of the bacterial cell wall. Other bacteria are responsible for recycling this molecule back into the cytoplasm. However, in case of *Bordetella* and *Neisseria gonorrhoeae* it is released into the environment. Tracheal cytotoxin is resnsible to paralyse the ciliated epithelial cells by the inhibition of DNA synthesis in epithelial cells and ultimately destroy them. One of the most important of the regulated toxins is adenylate cyclase toxin, which aid in the evasion of innate immunity also plays an important role in the pathogenesis process. The toxin is delivered to phagocytic immune cells upon contact[48]. Immune cell functions are then inhibited due to the accumulation of cyclic AMP. Recently discovered activities of adenylate cyclase toxin, including transmembrane pore formation and stimulation of calcium influx contribute to the intoxication of phagocytes.

Adherence mechanisms of *B. pertussis* involve a "filamentous hemagglutinin" (FHA), which is a fimbrial-like structure on the bacterial surface, and cell-bound pertussis toxin (PTx). Short range effects of soluble toxins play a role as well in invasion during the colonization stage. Various virulence factors associated with *Bordetella pertussis* include agglutinogens, fimbriae, P.69/pertactin, pertussis toxin, filamentous haemagglutinin, adenylate cyclase, tracheal cytotoxin, dermonecrotic toxin, lipopolysaccharide, tracheal colonisation factor, serum resistance factor, and effector molecules released by type III secretion system.

Epidemiology

Bordetella pertussis is transmitted by respiratory droplets and causes disease only in humans. Various diagnostic methods are available, including culture, serological methods, and the polymerase chain reaction. Serotyping of isolates to detect agglutinogens 2 and 3 is useful because serotype 1,2 are considered to be associated with higher mortality. The antibodies to these antigens (agglutinins) may be protective in both animals and humans. Immunisation using whole-cell vaccine is effective but it develops reactions in the body. Acellular vaccines containing one to five components are administered in various countries. Protective immunity to pertussis correlates with high levels of antibody to each of its component - pertactin, fimbriae and pertussis toxin. Although pertussis is relatively well controlled at present by extensive vaccination programs however evidences show that the circulation of *B. pertussis* throughout the world continues remain largely unabated.

Incomplete immunity in infants has been observed who have not received three doses of vaccine. Relatively, short-lived immunity results from vaccination. Although adolescent and adult pertussis is increasing significantly resulting in heavy medical costs and lost work, the most worrisome consequence is epidemiological[49]. Numerous studies have shown that adults and adolescents provide a reservoir of *B. pertussis* and are the major source of transmission to partially immunized infants and children

Transmission, Treatment and Prevention

Administration of the antibiotic Erythromycin at an early stage of the disease helps in ameliorating the symptoms and also helps in decreasing the rate of transmission of the disease to other individuals., A 14-day course of the antibiotic erythromycin is recommended in order to eradicate the bacteria from the upper respiratory tract. To eradicate detectable bacteria from the upper respiratory tract, a 14-day course of erythromycin is often recommended[50]. However, evidences show that the use of erythromycin has little effect in preventing secondary transmission of pertussis to close contacts of primary cases apart from prophylaxis in newborns[51].

Vaccine: Pertussis vaccine is administered along with diphtheria and tetanus toxoids (DPT) at 6 weeks and 4, 6 and 18 months of age, with a booster dose between 4 and 6 years. Despite more than 40 years of vaccination, Bordetella pertussis has remained an endemic disease, with epidemic outbreaks occurring every 3 to 5 years. In some vaccinated populations e.g. Australia, Canada, The Netherlands and US, the incidence of pertussis is found to be increasing. This is implicated due to strain variation or the polymorphism of the pathogen. Therefore, it is suggested that the inclusion of predominant protein variants in vaccines may improve their efficacy.

Pertussis outbreaks have occurred cyclically since the early 1980s, with large increases in incidence in 1983, 1986, 1990 and 1993 as documented by the Centers for Disease Control and Prevention[52].

In recent years, outbreaks due to pertussis have been reported from many parts of the world despite the widespread use of a vaccine[53]. Non-immunized or partially immunized children are at a greater risk of developing infections. Partially immunized individuals maintain the infectious agents and offer a risk for the community as a whole especially in developing countries. There is also a shift in the epidemiology of pertussis, with more cases being reported in older age groups, which is due to losing immunity to *Bordetella pertussis* a number of years after immunization.

A cellular vaccines, contain the antigens that stimulate immunity but lack other components of the bacterial cell wall. This was first approved for use in the United States in 1991. Better vaccines are now being proposed giving better immunity. Future vaccines will be safe enough to give as booster doses to adolescents and adults who are now considered as major reservoir of *B. pertussis* because of their waning immunity.

3.3.2.3 Tuberculosis

Tuberculosis (TB) is a contagious disease. Like the common cold, it spreads through the air. Only people who are sick with TB in their lungs are infectious. When an infected person cough, sneeze, talk or spit, they propel TB germs. These germs are also known as bacilli. Once released in air a person needs only to inhale a small number of these to be infected.

Pathogenesis

Tuberculosis is a chronic bacterial infection of the lungs and often characterized by slight fever, weight loss, sweating at night and chronic cough production with the release of blood streaked sputum. Infection of *Mycobacterium tuberculosis* begins by inhaling aerosolized organisms from an infected individual. TB is defined as a disease caused by members of the *M. tuberculosis* complex, which includes the tubercle bacillus (*M. tuberculosis*), *M. bovis, M. africanum, M. microti, M. canetti, M. caprae* and *M. pinnipedii*[54]. Aerosol transmission is the predominant route of infection, whereby a droplet of sputum may contain one to three bacilli (particle size, <5 pm) and soon gain access to alveoli through the airway. Subsequently, the bacteria are engulfed by alveolar macrophages, which presumably are equipped with multiple microbicidal mechanisms, including phagolysosome fusion and a respiratory burst to eliminate the infecting microorganisms from the host[55]. To establish infection successfully, the tubercle bacillus must survive in the alveolar macrophages by inhibiting phagosomal maturation and ultimately gain access to the lymphatics or the bloodstream. Once in the bloodstream, dissemination to the other parts of the body can be widespread. Delayed hypersensitivity to the tubercle bacilli develops leading to its sensitivity to tubercle-derived proteins. This has proved enormously useful in diagnosing tuberculosis infections. Use of purified protein derivatives (PPD) of tuberculin is the basis for estimating the latent tuberculosis infection in populations. This is an important tool to study the natural history of tuberculosis infection and frequently useful in evaluating patients with suspected tuberculosis disease.

Pathophysiology of tuberculosis includes following steps

1. Inhalation of bacilli
2. Containment in a granuloma
3. Breakdown of the granuloma in less immunocompetent individuals

Epidemiology

More than 2 billion people encompassing around one-third of the world population are estimated to be infected with tuberculosis[56]. The global incidence of TB peaked around 2003 and now appears to be declining slowly. In 2006 the World Health Organization (WHO) issued the following estimates[57]:

- The prevalence of active infection was 14.4 million, corresponding to a prevalence rate of 219/100,000 persons.
- The incidence of new cases was estimated to be 9.2 million, corresponding to an incidence rate of 139/100,000.
- Twelve of the 15 countries with the highest estimated TB incidence are in Africa, where the TB incidence rate was 363/100,000
- In 2006 there were 1.7 million deaths from TB worldwide thus showing a death rate of 25/100,000.

The epidemiology of tuberculosis varies substantially around the world. The highest rates (100/100,000 or higher) are observed in sub-Saharan Africa, India, China, and the islands of Southeast Asia and Micronesia. Intermediate rates of tuberculosis (26 to 100 cases/100,000) occur in Central and South America, Eastern Europe, and northern Africa. Low rates have been recorded which is less than 25 cases per 100,000 inhabitants in Unites States,Western Europe, Canada, Japan, and Australia. Current incidence rate reported by WHO in 2011 is shown in Fig. 3.9.

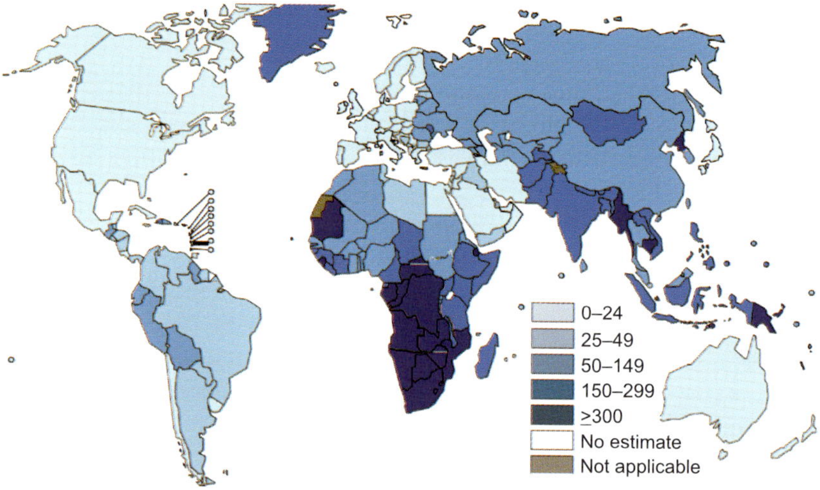

Estimated new TB cases (all forms) per 100,000 population

Fig. 3.9: Estimated TB incidence rates, 2011
(Adapted from WHO Global Tuberculosis Report, 2012)

Poverty, HIV and drug resistance are major contributors for the revival of global TB epidemic. Approximately 95 percent of TB cases occur in developing countries. Approximately 1 in 14 new TB cases occur in individuals who are infected with HIV; 85 percent of these cases occur in Africa. An estimated half million cases of multidrug resistant (MDR)-TB also occur annually in Africans; even higher rates of drug resistant infections occur in Eastern Europe.

Indian Scenario

India is classified along with the sub-Saharan African countries to be among those with a high burden of the disease. The average prevalence of all forms of tuberculosis in India is estimated to be 5.05 per thousand individuals. The prevalence of smear-positive cases reported as 2.27 per thousand and average annual incidence of smear-positive cases are estimated as 84 per 1,00,000 annually[58]. Epidemiologists forecast a rise of 20 per cent in incidence in the next 20 yr, for India. This will lead to a cumulative rise of 46 million cases of tuberculosis during that period. This could be as a consequence of HIV which is also considered as an epidemic for India.

Therefore, India's Revised National TB Control Program (RNTCP) is currently recording TB prevalence and death rates for the period 1990–2015. This has been specified by the United Nations Millennium Development Goals. Without the RNTCP, HIV would increase TB prevalence (by 1%), incidence (by 12%), and mortality rates (by 33%) between 1990 and 2015. With the RNTCP, we expect substantial reductions in prevalence (by 68%), incidence (by 41%), and mortality (by 39%) between 1990 and 2015[59]. In May 2012, India declared TB as one of the notifiable disease and data collection on TB using a web based system has been initiated: (http://pib.nic.in/newsite/erelease.aspx?relid=83486)

Treatment and Prevention

One of the strategies used to curb the disease is identifying unsuspected cases with a routine check up using DOT test on the skin and X-rays. Individuals with active disease are then treated, thus interrupting the spread of the disease. Drug therapy started in 1944 using streptomycin (SM) and para amino salicylic acid (PAS). Clinical trials carried out with individual and combination of drugs SM and PAS were carried out. The study showed that combination therapy was more effective. In 1952, a third drug, isoniazid (INH), was added to the mix, dramatically improving the effectiveness of treatment. However, the duration of the treatment was for 18-24 months. The ethambutol was introduced in 1960, replacing the PAS in the treatment schemes and reduced the duration to 18 months. With the introduction of rifampicin (RAM) in the combination, treatment was further shortened to 9 months. Finally, in 1980, pyrazinamide (PZA) is introduced into the therapeutic plan; the duration was further reduced to 6 months.

The antibiotics are classified into two groups according to their efficacy, potency and side effects:

- **First-line drugs:** isoniazid, rifampin, pyrazinamide, ethambutol or streptomycin
- **Second-line drugs:** Cycloserine, Ethionamide, ciprofloxacin. They are used in cases of resistant tuberculosis, or when the front-line drugs lead to side effects. Yet its use has poor prognosis due to Multi resistant TB.

Because of the long generation times of *M. tuberculosis* and its resistance to destruction by body defenses, drug treatment of tuberculosis must generally be continued for a minimum of six months to one or more years to cure the disease.

Development of resistance to antibiotics: The emergence of multidrug-resistant (MDR) Mycobacterium tuberculosis (MT) strains arise when the strains are resistant at least to isoniazid (INH) and rifampin (RMP). This is alarming and represents a worldwide health care problem. A dramatic increase in the resistance and spread of MDR strains has been observed in recent years[60]. Clinical isolates exhibit a diversity of resistance patterns. Isolates resistant to 3 or more anti-tuberculosis drugs became very common. These MDR strains carry multiple mutations in different resistance-related genes, and each mutation results from an independent mutational event. An MDR strain is thus the product of a multistep process, in which a progressive accumulation of genetic alterations occurs and results in the selection of a viable and fit bacterium.

During an infection process, the bacterial population results in a mixture of different subclones. Mixed populations have been detected confirming the dynamic evolution of subpopulations from a single founder strain. Markers are used to detect the wild type or mutated type for any marker; This leads to generation of different variants that become prevalent in diverse times of disease. The prevalence of each subpopulation could affect the outcome and the effectiveness of therapy[61]. Once the resistant strains start infecting others in the community and become increasingly responsible for new infections.

MDR TB status in India: Prevalence of multidrug-resistant TB (MDR TB) cases are on the rise in India, and proportions of new cases of MDR TB have been observed to vary from 1.1% to 5.3% in reported studies. The proportion of previously treated patients with MDR TB varied from 8% to 67%[62]. Studies conducted in different parts of India indicate an increasing trend of MDR TB cases. Extensively drug-resistant TB (XDR TB) is defined as TB caused by a *Mycobacterium tuberculosis* strain that is resistant to at least rifampin and isoniazid among the first-line anti-TB drugs (MDR TB) in addition to resistance to any fluoroquinolones and also show resistance to at least 1 of 3 injectable second-line drugs[63, 64]. Although India has high annual risk for TB cases and increasing prevalence of MDR TB cases, XDR TB has not yet been described in India however, a community-based, multicenter study, which includes all parts of the country and uses the full spectrum of drugs, is needed to describe the true prevalence of XDR TB in India[65].

Prevention: The vaccination agent used is a living attenuated mycobacterium known as Bacillen Calmette-Guerin (BCG) derived from *Mycobacterium bovis*. The vaccine is normally administered in a single dose, intradermally by a nurse or by any professional healthcare personnel. It is important to undergo a skin tuberculin test before administering the vaccine. Positive reaction to tuberculin is a contraindication to BCG. BCG uses shared antigens to stimulate the development of cross-immunity to *Mycobacterium tuberculosis* and *Mycobacterium leprae*. Worldwide efficacy rates of BCG vaccination have varied from 0-80% in studies of different patient population. Evidence of reproducible efficacy has been limited but recent meta-analyses have shown that the BCG vaccination[66]

- Is able to reduce the risk of tuberculosis (TB) at least by 50%
- It also minimizes severe, non-pulmonary forms like childhood tuberculous meningitis at least by 70%.

This protective effect of the vaccine may lasts for 10 years or so. The prolonged benefit of the vaccine is still doubtful. The efficacy of BCG vaccination when administered to adults is questionable. Therefore, older people remain at highest risk of mortality, main source of transmission and also remain unprotected[67]. However people may remain protected provided there is no history of residence in a high-risk area and contact with a tuberculosis patient. For all other patients, a tuberculin test is necessary before BCG vaccination

The BCG vaccine should not be given to patients:

- Having a past history of TB.
- Showing a positive response with pre-immunisation tuberculin test.
- A previous anaphylactic reaction to vaccine component.
- Compromised immunity due to treatment or disease.
- Any report of septic skin conditions.
- Acute illnesses with fever or systemic upset.
- HIV-positive individuals - the BCG vaccine is absolutely contra-indicated in symptomatic HIV-positive patients

New vaccines are being developed and are entering into clinical trials. The new candidates include live attenuated *M. tuberculosis,* recombinant BCG, DNA vaccines and fusion proteins with novel adjuvants. These are targeted to provide a long lasting immune response in heterogeneous populations[68]. A triple fusion protein was created by Dr. Claus Aagaard and Dr. Peter Andersen which was named as H56. This vaccine is a product of 2 proteins that the bacteria produces in the early stages of infection with another produced during later-stage of the infection. This vaccine is at present in clinical trials[69]. WHO proposes to reduce the global burden of TB by 2015 in line with the Millennium Development Goals (MDGs). The aims of this goal are as follows:

- access to high-quality health care for all people with TB
- reduce the socioeconomic burden associated with TB
- define measures to protect the vulnerable populations from TB, TB/HIV and drug-resistant TB
- development of new tools for prevention and enable their timely and effective use
- promote human rights in TB prevention, care and control

3.3.2.4 Legionellosis

Legionellosis is a potentially fatal infectious disease caused by Gram negative, aerobic bacteria belonging to the genus *Legionella*[70]. Over 90% of legionellosis cases are caused by *Legionella pneumophila*, a ubiquitous aquatic organism that thrives in temperatures between 25 and 45 °C, with an optimum around 35 °C.

Pathogenesis

The disease begins typically with symptoms very similar to a typical bacterial pneumonia. The symptoms are fever, chills, and a dry cough and may produce sputum. Some patients also develop muscle aches, headache, tiredness, loss of appetite and ataxia and occasionally may develop diarrhea and vomiting. Shortness of breath is a common phenomenon and may require oxygen therapy. Recovery is usually slow and weakness followed by fatigue last for couple of weeks. Since the symptoms of Legionnaires' resemble other types of pneumonia, therefore diagnosis is difficult to carry out only on the basis of symptoms or radiologic findings alone.

The *Legionella pneumophila* infection is normally acquired after inhaling an aerosol contaminated with Legionella bacteria. Origin of the infected particles is normally a contaminated water source. People of any age may suffer from Legionnaires' disease, but the illness most often affects middle-age and older persons. Individuals who smoke cigarettes and have a history of chronic lung disease are more susceptible to this infection. Immuno compromised patients are at elevated risk. The organisms lodge in and near the alveoli of the lung and are taken up by pulmonary macrophages. It invades and grows within alveolar macrophages, and indeed, intracellular legionellae display a remarkable capacity to avoid endosomal and lysosomal bactericidal activities and to establish in an unique replicative phagosome. Bacteria are not killed by the phagocytes, but instead multiply within them and are released upon cell death to infect other tissues. However, the infection remains confined to the lung in most cases.

Epidemiology

The organisms are widespread in natural waters. Potential sources of such contaminated water include cooling towers used in industrial cooling water systems as well as in large central air conditioning systems, coolers, hot water systems, showers etc. Various other water facilties that use public water supply are also potential source of contamination. According to an estimation, between 10,000 and 50,000 cases of Legionnaires' disease are reported from USA each year[71]. Hospital-acquired Legionella pneumonia has a fatality rate of 28%, and the principal source of infection in such cases is the drinking-water distribution system[72]. Interestingly, in natural waters, the organism is taken up and multiplies within amebas in a manner similar to its growth in macrophages. Some of the protozoa such as *Hartmannella vermiformis* and related protozoa have been shown to be able to support the growth of *L. pneumophila* in tap water of public distribution system. Besides that even other protozoans like *Acanthamoeba, Naegleria* and Tetrahymena can be infected by *L. pneumophila*. The organism is considerably more resistant to chlorine than the usual Gram-negative bacterial pathogens and thus survives well in hot water systems of buildings. Between 1995 and 2005 over 32,000 cases of Legionnaires' disease and more than 600 outbreaks were reported to the European Working Group for Legionella Infections (EWGLI)[73]. The data from developing countries are scanty on Legionella-related illness. It is therefore presumed to be under diagnosed worldwide. Direct person-to-person spread does not occur in this case.

Treatment and Prevention

Members of quinolones like levofloxacin, moxifloxacin, gemifloxacin or macrolides like azithromycin, clarithromycin, roxithromycin are generally prescribed as first line of drugs. The antibiotics used most frequently have been levofloxacin and azithromycin. Different drugs are administered according to the age groups like macrolides are used in all age groups while tetracyclines are prescribed only to patients above 12 years of age whereas quinolones are normally given to the individuals in the age group of 18 and above. Rifampicin is given in combination with a quinolone or macrolide. The drugs like tetracyclines and erythromycin gave better results compared to other antibiotics during outbreak. These drugs possess an inherent property of excellent intracellular penetration in the Legionella infected cells. The mortality rate was dependent on the time of administration of the antibiotic therapy including penicillins, cephalosporins, and aminoglycosides. Mortality rate can be reduced considerably in case the drug is administered at an early stage of diagnosis.

Regular monitoring of possible source of infection and timely intervention by the public health officials by recommending decontamination of the aerosol-producing devices, maintenance of cooling towers, plumbing systems, swimming pools and other public water supply sources can prevent any outbreak and spread of the disease.

3.3.3 Viral Infections of the Lower Respiratory Tract

Viral infections represent more than 90% of acute diseases in children and are considered the most important cause of lower respiratory tract illness commonly observed during winter time[74]. Despite significant advances in diagnostics, only around 35 to 87% of viral infections can be diagnosed. Therefore high rate of viral pathogens causing various viral infections remain undiagnosed[75].

A large variety of viruses are responsible for causing respiratory tract disease in children. They are predominantly categorized as respiratory syncytial virus, influenza viruses, parainfluenza viruses, rhinovirus, adenovirus, and the recently identified human metapneumovirus. Few other members included as pathogens are the members of paramyxovirus and orthomyxovirus. The former group includes measles (rubeola), parainfluenza and respiratory syncytial viruses (RSV) whereas the latter group includes influenza A and B.

A. Respiratory Syncytial Virus-(RSV)

RSV was first isolated from chimpanzees with colds, and it was soon found to cause colds in man as well. It is found to be associated with severe pulmonary infections in infants - especially Bronchiolitis. RSV is the prime cause of Bronchiolitis. The virus causes a localized infection of the respiratory tract and infants do not possess any maternal or passive protection.

B. Influenza

It is commonly known as the flu. This respiratory infection is caused by one of the influenza viruses that typically spread by air or by direct contact. It may surface as an epidemic and highest rate of incidence is seen during onset or outset of the winter months. A particularly widespread and severe epidemic is called a pandemic. One of the viruses such as Influenza A is responsible for the deadly influenza pandemics.

Influenza is spread primarily by aerosols, but occasionally through any contaminated inanimate object capable of transmitting the diseases. The primary infection involves the ciliated epithelial cells of the upper respiratory tract. The infection turns more acute as these cells begin necrosis that results to symptoms including fever, chills, muscular aching, and headache. The incubation period lasts from one to four days with an infection period lasting three to seven days. Mortality from a primary influenza infection is extremely rare. Flu deaths whenever occur are normally in the older age group above sixty or persons who suffer from immunodeficiency. Most deaths that involve the flu are attributed to secondary bacterial infections that appear after serious damage of the upper respiratory tract.

Human Metapneumovirus: Human metapneumovirus (hMPV) has recently been described as a causal agent of acute respiratory disease predominantly in children. This virus was first discovered in 2001 in The Netherlands in patients with respiratory infections. It was then classified in the *Pneumovirinae* subfamily of the *Paramyxoviridae* family[76]. hMPV has been found in many countries, including Australia, Canada, Finland, the United States, France, the United Kingdom , Spain, Ireland, and Japan

The clinical manifestations of hMPV may range from mild upper-airway disease to severe pneumonia. Other manifestations may include rhinorrhea, nasal congestion, pharyngeal erythema, myalgia, cough, and fever and, in more severe cases, wheezing, dysphonia, respiratory difficulty, bronchiolitis, pneumonia, and in rare case may turn into respiratory failure[77].

Pathogenesis and Symptoms

Annually, influenza outbreaks affect millions of people worldwide. The flu virus is in the family Myxovirus. Flu particles are highly pleomorphic showing varied shape of capsid leading into various subtypes. The most common shape is icosahedral, but helical viruses are also present. The size of icosahedral capsid ranges from 80 to 120 nm in diameter whereas the helical virus may retain the same diameter, but can extend up to 2000 nm in length. All influenza viruses have lipid envelope that contain two types of spikes, haemagglutinin (HA) and neuraminidase (NA). The genome of influenza is found to be segmented and possess a negative-sense single stranded RNA. There are three subtypes of the influenza virus A, B, and C. Subtype A is known to infect horses, pigs, birds and humans. The primary reservoir is considered to be pigs and birds. Flu A is the most prevalent subtype in the human population. The influenza B subtype may also infect humans and birds but the infection is considered to be less severe than flu A subtype. Finally, flu C subtype is is found in humans alone and not considered as a disease causing organism.

Influenza is categorized as a highly infectious respiratory disease. The disease is caused by certain strains of the influenza virus. Influenza victims are also susceptible to potentially life-threatening secondary infections.

Epidemiology

The disease may strike any time between the months of October and May. It normally lasts from three to five days however following the infection the patient normally develops fatigue for two or three weeks. The disease may also lead to complications like pneumonia, nerve and brain damage and even death however such complications are rare. The viral infection normally attacks respiratory system including nose, throat, bronchial tubes and lungs.

Influenza causes frequent epidemics and periodic pandemics, and is a major public health problem. Pandemics of influenza experienced in 1918 was swine influenza, Asian influenza in 1957, Hong Kong influenza in 1968 and Russian influenza in1977 were due to major antigenic variation of the type A influenza virus. Frequent epidemics may follow after pandemics as a result of minor antigenic variation of the pandemic virus strains. Minor antigenic variant strains of type A are H1N1, H3N2 and type B influenza viruses are currently circulating globally that may be responsible in causing frequent epidemics. The World Health Organization (WHO) has established a network of National Influenza Centres (Flunet) all over the world to study the epidemiology and ecology of influenza. Flunet has also set up collaborating centres for updating the influenza vaccines and other research activities. National Influenza Centres in India have been investigating and reports include pandemics and epidemics caused by global influenza virus strains during the past 50 years. There is a need to expand influenza surveillance system in India.

Treatment and Prevention

Antiviral drugs have been developed directly targeting the viruses responsible for influenza infections. Anti-viral drugs work optimally when taken within a few days of the onset of symptoms. Certain drugs are also used in uninfected individuals to guard against infection. Four licensed influenza antiviral agents are available in the United States: amantadine, rimantadine, zanamivir, and oseltamivir[78]. These drugs are available through prescription only. These drugs fall into two categories: 1. M2-inhibitors (admantane derivatives) and 2. neuraminidase inhibitors as illustrated in the following Table 3.2

Table 3.2: Antiviral drugs to treat influenza

Class	Effective Against	Drug Name (INN)	Brand Name	Year Approved	Manufacturer
M2 inhibitors (adamantane derivatives)	Influenza A	Amantadine	Symmetrel	1976	Endo Pharmaceuticals
		Rimantadine	Flumadine	1994	Forest Laboratories
Neuraminidase inhibitors	Influenza A and B	Zanamivir	Relenza	1999	Glaxo SmithKline
		Oseltamivir	Tamiflu	1999	Hoffmann-La Roche

Note: Neuraminidase inhibitors are approved for prophylaxis use in children and adults

3.3.4 Fungal Infections of the Lower Respiratory Tract

Coccidioidomycosis and Histoplasmosis are two important fungal diseases that are also referred as mycoses. Coccidioidomycosis caused by *Coccidioides immitis* is a fungus unique to certain arid regions of the western hemisphere. Histoplasmosis is caused by *Histoplasma capsulatum*. Coccidiomycosis is endemic in certain parts of Arizona, California, Nevada, New Mexico, Texas, Utah and northwestern Mexico.

3.3.4.1 Coccidioidomycosis (Valley Fever)

Pathogenesis

Coccidioides immitis is a dimorphic fungus normally inhabit soil and when conditions are favourable for growth, they appear as a mycelium in the soil and reproduces as arthroconidia. The mature arthroconidia gets transmitted through air to a host where it developes a thick-walled structure known as spherule form. These spherules are filled with endospores in the host organism (Fig. 3.10). The infected person then develops symptoms like flu. Common symptoms are fever, cough, chest pain and loss of appetite and weight. In some occasion, a particularly susceptible individual may develop a serious or even fatal illness from valley fever. Serious complications include severe pneumonia, lung nodules, and the disease may get disseminated to other organs of the body. The disseminated form of valley fever can cause skin ulcers, abscesses, bone lesions. The patients may also suffer from severe joint pain, heart inflammation, urinary tract problems, meningitis, and often leads to death. Some patients fail to recover and succumb to the secondary manifestation of the disease like chronic pulmonary infection or may develop widespread disseminated infection affecting meninges, soft tissues, joints, and bone. Severe pulmonary disease may develop in HIV-infected persons[79]. The disease shows close resemblance to tuberculosis. Due to the destruction of lung tissue the patient develops cavities in the lungs. The fungal infection can be demonstrated by microscopic detection of diagnostic cells in body fluids, exudates, sputum and biopsy-tissue. The molecular detection procedure involves amplification of genomic DNA using specific nucleotide primers of *C.immitis*. It can also be detected in culture by morphological identification. They can also be detected using molecular probes that hybridize with *C.immitis* RNA. An indirect demonstration of fungal infection can be achieved also by serologic analysis detecting fungal antigen or host antibody produced against the fungus.

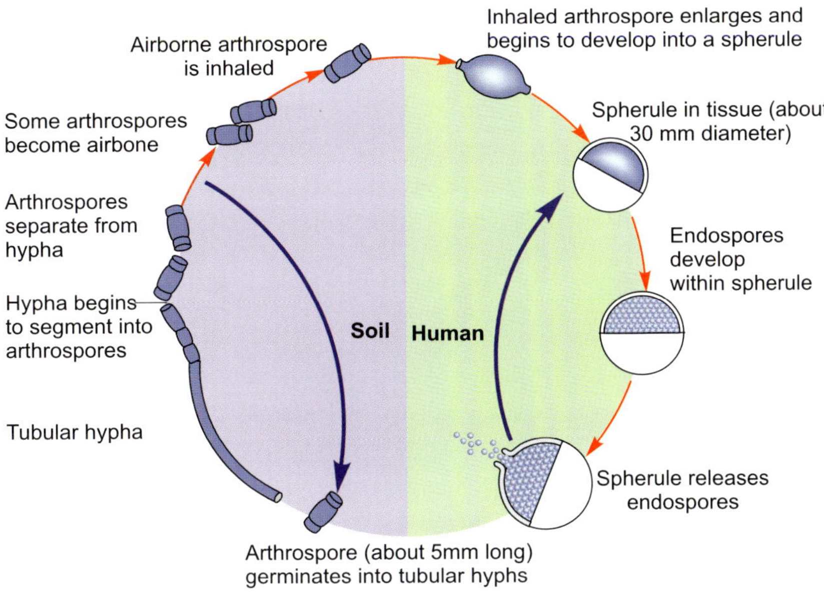

Fig. 3.10: The life cycle of *Coccidioides immitis*

Epidemiology

C. immitis/posadasii is a soil inhabiting fungus endemic in south-western U.S.A., northern Mexico and some areas in South America. More than 20,000 cases of valley fever are reported every year in the United States. Several cases have now been diagnosed in Australia. However it has been observed that the patients have a history of travel to endemic areas. People become infected with *C. immitis* by inhaling the airborne spores of the fungus, easily dispersed in dust from soil. The infections occur only during the hot, dusty seasons when its spores are airborne. It is dormant during long dry spells and then develops as a mold with long filaments that break off into airborne spores when the rains come. The spores, known as arthroconidia, are swept into the air by disruption of the soil[80] (Fig. 3.11B). Infection is caused by inhalation of the particles. The disease is not transmitted from person to person. People who are more prone to the disease include African Americans, women in third trimester of pregnancy, people with weak immune systems, including those with an organ transplant or who have HIV/AIDS. The endemic areas in USA are represented in Fig. 3.11A.

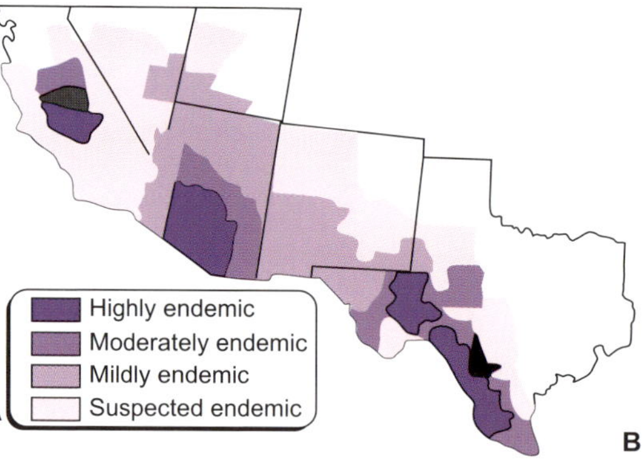

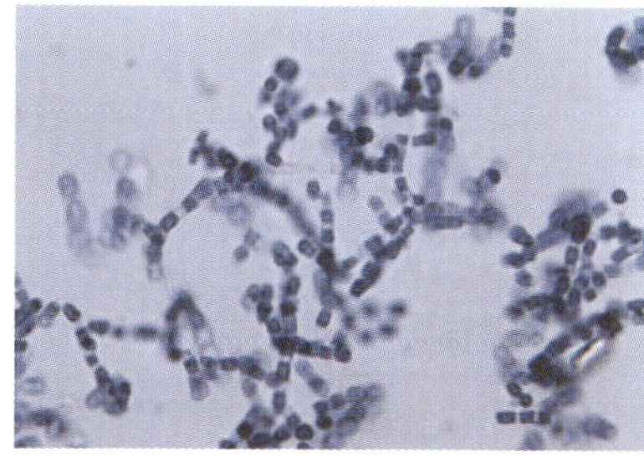

Fig. 3.11(A): Endemic areas for valley fever in USA (***Source:*** *CDC*), and (B) Arthroconidia of *Coccidioides immitis* showing septate hyphae *(**Source:** CDC)*

Treatment

There are no published reports on optimal antifungal therapy for coccidioidomycosis. Mild cases often do not require treatment. Oral Fluconazole and intravenous Amphotericin B are used in progressive or disseminated disease, or in patients who are immunocompromised. Other drugs that are also in use are itraconazole or ketoconazole. Posaconazole and voriconazole have also been used successfully[81].

3.3.4.2 Histoplasmosis

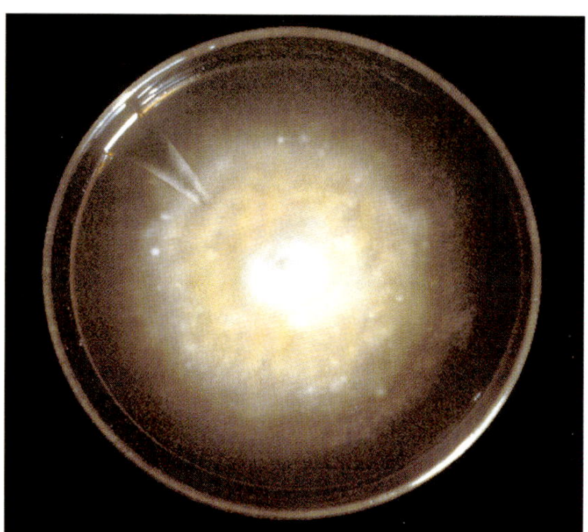

Fig. 3.12: Sabouraud's dextrose agar plate culture of soil isolate of *Histoplasma capsulatum* showing typical fuzzy appearance of mold colony.

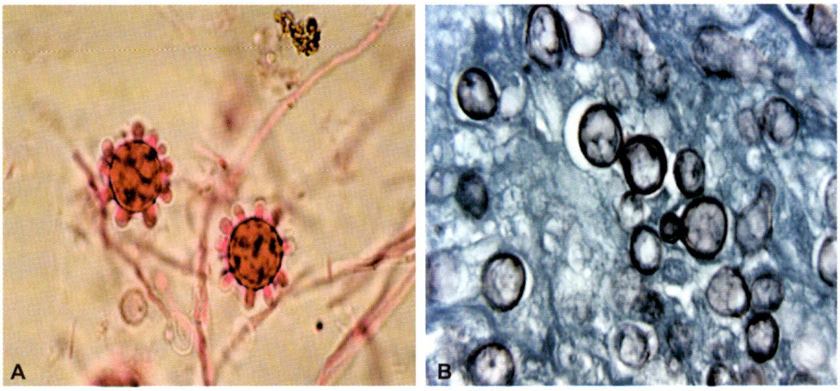

Fig. 3.13(A): Macroconidia of *H. capsulatum*, (12 μm) showing infectious microconidia budding off from hyphae and **(B)** Yeast like morphology of *H. capsulatum* (2–3 μm) growing in a lung tissue specimen ***Source:*** (faculty.ccbcmd.edu)

Pathogenesis

The disease resembles tuberculosis showing very similar lung lesions on X-ray plates. The causal organism of the disease is the fungus *Histoplasma capsulatum*. Symptoms of this infection may vary but the disease primarily affects the lungs[82]. At some occasions other organs are also affected due to the disseminated nature of histoplasmosis. This may be fatal if left untreated. Histoplasmosis is also quite common among AIDS patients because of their depressed immune system. *H. capsulatum* grows in soil and also in materials contaminated with bird or bat droppings. The fungus has been found in poultry litters, in caves and areas harboring bats, and in bird roosts. The fungus is known to exist in thermally dimorphic forms. It grows as a brownish mycelium in the environment at normal temperature however (Fig. 3.12) at 37°C it appears like yeast. The infection occurs with the help of microconidia. Once they are inhaled, settles in the alveolar spaces that eventually germinate and then transform into budding yeast cells (Fig. 3.13 a and b). Histoplasmosis is not contagious, but is contracted by inhalation of the spores from contaminated soil. Histoplasmosis can be diagnosed from samples like sputum, blood, or tissues from the infected organs. It can also be diagnosed by antigen detection system from blood or urine samples by ELISA or they can also be detected by PCR based techniques. It can also be diagnosed by a test for antibodies against *Histoplasma* in the blood.

Epidemiology

Histoplasma capsulatum is found throughout the world. It is endemic in certain areas of the United States, particularly in states bordering the Ohio River valley and the lower Mississippi River.

In India, so far no report is available of the histoplasmosis infection caused by *H. capsulatum*. Furthermore there is no authentic report on the isolation of *H. capsulatum* from the soil of this country. However, it is possible that cases of histoplasmosis may be seen in our country in patients who have resided or travelled in those parts of the world where this disease is endemic.

Treatment and Prevention

Antifungal medications are used to treat severe cases of acute histoplasmosis and all cases of chronic and disseminated disease. Mild disease usually resolves without treatment. Past infection may offer some protection against establishment of the disease due to reinfection. The introduction of azoles revolutionized the treatment for coccidioidomycosis. Some of the azole medications used in the treatment of Histoplasmosis includes Ketoconazole, Apoketoconazole and Nizoral.

3.3.4.3 Pneumocystis Pneumonia (PCP)

Pathogenesis

Commonly known as fungal pneumonia, caused by *Pneumocystis jirovecii* formerly known as *P. carinii*. This is known to be an opportunistic pathogen in HIV infected individuals. In the human lung, the organism is found mostly in interstitial, fibrous tissue of the lungs leading to a marked thickening of the alveolar septa and alveoli due to the infection. This may lead to a state of significant hypoxia that can be fatal if not treated at an early stage of the infection. If the infection prevails this may stimulate hyper-ventilatory effort, thereby causing breathlessness. The diagnosis is mainly done from the sputum samples or bronchio-alveolar lavage. The cells can be stained with toluidine blue, silver stain, periodic-acid schiff stain (Fig. 3.14 b). The characteristic cysts can also be detected by an immunofluorescence assay. In the lung, two major forms of the organism can be identified by microscopy. They are traditionally called cystic form (cysts) and trophic form (trophozoites). Trophic forms normally undergo asexual replication by typical binary fission. The trophic forms undergo another replicative cycle leading to the formation of the cyst form. A diploid zygote is produced by conjugation. This zygote then undergoes meiosis and subsequent mitosis, resulting in the formation of precyst, early cyst and finally mature cyst. As the mature cyst ruptures, 8 intracystic spores are released. These spores finally develop into trophic forms by the process of excystation (Fig. 3.14A).

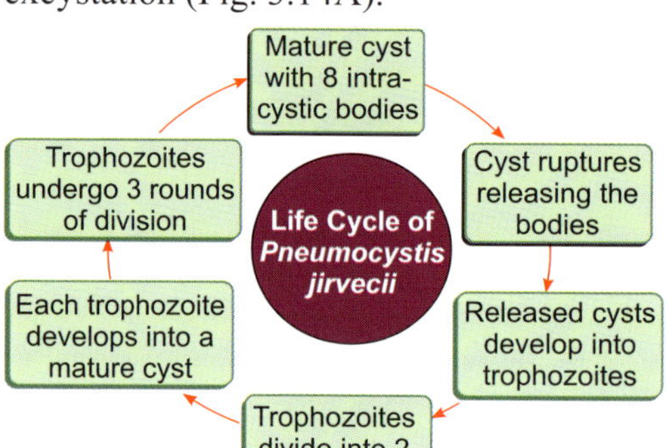

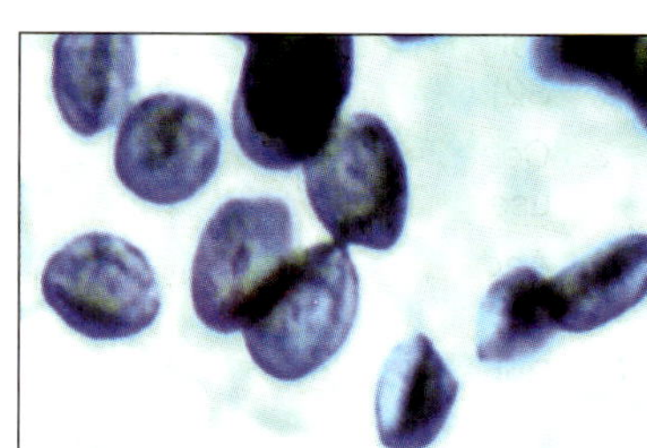

Fig. 3.14(A): Life cycle of Pneumocystis, and **(B)** *Pneumocystis jirovecii* cysts from bronchoalveolar lavage, stained with Toluidine blue ***Source:*** en.wikipedia.org

Some interesting observation can be made by examining bronchoalveolar lavage fluid (BALF) samples from patients with PCP. At low magnification, frequently cells are detected that are not of human origin. By using molecular and immunological methods, these cells have been identified as belonging to the trichomonad group of protozoa. These microorganisms multiply in alveoli and are found in BALF without oral or upper-respiratory-tract-fluid contamination. Furthermore, there is a strong correlation between the abundance of *Pneumocystis* organisms and trichomonad populations. It is established that *Trichomonas tenax*, a commensal of the human oral cavity, occurs in bronchial secretions during chronic respiratory conditions suggesting that trichomonads have a potential pathogenic effect on the alveolar epithelium during the course of PCP disease[83].

Epidemiology

The disease Pneumocystis pneumonia is relatively rare in individuals with normal immune systems. However, it is found to be quite common among severely malnourished children, the elderly, and in persons living with HIV/AIDS due to their low immunity. It is also prevalent in patients who have undergone solid organ transplantation or bone marrow transplantation or have undergone any form of surgery. Other clinical factors such as sex, race or ethnicity, and HIV transmission category have been also considered as risk factors for PCP. In sub-Saharan Africa, the effect of disease from PCP in infants and children is high. The incidence rate also remains high in adults than previously recognized. Colonization rates among both HIV-infected and non–HIV-infected populations may also be substantial.

Treatment and Prevention

The drug of choice for treatment of PCP is currently trimethoprimsulfamethoxazole. Other alternatives are clindamycin or intravenous pentamidine. The use of combination antiretroviral therapy reduced the rates of PCP among adults by 3.4% per year after 1992.

REFERENCES

1. Garcia-Rodriguez JA, Fresnadillo Martinez MJ. (2002) Dynamics of nasopharyngeal colonization by potential respiratory pathogens. J Antimicrob Chemother. 50(suppl S2):59-73.
2. Cooper RJ, Hoffman JR, Bartlett JG, et al (2001) Principles of appropriate antibiotic use for acute pharyngitis in adults: Background. Ann Intern Med. 134: 509-517.
3. Aagaard E, Gonzales R (2004) Management of acute bronchitis in healthy adults. Infect Dis Clin N Am. 18: 919-937.
4. Baltimore RS (February 2010). “Re-evaluation of antibiotic treatment of streptococcal pharyngitis”. Curr. Opin. Pediatr. 22 (1): 77–82.
5. Hayes CS, Williamson H (2001). Management of Group A beta-hemolytic streptococcal pharyngitis Surg. 72:560-6.
9. Dyne P and McCartan K (2005). “Pediatrics, Scarlet Fever”. eMedicine. emerg/402).

10. Faé KC, da Silva DD, Oshiro SE, et al. (2006). Mimicry in recognition of cardiac myosin peptides by heart-intralesional T cell clones from rheumatic heart diseas”. J. Immunol. 176 (9): 5662–70.

11. Porth, Carol (2007). Essentials of pathophysiology: concepts of altered health states. Hagerstown, MD: Lippincott Williams & Wilkins. ISBN 0-7817-7087-4.

12. Atkinson W, Hamborsky J, McIntyre L, Wolfe S, eds. (2007). Diphtheria. in: Epidemiology and Prevention of Vaccine-Preventable Diseases (The Pink Book) (10 ed.). Washington DC: Public Health Foundation. pp. 59–70.

13. “Diphtheria in the Former Soviet Union: Reemergence of a Pandemic Disease”. CDC, Emerging Infectious Diseases. 1998-10-01).

14. World Health Organization. (2006) Diphtheria vaccine—WHO position paper. Wkly Epidemiol Rec.81:24–31

15. Chen RT, Broome CV, Weinstein RA, Weaver R, Tsai TF.(1985) Diphtheria in the United States, l971–l981. Am J Public Health.75:1393–7).

16. Central Bureau of Health Intelligence. National health profile 2005. In: Health status indicators: Director General of Health Services, Government of India; Ministry of Health & Family Welfare Nirman Bhawan, New Delhi, 2006 Available from http://cbhidghs.nic.in/CBHI%20Book/chapter3.pdf

17. Bitragunta S, Murhekar MV, Hutin YJ, Penumur PP, Gupte MD. Persistence of diphtheria, Hyderabad, India, 2003-2006. Emerg Infect Dis. 2008; 14:1144-6.

18. Lodha, R., Dash, N., Kapil, A. & Kabra, S. K. (2000). Diphtheria in urban slums in north India. Lancet 355, 204.

19. Diphtheria Mechanism.” The Medical News. Web. 07 Oct. 2010. <http://www.news-medical.net/health/Diphtheria-Mechanism.aspx).

20. Paradise JL, Rockette HE, Colborn DK, et al. (1997) Otitis media in 2253 Pittsburgh-area infants: prevalence and risk factors during the first two years of life. Pediatrics. 99:318-33.

21. Erramouspe J, Heyneman CA. (2000) Treatment and prevention of otitis media. Ann Pharmacother. 34:1452-68.

22. Heikkinen T, Järvinen A (2003). “The common cold”. Lancet 361 (9351): 51–9.

23. Eccles R (2005). Understanding the symptoms of the common cold and influenza. Lancet Infect Dis 5 :718–25.

24. Abraham, G., and R.J. Colonno. 1984. Many rhinovirus serotypes share the same cellular receptor. J. Virol. 51:340-345.

25. Winther, B. “Rhinovirus Infections in the Upper Airway”, Proceedings of the American Thoracic Society, Vol. 8, No. 1 (2011), pp. 79-89.

26. Aronson MD, Weiss ST, Ben RL, Komaroff AL (July 1982). “Association between cigarette smoking and acute respiratory tract illness in young adults”. JAMA 248: 181–3.

27. Cohen S, Doyle WJ, Alper CM, Janicki-Deverts D, Turner RB (January 2009). Sleep habits and susceptibility to the common cold”. Arch. Intern. Med. 169: 62–7.

28. Eccles R (2002).Acute cooling of the body surface and the common cold. Rhinology 40: 109–14.

29. Mourtzoukou, E.G. Falagas, M.E. (2007) Exposure to cold and respiratory tract infections. Int. J. Tuberculosis & Lung Dis.. 11: 938-943.

30. Gwaltney JM, Druce HM.(1997) Efficacy of brompheniramine maleate for the treatment of rhinovirus colds. Clin Infect Dis. 25:1188–1194.

31. Watson, D. A., D. M. Musher, J. W. Jacobson, and J. Verhoef. 1993. A brief history of the pneumococcus in biomedical research: a panoply of scientific discovery. Clin. Infect. Dis. 17:913–924.

32. Avery, O. T., C. M. Macleod, and M. McCarty (1944) Studies on the chemical nature of the substance inducing transformation of pneumococcal serotypes: induction of transformation by a desoxyribonucleic acid fraction isolated from pneumococcus type III. J. Exp. Med. 79:137–157.

33. Baker, P. J. 1990. Regulation of the magnitude of the antibody response to bacterial polysaccharide antigens by thymus-derived lymphocytes. Infect. Immun. 58:3465–3468.

34. Felton, M. D., G. Kauffmam, B. Prescott, and B. Ottinger (1955) Studies on the mechanism of immunological paralysis induced in mice by pneumococcal polysaccharides. J. Immunol. 74:17–26.

35. D. Martı´nez-Herna´ndez, and E. Bouza. 1992. Susceptibility of Streptococcus pneumoniae to penicillin: a prospective microbiological and clinical study. J.Infect. Dis. 14:427–435.

36. Gillespie, S. H. 1989. Aspects of pneumococcal infection including bacterial virulence, host response and vaccination. J. Med. Microbiol. 28:237–248.

37. Gray, B. M., G. M. Converse III, and H. C. Dillon, Jr. (1979) Serotypes of Streptococcus pneumoniae causing disease. J. Infect. Dis. 140:979–983.

38. Black R. E., Morris S. S., Bryce J. (2003) Where and Why Are 10 Million Children Dying Every Year? The Lancet, 361:2226–34.

39. Coles CL, Kanungo R, Rahmathullah L, Thulasiraj RD, Katz J, Santosham M, et al. (2001) Pneumococcal nasopharyngeal colonization in young South Indian infants. Pediatr Infect Dis J. 20:289-95.

40. Ehrenworth L, Baer H (1956) The pathogenicity of Klebsiella pneumoniae for mice: The relationship to the quantity and rate of production of type specific capsular polysaccharide. J Bacteriol. 72:713-717.

41. Highsmith A.K and William R. Jarvis W.R. (1985) Klebsiella pneumoniae: Selected Virulence Factors That Contribute to Pathogenicity. Infection Control, 6: 75-77.

42. Yu VL, Hansen DS, Ko WC, Sagnimeni A, Klugman KP, von Gottberg A, Goossens H, Wagener MM, Benedi VJ; International Klebseilla Study Group (2007) Virulence characteristics of Klebsiella and clinical manifestations of K. pneumoniae bloodstream infections. Emerg Infect Dis.13:986-93.

43. Fung CP, Chang FY, Lee SC, Hu BS, Kuo BI, Liu CY, A global emerging disease of Klebsiella pneumoniae liver abscess: is serotype K1 an important factor for complicated endophthalmitis? Gut. 2002;50:420–4.

44. Wang JH, Liu YC, Lee SS, Yen MY, Chen YS, Wang JH, Primary liver abscess due to Klebsiella pneumoniae in Taiwan. Clin Infect Dis. 1998;26:1434–8.

45. Foy HM (1993) Infections caused by Mycoplasma pneumoniae and possible carrier state in different populations of patients. Clin Infect Dis. Suppl 1:S37-46.

46. Wright, SW. (1998) Pertussis infection in adults. South Med J. 91:702.

47. Guiso N (2009) Bordetella pertussis and pertussis vaccines. Clin Infect Dis. 49:1565-9.

48. Hewlett EL, Donato GM, Gray MC (2006). Macrophage cytotoxicity produced by adenylate cyclase toxin from Bordetella pertussis: more than just making cyclic AMP. Mol. Microbiol. 59 (2): 447–59.

49. Cherry, J. D., and U. Heininger. (2004) Pertussis and other Bordetella infections, p.1588–1608. In R. D. Feigin, J. D. Cherry, G. J. Demmler, and S. Kaplan (ed.), Textbook of pediatric infectious diseases, 5th ed. The W. B. Saunders Co, Philadelphia, Pa.149, 645

50. Kerr JR, Preston NW (2001) Current pharmacotherapy of pertussis. Expert Opin Pharmacother 2:1275–1282.

51. von Ko¨nig W (2005) The Pediatric Infectious Disease Journal Vol. 24, S66–S68. Use of Antibiotics in Pertussis.
52. CDC (1995). Pertussis – United States, January1992 – June 1995. Morb Mortal Wkly Rep 44, 525–529.
53. CDC (2006). Preventing tetanus, diphtheria, and pertussis among adults: use of tetanus toxoid, reduced diphtheria toxoid and acellular pertussis vaccine. Morb Mortal Wkly Rep 55, 1–33.
54. van Soolingen, D, Hoogenboezem, T, de Haas, PE, et al.(1997) A novel Pathogenic taxon of the Mycobacterium tuberculosis complex, Canetti: characterization of an exceptional isolate from Africa. Int J Syst Bacteriol 47:1236
55. Verhoef, J. (1991) Host-pathogen relationships in respiratory tract infections. Clin. Ther. 13:172-180.
56. Lönnroth, K, Raviglione, M (2008) Global epidemiology of tuberculosis:prospects for control. Semin Respir Crit Care Med. 29:481
57. WHO. Global Tuberculosis control. WHO/HTM/TB/2008.393. Geneva: World Health Organization; 2008. Available online at http://www.who.int/tb/publications/global_report/2008/en/index.html
58. Chakraborty AK . (2004) Epidemiology of tuberculosis: Current status in India. Indian J Med Res 120:248-76.
59. Williams BG, Granich R, Chauhan LS, Dharmshaktu NS, Dye C. (2005) The impact of HIV/AIDS on the control of tuberculosis in India. Proc Natl Acad Sci. US A. 102:9619-24.
60. World Health Organization. 2000. Anti-tuberculosis drug resistance in the world. Prevalence and trends. WHO/CDS/TB/2000/.278 The WHO/IUATLD Global Project on Anti-Tuberculosis Drug Resistance Surveillance. Report 2. World Health Organization, Geneva, Switzerland.
61. Meacci F, Orrù G, Iona E, Giannoni F, Piersimoni C, Pozzi G, Fattorini L, Oggioni MR. (2005) Drug resistance evolution of a Mycobacterium tuberculosis strain from a noncompliant patient. J Clin Microbiol.43:3114-20.
62. Prasad R. Current MDR status (2005) Indian J Tuberc. 52:121–31.
63. World Health Organization (2006) Global tuberculosis control: WHO report. Geneva: WHO/HTM/TB/2006.362.
64. Shah NS, Wright A, Bai GH, Barerra L, Boulahbal F, Casabona N, et al. (2007) Worldwide emergence of extensively drug-resistant tuberculosis. Emerg Infect Dis.13:380–7.
65. Mondal R and Jain A. (2007) Extensively Drug-Resistant Mycobacterium tuberculosis, India. Emerg Infect Dis.13:1429–1431.
66. Colditz GA, Brewer TF, Berkey CS, et al (1994) Efficacy of BCG vaccine in the prevention of tuberculosis. Meta-analysis of the published literature. JAMA. 271: 698-702.
67. Chan ED, Iseman MD (2002) Current medical treatment for tuberculosis. BMJ 325:1282-6.
68. Stedwell RE, Allen KM, Binder LS (1992) Hypokalemic paralyses: a review of the etiologies, pathophysiology, presentation, and therapy. Am J Emerg Med. 110:143-8
69. Aagaard C, Hoang T, Dietrich J, Cardona PJ, Izzo A, Dolganov G, Schoolnik GK, Cassidy JP, Billeskov R, Andersen P (2011) A multistage tuberculosis vaccine that confers efficient protection before and after exposure. Nat Med. 217:189-94.
70. Swanson M, Heuner K (2008). Legionella: Molecular Microbiology. Caister Academic Pr. ISBN 1-904455-26-3.

71. http://www.osha.gov/SLTC/legionnaires disease/index.html Safety and Health Topics: Legionnaires' Disease.

72. Stout JE, Muder, RR et al (2007) Role of Environmental Surveillance in Determining the Risk of Hospital-Acquired Legionellosis: A National Surveillance Study With Clinical Correlations. Infect Control Hosp Epidemiol 28:818-824.

73. Swanson M, Heuner K (2008). Legionella: Molecular Microbiology. Caister Academic Pr. ISBN 1-904455-26-3.

74. van Woensel, J. B., W. M. van Aalderen, and J. L. Kimpen. 2003. Viral lower respiratory tract infection in infants and young children. BMJ 327:36–40.

75. Jennings, L. C., T. P. Anderson, A. M. Werno, K. A. Beynon, and D. R. Murdoch. (2004) Viral etiology of acute respiratory tract infections in children presenting to hospital: role of polymerase chain reaction and demonstration of multiple infections. Pediatr. Infect. Dis. J. 23:1003–1007.

76. van den Hoogen BG, de Jong JC, Groen J, Kuiken T, de Groot R, Fouchier RA (2001). A newly discovered human pneumovirus isolated from young children with respiratory tract disease. Nat Med. 7:719–24.

77. Crowe, J. E., Jr. 2004. Human metapneumovirus as a major cause of human respiratory tract disease. Pediatr. Infect. Dis. J. 23(Suppl. 11):S215–S221.

78. Smith, Nicole M.; Joseph S. Bresee, David K. Shay, Timothy M. Uyeki, Nancy J. Cox, Raymond A. Strikas (2006). "Prevention and Control of Influenza: Recommendations of the Advisory Committee on Immunization Practices (ACIP)". Morbidity and Mortality Weekly Report. Centers for Disease Control and Prevention.

79. Ampel N (2005) Coccidioidomycosis in persons infected with HIV type 1. Clin Infect Dis 41: 1174–8

80. Schneider E, Hajjeh RA, Spiegel RA, et al. (1997). A coccidioidomycosis outbreak following the Northridge, Calif, earthquake. JAMA 277 (11): 904–8.

81. Barron MA, Madinger NE: Opportunistic Fungal Infections, Part 3: Cryptococcosis, Histoplasmosis, Coccidioidomycosis, and Emerging Mould Infections. Infections in Medicine 2008 UBM Medica LLC, London, UK, 2008

82. Ryan KJ, Ray CG, ed (2004). Sherris Medical Microbiology (4th ed.). McGraw Hill. pp. 674–6. ISBN 0-8385-8529-9

83. Duboucher, C. et al. (2006).Molecular identification of Tritrichomonas foetus-like organisms as coinfecting agents of human Pneumocystis pneumonia. J. Clin. Microbiol. 44, 1165–1168

4

Gastrointestinal Tract Infections

Chapter Outline

4.1 Upper Alimentary tract infections
4.2 Lower Alimentary tract infections

4.1 Upper Alimentary Tract Infections

4.1.1 Oral Bacterial Infections- Pathogenesis, Epidemiology, Treatment and Prevention

The upper gastrointestinal tract consists of the mouth cavity, esophagus, stomach, and duodenum. The alimentary tract is a major route for the entry of the pathogenic microorganisms into the body. The oral cavity contains some of the most diverse microflora in the entire human body. In addition to bacterial organisms, oral microorganisms may include fungal, protozoal, and viral species. There exist hundred types of organisms out of which only twenty two are identified as predominant ones1. Microorganisms in the microenvironment of the oral cavity adhere to different regions of the mouth like the teeth, the gingival sulcus, the tongue, and the buccal mucosa. Each site offers unique niche for the establishment of the microorganisms typical for the site. The normal flora in healthy individuals maintains a uniform pattern. However, when a local or systemic disease progresses, an alteration in the pattern of the microorganisms is observed when atypical organisms begin to predominate and some normal organisms with a benign nature, such as Candida albicans may turn pathogenic. The microenvironment of the oral cavity is subject to change alongwith the age of the patient, the eruption or loss of teeth, and the appearance of disease states like dental caries or any periodontal disease. Systemic changes such as pregnancy or any drug intake are likely to alter the number and proportion of the microflora. These changes are also subject to alterations in the flow and composition of salivary fluid and in the levels and activity of defense components for example immunoglobulins and cytokines present in the saliva.

4.1.1.1 Dental Plaque

The formation of dental plaque initiates when the microorganisms like Streptococci or actinomycetes start adhering to the tooth pellicle and form a thin film of proteinaceous material that arises out of saliva that finally get adsorbed on the tooth. The microorganism *Streptococcus mutans* derives sucrose from the dietary contents and soon produces a sticky extracellular polysaccharide consisting of glucan. This polysachharide enables the organism to bind together and provides attachment sites for other species to adhere. Due to the metabolic activity of the early aerobic colonizers, depletion of oxygen takes place locally due to consumption of oxygen thus creating an ideal environment for the anaerobic microorganisms to follow. This leads to drastic drop in pH due to the production of lactic acids and other acids. The anaerobic bacteria colonizing the plaque belong to the genus *Veillonella* and lactobacilli predominantly. As the plaque formation progresses, it leads to the formation of a sponge like structure that gets soaked in acid. Acid then slowly diffuses into the tooth from overlying cariogenic plaque, penetrates the enamel and finally attacks the dentin that leads to the destruction of root canal lined with nerves and blood vessels.

Prevention of Dental Caries

Dental caries increases with age. The formation of dental caries starts right after the eruption of tooth in susceptible children then increases further with age. Studies have demonstrated that some populations show lower incidence of dental caries than other populations receiving the same diet. Three important control measures can reduce the incidence and also prevent dental caries:

1. **Fluoridation of water:** The most effective way to prevent coronal and root dental caries lies in the Fluoridation of water supply. Fluoride when applied to tooth surfaces either in the form of mouthwashes, gels or toothpastes provides antibacterial action. Therefore, use of this supplemental treatment is helpful for people especially prone to dental caries.
2. **Restricting sugar consumption:** Dental caries can be reduced substantially if sweet intake is controlled by reducing the frequency of consuming sweets in between the meals.
3. **Improving oral hygiene:** Excercising oral hygiene procedures like personal plaque removal by tooth brushing after every major meals. Even flossing helps to minimize the dental caries. The professional prophylaxis that is often carried out by a dentist during a periodic dental examination helps in reducing the formation of plaques considerably.

4.1.1.2 Peridontitis/Gingivitis

Gingivitis results into wounds that affect the gums. Periodontitis damages the bone and connective tissue that supports the teeth.

Pathogenesis

Peridontitis is a specific pathological condition that developes in the oral cavity where inflammation is observed in the periodontal attachment of the teeth and the alveolar bone (Fig. 4.1). In case the formation of the plaque continues for a long period may result in gum bleeding when it is referred as periodontal disease. In case the plaque is not removed periodically, it transforms into hard deposit of minerals, called as tartar and accumulates in the base of the tooth. The Periodontal disease is a chronic inflammatory disease of the gum and tissues. The gum and tissue surrounds and supports the teeth. If the disease is not treated in time, periodontal disease can lead to tooth loss. Further if they are left untreated can create vulnerable foci for infection in the oral cavity that can affect many other vital systems of the body such as cardiovascular and renal systems. Foci of infection in the oral cavity arising from chronic periodontitis or chronic periapical abscesses (ie, inflammation and abscess of the tissue attached to the apex of the root) may lead to subacute bacterial endocarditis (BE) and glomerulonephritis (GN).

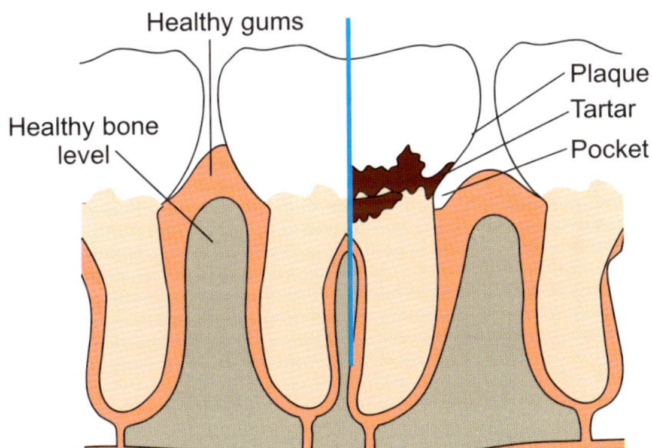

Fig. 4.1: Diagram of healthy gums vs. gums with periodontal disease

Bacteria causing Peridontal disease: More than 350 species of microorganisms have been found in a healthy mouth. Periodontal infections are caused by less than 5% of these species. Among the bacteria most implicated in periodontal disease and bone loss are the following:

Actinobacillus actinomycetemcomitans and *Porphyromonas gingivalis*: Both these bacteria are causal organism for an aggressive periodontal disease. Risk of serious gum disease increases when a person is infected with *P. gingivalis*. Virulent strains of these bacteria are responsible for periodontal disease. *P. gingivalis* produces a destructive enzyme referred as arginine-specific cysteine proteinase that is responsible to disrupt the immune system and leads to subsequent periodontal connective tissue destruction.

Presence of *Bacteroides forsythus* is also linked to the periodontal disease. Other bacteria found to be associated with periodontal disease are *Treponema denticola*, *T. socranskii* and *P. intermedia*. These bacteria, together with *P. gingivalis*, are frequently present at the same sites, and are associated with deep periodontal pockets. Some bacteria like few Streptococcal species are responsible only for gingivitis and do not participate in plaque development.

While gingivitis is nearly universal among children and adolescents, periodontitis typically occurs as people get older and more often encountered after 35 years of age. There are different forms of periodontal disease. They include:

Chronic Periodontitis. Chronic periodontitis is the most common type of periodontitis. It may begin at an adolescent stage however the clinical manifestation of the disease takes place usually in their mid-30s.

Aggressive Periodontitis. Aggressive periodontitis is a subtype of chronic periodontitis that can occur in childhood. It can lead to severe bone loss by the time patient reach their early 20s.

Disease-Related Periodontitis. Periodontitis can also be associated with a number of systemic diseases, including type 1 diabetes, Down syndrome, AIDS, and several rare disorders of white blood cells.

Necrotizing Periodontal Disease. Necrotizing periodontal disease is an acute infection of the gum tissue. This disease is characterized by bleeding gums and bad breath and onset of severe pain in the gum. If left untreated, necrotizing periodontal disease can spread throughout the facial areas like cheeks, jaw and often cause extensive damage. Necrotizing periodontal disease has been found to be usually associated with systemic health conditions such as HIV or malnutrition.

Treatment and Prevention

Practice of good oral hygiene is the best means of prevention. Scaling, polishing, and sometimes curettage are used to manage periodontal disease. Treatment Phases include the following:

- Initial cleaning, scaling, and curettage
- Surgery in serius cases for reducing deep pockets that remain underneath the gums evenafter extensive cleaning sessions
- Administration of low dose oral or topical antibiotics
- Maintenance of the teeth and gums

4.1.2 Viral Diseases of the Mouth and Salivary Glands-Pathogenesis, Epidemiology, Treatment and Prevention

4.1.2.1 Herpes Simplex Virus Infection

Symptoms of the infection and treatment of HSV is described in Chapter 2 in the section 2.3.2.1

The following section deals with the mode of action of the virus and its epidemiology

Pathogenesis

The HSV virus is a medium-sized, enveloped virus having double stranded, linear DNA. The virus can be readily cultivated in a variety of tissue cell cultures and is detected by using specific anti-HSV antibody.

Structure and mode of entry of HSV

The structure of herpes viruses consists of a relatively large double-stranded, linear DNA genome surrounded by an icosahedral protein coat known as capsid. This is finally wrapped in a lipid bilayer called the envelope (Fig. 4.2). The envelope is joined to the capsid by means of a tegument. This makes the complete particle known as the virion[2]. The process of infection begins when the virions bind to the heparan sulfate moieties present on host cell surfaces. This initial attachment is essential to trigger a cascade of molecular interactions involving multiple viral and host cell proteins and receptors that leads to penetration of the viral nucleocapsid and tegument proteins into the cytoplasm of the host cell. The nucleocapsid is then transported to the nuclear membrane and the viral DNA is released ready for replication in the nucleus. Both Herpes simplex type-1 (HSV-1) and type-2 (HSV-2) viruses are closely related and they are members of the alphaherpesvirus subfamily. Both the members are responsible for highly prevalent infections among humans[3].The viral envelope contains a set of glycoproteins. During initial interactions with the host, one of the glycoprotein C (gC) binds to a cell surface particle called heparan sulfate. In the second stage,a second glycoprotein, glycoprotein D (gD), binds specifically to at least one of three known entry receptors. These include herpesvirus entry mediator (HVEM), nectin-1 and 3-O sulfated heparan sulfate. The receptor provides a strong, fixed attachment to the host cell. These interactions bring the membrane surfaces into mutual proximity. This process allows other glycoproteins encased in the viral envelope to interact with other cell surface molecules. Once bound to the HVEM, gD normally alters its conformation and interacts with viral glycoproteins H (gH) and L (gL) and results in the formation of a complex. The interaction of these membrane proteins results in the hemifusion state. Thus gB interaction with the gH/gL complex creates an entry pore for the viral capsid[4]. Following infection of a cell, a scries of herpes virus proteins, defined as immediate-early,early and late, are generated (Fig. 4.3). These stages of lytic infection like late lytic stage can be clearly distinguished from the latency stage. It is quite apparent that in case of HSV-1, no protein products are detected during latency whereas the protein products are detected during the lytic cycle. The early proteins transcribed are used in the regulation of genetic replication of the virus. On entering the cell, α-TIF protein joins the viral particle and aids in immediate-earlytranscription. The virion shuts off the host protein synthesis by degrading host mRNA thereby helps in viral replication and regulates gene expression of viral proteins[5]. The viral genome eventually travels to the nucleus but the coat protein remains in the host cytoplasm.

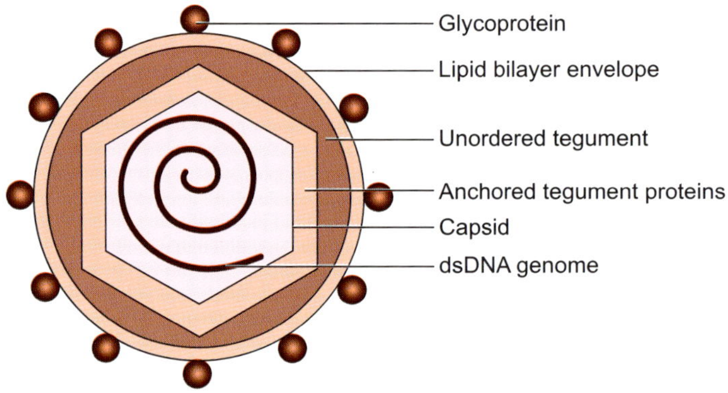

Fig. 4.2: Structure of HSV

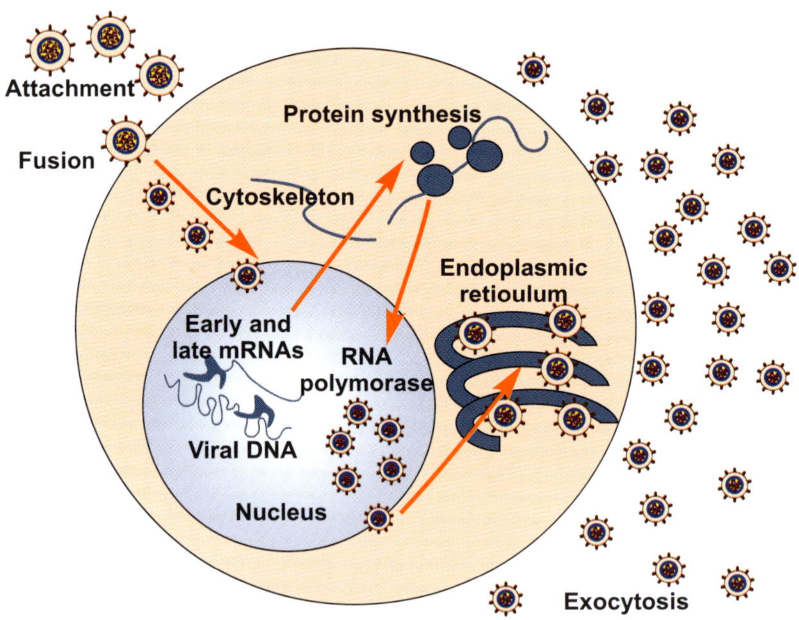

Fig. 4.3: A simplified diagram of HSV replication

The late proteins that are synthesized are responsible for the formation of the capsid and the receptors on the surface of the virus. The complete process of packaging of the viral particles including the genome, core and the capsid takes place in the nucleus of the host cell. The concatemers of the viral genome are separated by cleavage and are packaged into pre-formed capsids. HSV-1 further undergoes a process of primary followed by secondary envelopment. The primary envelope is acquired by the process of budding into the inner nuclear membrane of the cell. This is followed by a fusion with the outer nuclear membrane that releases a naked capsid into the cytoplasm. The virus finally acquires its final envelope by budding into cytoplasmic vesicles.[6]

An interesting aspect of herpes simplex virus (HSV) can be observed by its ability to enter a quiescent state and establish a lifelong latent infection in sensory neurons that are present at the site of primary infection. After a productive infection by HSV at the site of inoculation, the virus is able to disseminate to and makes an entry ino the sensory neurons, where it establishes a latent infection. Latent infection forms a reservoir of virus and leads to recurrent infection, disease, and transmission to other individuals[7].

Epidemiology

Oral and Genital Herpes are the two most common manifestation of the disease caused by these viruses. Unfortunately, the data on the extent of the virus's infiltration into the human population is not conclusive however number of infected people worldwide is estimated to be at approximately 86 million people as of 1999[8]. Due to patient's ignorance of their infection along with the occurrence of atypical symptomology and asymptomatic viral shedding of HSV has contributed to the increases in transmission of HSV. Since the disease is not considered as life threatening, therefore in the past HSV did not receive much attention for eradication because it was not considered an "emergency" disease. Rare incidences of Neonatal Herpes infection and the rare incidences of HSV induced encephalitis indicate that the disease needs immediate attention and more so because of increasing association of HIV transmission within HSV infected patients. It was found that in approximately 20% of children younger than 5 years of age reported positive for antibodies against HSV-1. The titre further rises in a linear fashion through the age till 70 years[9]. During pregnancy, HSV infection is found to be associated with spontaneous abortion, prematurity and congenital neonatal herpes. The most serious consequence of genital HSV infection is neonatal herpes. This results in perinatal transmission from mother to infant. The greatest risk of infection is from contact with lesions or saliva from patients within a few days of disease onset. During this stage, the disease remains highly contagious because at this time large number of virions are present. HSV can infect almost any part of the body tissue.

4.1.2.2 Mumps Virus

The pathogenesis, epidemiology and prevention of the disease is described in the earlier Section 2.3.3 of Chapter 2

The following section describes the structure of the Mumps virus

The virus appears like a spherical particle consisting of concentric layers of fatty lipids, large protein molecules and nucleic acid. The virus posesses large 'spikes' on its surface. This spikes are proteinaceous in nature and help the virus to gain entry to host cells. The core of the virus particle consists of a single, long molecule of RNA wrapped up in protein which is released into the cell.

MuV is an enveloped, nonsegmented, negative-sense, singlestranded RNA virus. This virus is classified as a member of the family *Paramyxoviridae*, subfamily *Paramyxovirinae* and genus *Rubulavirus*. Mumps virus (MuV) resembles other members of the Paramyxoviridae in morphology. The particle appears roughly spherical or pleiomorphic in shape. The sizes of Paramyxovirus particles may vary from 120 – 450 nm in diameter. The outer surfaces of virions consist of host-derived membrane, enriched with viral integral membrane glycoproteins

An electron microscopic view of the virus particle reveals the organization and the assembly of the virus particle. The viral envelope appearing like 'spikes' are made up of various layers of densely packed proteins (Fig. 4.5). The core of the paramyxovirus particle contains the RNA genomes that is bound to nucleocapsid (N or NP) proteins to form helical structures called ribonucleoproteins (RNPs)[10]. The viral matrix protein known as M proteins lies directly below the viral membrane and form a bridge between viral glycoproteins and the ribonuclear proteins (RNPs). Paramyxoviruses form particles that are mainly spherical, but at times they appear as filamentous presenting considerable variation in size and shape. (Fig. 4.4). The paramyxoviruses encode two glycoproteins: a fusion (F) protein, and an attachment [HN (hemagglutinin-neuraminidase), H (hemagglutinin), or G (glycoprotein)

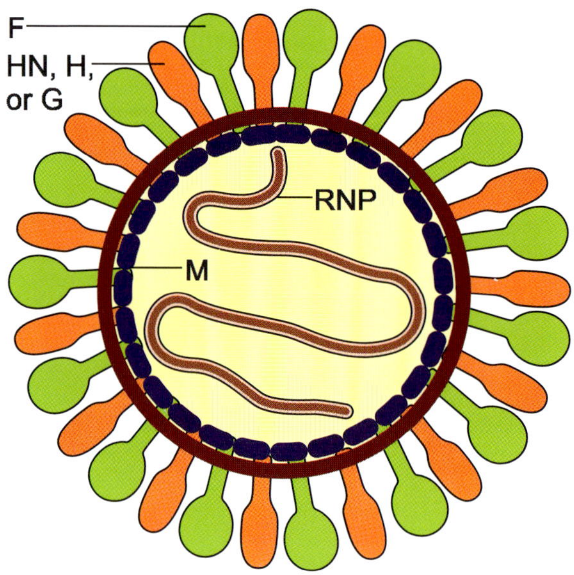

Fig. 4.4: Paramyxovirus virions: Schematic representation of a spherical paramyxovirus virion illustrating M protein lining the inner surface of the lipid envelope, the viral glycoproteins (HN, H or G and F) decorating the virion surface, and the helical RNP enclosed within.

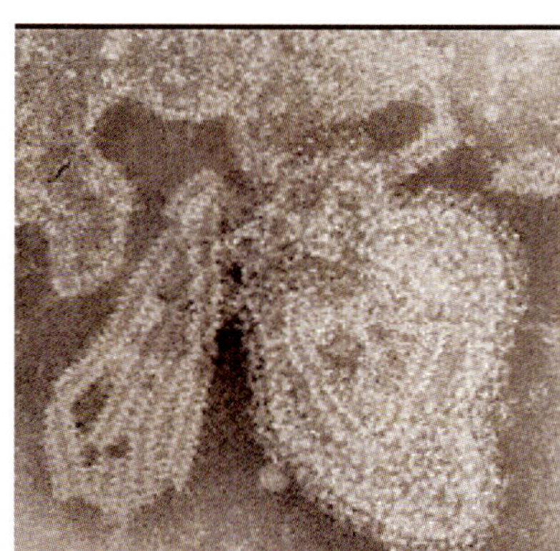

Fig. 4.5: Transmission electron micrograph of virus particles ***Source:*** en.wikipedia.org

4.1.3 Infections of Stomach-causative Agent, Pathogenesis, Epidemiology, Treatment and Prevention

Helicobacter pylori have been formally recognized and documented as a bacterialcarcinogen. This gram-negative bacterium is also considered as one of the most successful human pathogen that is able to colonize almost fifty percent of world population. The bacteria remain colonized and may persist lifelong in case the individual is left untreated. *H. pylori* infection remains a key factorin the etiology of various gastrointestinal diseases. The manifestation of the disease ranges from chronic active gastritis, asymptomatic state without clinical symptoms to peptic ulceration, gastric adenocarcinoma, and gastric mucosa-associated lymphoid tissue lymphoma (Table 4.1).

Table 4.1: The correlations between the pattern of *H. pylori* colonization, inflammation, acid secretion, gastric and duodenal histology, and clinical outcome

Infection pattern	Histopathological changes	Colonization pattern	Acid secretion	Clinical manifestation
Pangastritis	Mucosal atrophy marked inflammation is distributed throughout the stomach,	Throughout the stomach	Reduced amount	Gastric ulcer and prone to gastric carcinoma
Antral predominant gastritis	higher density of lymphoid aggregates and follicles Chronic inflammation	Primarily colonizes pyloric antrum	Large amount	Duodenl ulcers

Most of the duodenal and stomach ulcers have been found to be as a result of *H. pylori* infections. In few cases this bacteria may also cause non-ulcer dyspepsia. Large number of individuals infected with *H. pylori* remain mostly asymptomatic. The visible clinical symptoms of the infection during pathogenesis are inflammation in the stomach, upper or lower abdominal pain, stomach ache, regurgitation, acid reflux, nausea, flatulence, and belching. A symptomatic individual should be attended by a medical practioner immediately to avoid serious ailments[11].

Detection and Symptoms

H. pylori a member of microaerophilic gram negative bacilli. The organism has adapted to a unique niche as it thrives in the acidic environment of the stomach and also adheres to the mucosal layer. The *H. pylori* infection can be identified with the help of various diagnostic tests like antigen tests done on samples of stool, blood antibody tests, and carbon-urea breath tests. However, examining the biopsy samples during endoscopy is considered to be most reliable test for detection of this organism. Normally histological examination of the biopsy sample is carried out along with a microbial culture for confirmation.

The clinical symptoms of peptic ulcer are:

(*a*) Severe abdominal pain during meals
(*b*) Chronic inflammation
(*c*) Feeling of abdominal fullness followed by bloating
(*d*) Nausea accompanied by copious vomiting
(*e*) Loss of appetite resulting into drastic weight loss
(*f*) Bleeding through stool due to gastric ulcer
(*g*) Duodenal or gastric perforations appear in severe cases

Pathogenesis

H. pylori Bacteria causing peptic ulcers on the lining of the stomach or duodenum (Fig. 4.6)

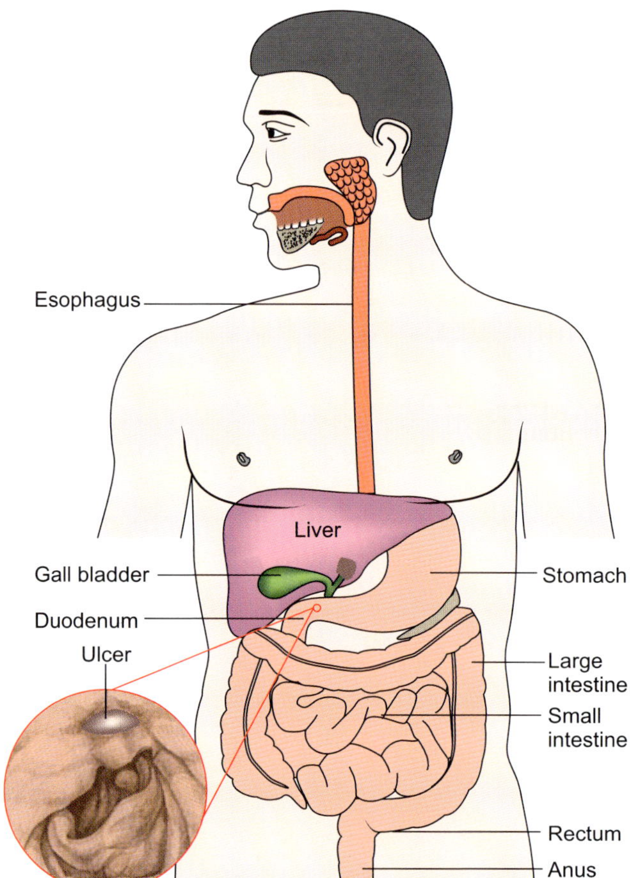

Fig. 4.6: Schematic representation showing the location of peptic ulcer in the Gastrointestinal tract

Habitat: H. pylori has found an unique habitat as it harbours in the protective mucous layer of the stomach. These bacteria possess long thread like structures protruding from them with the help of which the organism attach to the underlying stomach cells. The mucous layer of the stomach gives protection to *H. pylori*. For survival in the harsh, acidic environment of the stomach *H. pylori* is known to secrete an enzyme called urease, which converts the chemical urea to ammonia. The production of ammonia around *H. pylori* neutralizes the acidity of the stomach, making the environment suitable for its existence. In addition, the helical shape of *H. pylori* allows it to burrow into the mucus layer. The mucous layer is less acidic than the inside space, or lumen, of the stomach. *H. pylori* can also attach to the cells that line the inner surface of the stomach.

Studies reveal that an increased number of plasma cells accumulate in the gastric mucosa where *H. pylori* resides revealing that the body presents an active humoral immune response against the bacteria. The anti- *H. pylori* antibodies produced are primarily of the IgG type however this issue is still debatable.

Structure: H. pylori exist as a helix shaped Gram-negative bacterium and is about 3 micrometres long with a diameter of about 0.5 micrometres (Fig.4.7). The organism is microaerophilic in nature. The organism harbours a unique hydrogenase enzyme to harvest energy by oxidizing molecular hydrogen (H2) produced by other intestinal bacteria. The organism also produces three important enzymes- oxidase, catalase, and urease. The bacteria are also capable of forming biofilms[12]. During the formation of biofilm the bacteria can convert from spiral shape to a possibly viable but nonculturable coccoid form favoring its survival. *H. pylori* possesses *a* family of five major outer membrane protein (OMP)[13]. The largest family includes putativeadhesins. The other four families are described as porins, iron transporters, flagellum-associated proteins and some proteins of unknown function. As expected in a typical Gram-negative bacterium, the outer membrane of *H. pylori* consists of phospholipids and lipopolysaccharide (LPS). The O antigen of LPS is found to be fucosylated and thus mimics Lewis blood group antigens found on the gastric epithelium. The outer membrane also contains cholesterol glucosides. *H. pylori* display four to six lophotrichous flagella making the organism highly motile owing to flagella[14]. The characteristic sheathed flagellar filaments of Helicobacter are composed of two copolymerized flagellins, FlaA and FlaB

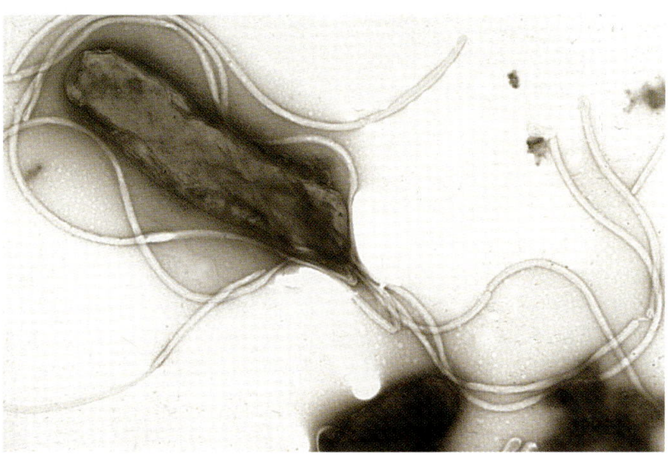

Fig. 4.7: Microscopic view of *Helicobacter pylori*
Source: (http://en.wikipedia.org/wiki/Helicobacter_pylori)

Gastric cancer patients can be diagnosed by a non-invasive method by subjecting their serum samples to Pepsinogen testing. Pepsinogens are considered to the proteolytic stomach enzymes and are known as noninvasive biomarkers of gastric atrophy. Pepsinogen I is normally present in the gastric corpus and pepsinogen II is present in the gastric corpus and antrum. Low serum levels of pepsinogen I or low pepsinogen I:II ratio is found to be associated with gastric atrophy. This can be further confirmed by endoscopy.

Disease outcome: A complex interplay between the host and the bacterium contributes in the final outcome of the disease. Polymorphisms observed in the host genes that function as immune mediators often contribute in the severity of the disease. Bacterial virulence factors such as the cytotoxin-associated gene pathogenicity island-encoded protein CagA and the vacuolating cytotoxin VacA contribute in the process of colonization of the gastric mucosa and subsequently modulate the host's immune system.

H. pylori infection is designated as a major cause of gastric cancer. The disease may involve any area of the stomach except the area where it joins the esophagus. In 1994, the International Agency for Research on Cancer classified *H. pylori* as a carcinogen, or cancer-causing agent. Since then, it has been accepted that colonization of the stomach with *H. pylori* is an important cause of gastric cancer and of gastric mucosa-associated lymphoid tissue (MALT) lymphoma.

Role of virulence factors: Various pathogenic factors have been identified contributing into the pathogenesis of disease. Two important candidate genes are cag (cytotoxin-associated antigen),polymorphic variation in the VacA vacuolating exotoxin and the blood group antigen binding adhesin BabA. These genes are found to be located in the pathogenicity island of the genome. The protein Cag encoded by cag gene is one of the important fator in the disease related outcome particularly during gastric cancer. It was suggested that there is a need to consider the geographical contributions to disease outcome. A study reveal that in the population belonging to western countries, the presence of CagA is associated with an increased risk of severe disease outcome however the same is not true in case of East Asian populations where the the presence of potent CagA does not pose an increased risk of *H. pylori*-related diseases.

Epidemiology

By simulating the genetic diversity data available the history of this organism can be traced. The study indicated that the bacteria originated from East Africa which is reported to be the birthplace of modern humans[16]. Their results indicate modern humans were already infected by *H. pylori* before their migrations out of Africa. The bacterim was therefore harboured by humans around 58,000 years ago before it spread to other continents. Therefore, *H. pylori* is considered to be a common bacterium, and approximately 50 percent of the world's population has been estimated to be infected. Humans remain the principal reservoir for this organism. The prevalence of *H. pylori* infection varies widely by geographic area, age, race and ethnicity. A person contracts the *H. pylori* infection through contaminated items like food, fluid, or perhaps from contaminated utensils.

This causes world wide infection. The rate of infection is found to be higher in developing than in developed countries and most of the infections occur during childhood. The infection rate decrease with improvements in hygiene practices. The rate of infection is again high in areas with dense population and living in poor sanitary conditions. The disease may manifest in all age groups with older people being more vulnerable. The disease may remain asymptomatic in many infected individuals however the infection remains localized to the gastric area, and persists unless specific treatment is given.

The person-to-person transmission is recorded among institutionalized children and adults. The incidence rate *H. pylori* is also found to be high within families. This supports the detection of *H. pylori* DNA in vomitus, saliva, dental plaque, gastric juice, and feces. Recent studies in the United States have linked clinical *H. pylori* infection with consumption of *H. pylori*-contaminated well water.

Treatment and Prevention

The standard therapy given to the patients normally consists of a combination of a proton pump inhibitor such as Omeprazole and the antibiotics Clarithromycin and Amoxicillin, administered for a week. Variations of the triple therapy have been developed over the years, such as using a different proton pump inhibitor, as with Pantoprazole or Rabeprazole, or replacing amoxicillin with Tinidazole in case the patient shows allergic reaction to Penicillin. The triple therapy has revolutionized the treatment of peptic ulcers and is an excellent therapy to cure ulcer patients. The combination drugs help to kill the bacteria, reduce stomach acid, and therefore able to protect the stomach lining. Antibiotics are used to kill the bacteria. Proton pump inhibitors suppress acid production by suppressing the action of the pump in releasing the acid into the stomach. Proton pump inhibitors were earlier prescribed for many years for treatment of ulcers. But when used in isolation, these drugs do not eradicate *H. pylori* and therefore do not cure *H. pylori*-related ulcers. However, an increasing number of infected individuals are found to harbour antibiotic-resistant bacteria. As a result, initial treatment failure may occur where an additional round of antibiotic therapy may be required. An alternative therapy known as quadruple therapy has been developed where a bismuthcolloid, such as bismuth subsalicylate is added in the treatment[17].

Recent studies suggest that the activity of *H. pylori* could be suppressed via dietary supplements[18]. The study recommended that consumption of around 70 g of broccoli sprouts daily for two months may reduce substantially the number of colonies of *H. pylori* bacteria in the stomach by 40% in humans. This treatment is able to enhance the protection of gastric mucosa against *H. pylori*. However, it has been found to be ineffective on related gastric cancers.

4.2 Lower Alimentary Tract Infections

4.2.1 Infectious Intestinal Diseases

Infectious diarrhea is a major cause of death worldwide. In developing countries, one out of five children dies of diarrhea before the age of five. Each year, more than five million people die of diarrhea, out of which 80% at one year of age. Infection of the intestinal tract is caused by an array of bacterial, parasitic, and viral pathogens. The list of the pathogens is ever increasing. It can profoundly disrupt intestinal function and most of the time causing overt dehydrating diarrhea. Diarrhea is a syndrome that is frequently not differentiated clinically by specific etiologic agent.

The ingestion of potentially contaminated food and drinks provides an easy portal of entry for infectious organisms that may affect the colon, or large intestine. The gastrointestinal system has developed several mechanisms to protect its organs from infection. However, intestinal infections remain a common malady.Although global mortality from diarrhea has declined in recent years, from approximately 4.6 million deaths during the mid-1980s to the current estimate of 1.6–2.1 million till 2000, most of these deaths occur in children in developing countries under the age of 5 years. Most of the diseases spread through fecal-oral contamination of water sources. This is supported by the high incidence rate of these infectious diseases observed in population living in poor hygienic conditions.

Microbes Involved in Food Poisoning- Pathogenesis, Epidemiology, Treatment and Prevention

Classification of Food Poisoning

I. ***Based on symptoms and duration of onset:***

(*a*) Nausea and vomiting reported within six hours of infection (*Staphylococcus aureus, Bacillus cereus*)

(*b*) Abdominal cramps and diarrhoea develop within 8-16 hours (*Clostridium perfringens, Bacillus cereus*)

(*c*) Abdominal cramps, diarrhea followed byfeverdevelop within 16-48 hours (Salmonella, Shigella, *Vibrio parahemolyticus,* Enteroinvasive *E.coli, Campylobacter jejuni*)

(*d*) Abdominal cramps and watery diarrhoea within 16-72 hours (Enterotoxigenic *E.coli, Vibrio cholera* O1, O139, *Vibrio parahemolyticus*, Non agglutinable vibrios (NAG), Norwalk virus)

(*e*) Fever followed by abdominal cramps within 16-48 hours (*Yersinia enterocolitica*)

(*f*) Bloody diarrhoea without any fever within 72-120 hours of infection (Enterohemorrhagic *E.coli* O157:H7)

(*g*) Nausea, vomiting, diarrhoea and paralysis when developing within 18-36 hours (*Clostridium botulinum*)

II. ***Infectious diseases classified based on their mode of pathogenesis:***

(*a*) Food intoxications caused due to the ingestion of preformed bacterial toxins. (*Staphylococcus aureus, Bacillus cereus, Clostridium botulinum, Clostridium perfringens*)

(*b*) Food intoxications caused by noninvasive bacteria that secrete toxins only when adhere to the intestinal wall (Enterotoxigenic *E.coli, Vibrio cholerae, Campylobacter jejuni*) (*c*) Food intoxications that follow an intracellular invasion of the intestinal epithelial cells.v(*Shigella, Salmonella*)

(*d*) d. Diseases caused by bacteria that enter the blood stream via the intestinal tract. (*Salmonella typhi, Listeria monocytogenes*)

Some of the most common microorganisms that cause food poisoning include the following:

A. Salmonella

Salmonella is considered to be the leading cause of foodborne illness. Salmonella can be classified into different types including those that cause typhoid fever and those that cause gastroenteritis. Enteric fever can extend for a longer period and causes systemic illness that results from infection with the exclusively human pathogens, *S. typhi* and *S. paratyphi*. Clinical manifestations of the disease include fever, abdominal pain, transient diarrhea or constipation, and occasionally develop a maculopapular rash. The symptoms show considerable variation depending on the virulence of the strain of salmonella and the number of infecting organisms.

Clinical Syndromes

- **Enteritis (acute gastroenteritis)**
- **Enteric fever** (prototype is **typhoid fever** and less severe paratyphoid fever)
- **Septicemia** (particularly *S. choleraesuis, S. typhi*, and *S. paratyphi*)
- **Asymptomatic carriage** (gall bladder is the reservoir for *Salmonella typhi*)

Salmonellae are facultative anaerobic gram-negative bacilli and usually enter the body via the gastrointestinal tract. They can persist in the intestine for long periods of time. Salmonellae can act as both commensals and pathogens and can be isolated also from the GI tracts of domestic and wild animals, including insects, reptiles, birds, and mammals.

Poultry are commonly infected with a wide variety of salmonella serovars and infection is generally subclinical. Poultry farming is often affected by reduction in productivity due to infections with large number of bacteria. Salmonellosis in chicken is of great concern as it causes great economic losses due to mortality and reduction in egg production[19]. Under the new naming system, the Salmonella serovar is divided into biovar Gallinarum and Pullorum, which are identified to cause Fowl typhoid and Pullorum disease respectively. While fowl typhoid is a disease of mature birds, pullorum causes mortality of embryos and chicks. Infection with these pathogens is responsible for considerable economic losses in poultry production.

Pathogenesis

Salmonella strains are Gram-negative, facultatively anaerobic, rod-shaped enterobacteria. They are non lactose fermenters and utilize citrate as their sole carbon source and produce H_2S. Salmonella virulence requires the coordinated expression of complexarrays of virulence factors that allow the bacterium to evade the host's immune system. All Salmonella serotypes share the ability to invade the host. They are capable of secreting effector molecules that are delivered into the intestinal epithelium using their secretion apparatus.

All Salmonella infections begin with the ingestion of organisms in contaminated food or water. Conditions that increase gastric pH reduce the Salmonella infectious dose, which suggests that gastric acidity represents a significant initial barrier to infection. Interestingly, salmonellae exhibit an adaptive acid-tolerance response on exposure to low pH, possibly promoting survival in acidic host environment such as the stomach. After entering the small bowel, salmonellae must traverse the intestinal mucus layer before encountering and adhering to cells of the intestinal epithelium. Salmonellae express several fimbriae that contribute to their ability to adhere to intestinal epithelial cells and contribute in biofilm formation[20].The invasion process is mediated by Type three secretion system (TTSS).Invasion of epithelial cells in vivo is observed within 10 to 15 min after introduction of *S. typhi* into the intestinal lumen. The infectious dose is about 10^6 colony forming units. Within 4 h after infection the majority of bacteria tend to get located in the lamina propria from the epithelial cells. This progress can be monitored using electron microscopy. The progression of the disease takes place as depicted in Fig. 4.8.

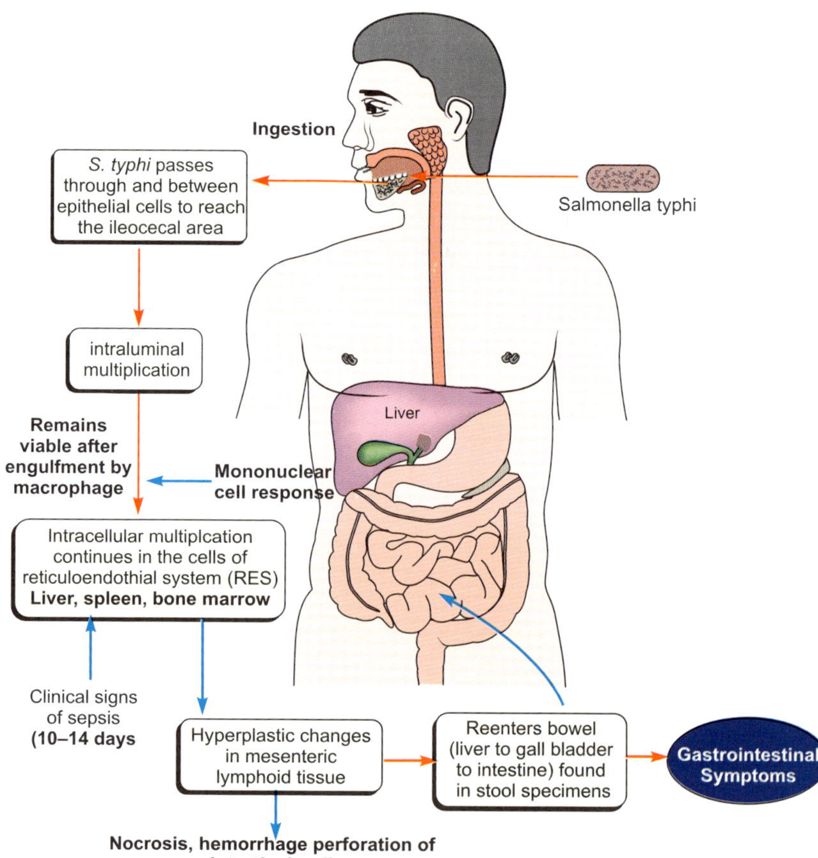

Fig. 4.8: Clinical progression of enteric fever (*Typhoid fever*)

Epidemiology

Typhoid fever is more common in children and young adults than in older patients[21]. Worldwide, typhoid fever is most prevalent in impoverished areas that are overcrowded with poor access to sanitation. Non-epidemic incidence estimates suggest that south-central Asia, Southeast Asia, and southern Africa are regions with high incidence of *S. typhi* infection. The reported cases are around 100 cases per 100,000 person years[22]. Other regions of Asia and Africa, Latin America, the Caribbean, and Oceania have a medium incidence of 10 to 100 cases per 100,000 person years.

A bulletin from WHO (2008) revealed that many countries in South Asia and possibly South-east Asia have a high burden of typhoid fever(http://www.who.int/bulletin/volumes/86/4/06-039818/en /index.html). Recent data from Bangladesh also suggest a very high burden of the disease. WHO findings highlight the complex geographical variation in the disease incidence, age groups affected and level of antimicrobial resistance, all of which need to be considered while implementing typhoid vaccines as tools to assist in the control of typhoid fever. Development of antimicrobial resistance in salmonella led to complications when a vaccine was introduced against typhoid in Vietnam population.

Treatment and Prevention

Salmonella infections usually resolve in 5-7 days and often do not require treatment other than oral fluids. In patients with severe diarrhea may require rehydration with intravenous fluids. Antibiotics, such as ampicillin, trimethoprim-sulfamethoxazole, or ciprofloxacins should be avoided as long as the infection does not spread from the intestine. Indiscrimante use of antibiotics to promote the growth of poultry animals has led Salmonella bacteria developing resistance to antibiotics.

There is no effective vaccine to prevent salmonellosis. Foods of animal origin may be contaminated with Salmonella therefore people should avoid consuming raw or undercooked eggs, poultry, or meat.

Cross-contamination of foods also poses a great risk. Uncooked meats should be kept separate from produce, cooked foods, and ready-to-eat foods.

Several antibiotics are effective for the treatment of typhoid fever. Chloramphenicol was the original drug of choice for many years. Sine the drug has serious side effect, chloramphenicol has been replaced by other effective antibiotics. Ampicillin (Omnipen, Polycillin, Principen) and trimethoprim-sulfamethoxazole (Bactrim, Septra) are frequently prescribed antibiotics although resistance has been reported in recent years. Additional antibiotics are required in case of a relapse.

Safe and effective vaccines have been developed now against typhoid fever. But they have restricted use and are used primarily by affluent travelers visiting developing countries. These are not made available to poor families residing in the developing countries, who account for almost all the global burden of typhoid morbidity and mortality. Despite a recommendation by the World Health Organization for use of Vi vaccines in developing countries, its use has been restricted. Since data is still not sufficient to prove its efficacy in protecting pre-school age children. This age group remains at exceptionally high risk in the areas of poor urban settings. Doubts remain about Vi's ability to confer "herd protection," or protection of unvaccinated residents living in highly vaccinated areas.(http://www.sciencedaily.com/releases/2009/07/090722191212.htm)

B. Campylobacter

Campylobacter Jejuni

These are small, curved-spiral gram-negative bacilli with polar flagella. *Campylobacter jejuni* appear in comma, S-shaped or appear as"gull-wings/sea-gull"forms. Campylobacter predominantly harbour in reproductive and alimentary tracts of some animals. Transmission to humans may occur via a fecal-oral route. The source of 50-70% of all campylobacter infections mostly originate from farm animals, birds, dogs, and processed poultry, with chicken preparation. The organism is transmitted to man through milk, meat products and contaminated water. Undercooked poultry and unpasteurized dairy are most often implicated as potent source of *C. Jejuni* infection (Fig. 4.9)

(A)

(B)

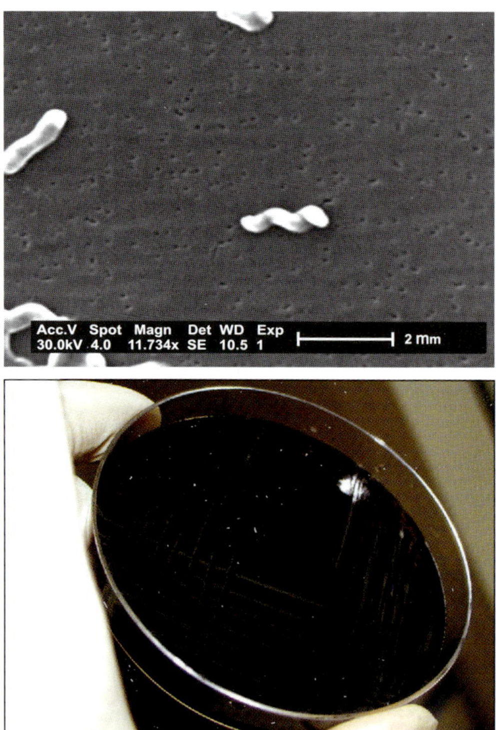

Fig. 4.9A: Scanning electron micrograph depicting a number of *Campylobacter jejuni* bacteria. **B.** Charcoal-based selective medium agar (CSM) for isolation of *Campylobacter jejuni* (***Source:*** http://en.wikipedia.org/wiki/File:Campylobacter_jejuni_5778_lores.jpg)

Pathogenesis

As few as 500 organisms can cause enteritis. The organism is of invasive nature but generally less severe than Shigella. Campylobacter produces adenylate cyclase-activating toxins as produced by *E.coli* LT and cholera. Incubation period of the organism to cause disease ranges from 2 to11 days.

Clinical Features

Patients present with abdominal pain and cramps, diarrhea, malaise, headache, and usually shows symptoms of fever. Typically the patient developes watery diarrhoea, but in severe cases bloody diarrhea may occur. Diarrhea normally last for 2-7 days. However, the patient may shed the organisms through stool for up to 2 months. Bacteremia is observed in a small minority of cases. The disease is usually self-limiting.

Treatment

The infection typically goes away on its own and is not usually treated with antibiotics. Severe symptoms may respond to treatment with antibiotics such as ciprofloxacin and azithromycin. Fluid and electrolyte replacement may be required for serious cases.

C. Yersinia Enterocolitica

It is described as gram negative psychrophilic rod shaped bacterium that is motile only at temperature below 30°C. *Yersinia enterocolitica* is widely distributed in environment and have been isolated frequently from soil, water and animals. The major animal reservoir for *Y.enterocolitica* strains causing human illness is pigs. They are also found in few other animals including rodents, rabbits, sheep, cattle, horses, dogs, and cats. Human pathogenic strains are usually confined to the intestinal tract and lead to either enteritis or diarrhea[23]. Predominant serogroups involved in human illness are designated as O:3, O:8, O:9, and O:5.

Infection is most often acquired due to consumption of contaminated food, raw or undercooked pork products. Transmission of the disease can also occur through contaminated unpasteurized milk or untreated water. This organism can survive and grow when food products are stored in refrigerators

Pathogenesis

Strains that cause human yersiniosis carry a plasmid that is associated with a number of virulence traits. Ingested bacteria adhere and invade M cells or epithelial cells. They exhibit resistance to complement and phagocytosis. They produce hear-stable enterotoxin (ST) only at temperatures below 30°C. The role of ST in the disease process remains uncertain. Incubation period is for 4-7 days. On the basis of phenotypic characteristics,*Y.enterocoliticais* considered to be a heterogeneous group of strains, which are classified by biotyping into 6 biogroups.When they are classified on the basis of serotyping, they are represented as 57 O serogroups, on the basis of their O (lipopolysaccharide or LPS) surface antigen. Five of the 6 biogroups (1B and 2–5) are regarded as pathogens[24]. The virulence factors characterized in *Y. enterocolitica* are located within the chromosome and also on a 70 kb virulence plasmid designated pYV which is only detected in virulent strains. These virulence properties, particularly the plasmid-encoded genes, guide the invading Yersinia pathogen and enable bacteria to survive inside the human host[25]. All pathogenic *Y. enterocolitica* strains, both low and high- pathogenicity isolates, show a thermally responsive adaptation process which aids the transition from the environment to the adverse conditions inside of the human host. The rapidity by which the organism adapts to the change in the environmental conditions for their survival, determines the clinical outcome, as well as the incubation period. Thus, regulation of virulence genes plays a key role in the successful accommodation for the increase in temperature to 37° C when infecting eukaryotic tissue. Virulent yersiniae traverse the intestinal lumen, attach to and penetrate the mucus barrier overlying the mucosal epithelial cells, and eventually adhere to intestinal cells.

Clinical Features

The typical gastroenteritis caused by *Y.enterocolitica* is characterized by fever, abdominal pain, and often leads to bloody diarrhea. Illness generally lasts from 1 to 2 weeks but chronic cases may persist for up to a year. Other than gastroenteritis it may also cause pseudoappendicitis, mesenteric lymphadenitis, and terminal ileitis. *Y. enterocolitica* infections are sometimes followed by chronic inflammatory diseases such as arthritis[26].

Epidemiology

Yersinia enterocolitica and free-living protozoa share the same ecological niches, such as water, vegetables, and anthropogenic environments, *e.g.,* domestic refrigerators. The role of free-living protozoa *Acanthamoeba castellanii* in the ecology and epidemiology of food-borne pathogen such as Campylobacter has recently been described[27]. This interaction identifies a role of free-living protozoa in the ecology and epidemiology of *Y. enterocolitica*. Healthy pigs are frequently colonized with strains that cause human illness, such as serotype O:3 and serotype O:9. Nosocomial transmission has been reported and there have also been reports of transmission by transfusion of stored blood from asymptomatic donors.

Treatment

Despite antibiotic susceptibility patterns varying among serogroups, the organism is usually susceptible to aminoglycosides, cotrimoxazole, chloramphenicol, tetracycline, third generation cephalosporins and fluoroquinolones but is resistant to penicillin, ampicillin and first generation cephalosporins. The intrinsic resistance to these beta-lactam antibiotics is due to the production of two chromosomally encoded beta-lactamase genes, blaA and blaB, encoding for one class A enzyme showing constitutive expression and one inducible class C enzyme (AmpC-type), respectively. The differential expression and activities of these two enzymes determines the level and spectrum of resistance.

D. Listeriosis

Pathogenesis

The disease caused by *Listeria monocytogenes* has a serious manifestation in animals where it was found to be associated with abortions and encephalitis in sheep and cattle. The organism is widely distributed in the environment. It is able to tolerate adverse condition for a longer period of time. It can also multiply and grow at refrigeration temperatures, *Listeria* has since become recognized as an important foodborne pathogen. Both humans and animals are considered to be the carriers of *L. monocytogenes*. The organism can grow in the pH range of 4.4 to 9.6. It is also salt tolerant and relatively resistant to drying, but easily get destroyed by heat. The organism grows between 32°F and 113°F. The disease primarily affects pregnant women, newborn infants and immune compromised individuals. The symptoms of the disease can include mild illness with fever, headaches, nausea and vomiting. Among pregnant women, the organism may lead to intrauterine or cervical infections that may have a serious consequences resulting in spontaneous abortion or still-birth. Even infants who are born alive may develop meningitis at a later stage. During listeriosis, the organism is isolated from blood, cerebrospinal fluid or from sites that are sterile otherwise. Pathogenic listeria enters the host primarily through the intestine. The liver is considered to be first target organ after intestinal translocation. In the liver it actively multiplies until the infection is controlled by a cell mediated immune response.

The pathogenesis of this facultative intracellular pathogen includes phases where the bacteria successfully proliferate in the challenging environment in a variety of mammalian cell types. They can thrive in epithelial cells, endothelial cells, hepatocytes, dendritic cells, and macrophages. While invading the host cells, *L. monocytogenes* has to face adverse growth conditions, such as a rapid pH drop within the phagosome[28]. Following early escape from the primary phagosome by membrane lysis, the pathogen is subsequently released into the cytosol where ithas to encounter conditions of nutrient and iron starvation.

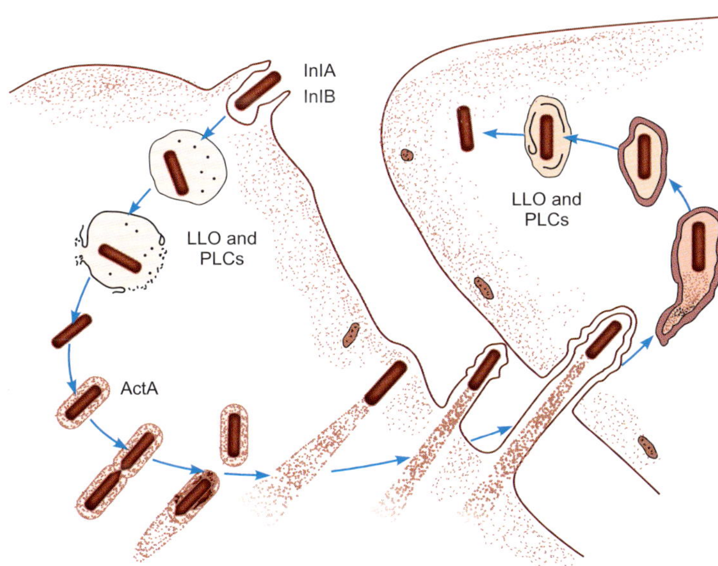

Fig. 4.10: Stages in the intracellular life-cycle of *Listeria monocytogenes*. A Cartoon depicting entry, escape from a vacuole, actin nucleation, actin-based motility, and cell-to-cell spread[26].

As the bacteria escapes from the phagocytic vacuole *Listeria* gets coated with a large population of actin filaments. These filaments rearrange to form a large "tail" of actin up to 5μm long. This complex then moves to the surface of the host cell and produces a large microvillus or, more accurately, a pseudopodal projection with the *Listeria* now at its tip. This pseudopod touches a neighboring cell that phagocytoses the apical end of it.

Mode of infection: The bacteria invade the intestinal epithelium by stimulating listerial internalins (Inl) that binds to the host's adhesion factor E-cadherin (Fig. 4.10). Subsequently, actin polymerization creates the phagocytic cup, an actin-based structure formed around foreign materials by phagocytes prior to endocytosis. Following internalization, the bacterium escapes from the vaculole/phagosome before fusion with a lysosome can occur. The bacterium employs three important virulence factors for their escape. They are listeriolysin O (LLO), phospholipase A (plcA) and phospholipase B (plcB)[30]. The secretion of both the virulence factors helps in the disruption of the vacuolar membrane and makes a passage for the bacterium to escape into the cytoplasm, where it may proliferate. Once inside the cytoplasm, *L. monocytogenes* exploits host actin for the second time using ActA proteins. Actin polymerization then propels the bacterium unidirectionally into the host cell membrane. This is considered to be the probable mode of cell to cell contact. The protrusion that is formed may then be internalized by a neighboring cell, forming a double-membrane vacuole from which the bacterium must escape using LLO and PlcB. This mode of direct cell-to-cell invasion involves a cellular mechanism known as paracytophagy[31].

Epidemiology

The mortality rate in diagnosed cases is reported as 20 to 25 percent. The incubation period extends from a few days to several weeks. Infected cases have been reported due to consumption of raw milk, soft cheeses made with raw milk, and also from raw or refrigerated ready-to-eat meat, poultry or fish products. Major outbreaks of invasive disease have been caused by a specific serotype of *L. monocytogenes* defined as *4b*. Studies have shown that *L. monocytogenes* exist in three separate lineages with lineageI mostly containing epidemic clones, lineageII containing clones associated with sporadic human disease and lineage III represent isolates not found in humans. In India, *L. monocytogenes* has been found to be one of the etiological factors in the causation of abortions and premature births in humans.[32,33]

Treatment and Prevention

Once diagnosed, listeriosis can be treated using antibiotics. So far no vaccine is available against listeriosis. However, listeriosis can easily be prevented following proper food handling practices. Preventive measures for listeriosis include maintaining good sanitation, consuming refrigerated ready-to-eat foods in time, pasteurizing milk and avoiding milk products made out of unpasteurized milk and cooking foods thoroughly. Recently, the Indian Council of Agricultural Research has initiated an outreach program on listeriosis for active surveillance on farm animals. Such a system will help in deriving proper estimates of disease burden among animal and human populations.

E. Shigellosis

Pathogenesis

Shigellosis is commonly known as acute bacillary dysentery. The disease produces inflammatory reactions and ulceration on the intestinal epithelium. The disease manifests as bloody or mucoid diarrhoea. Four important species that can cause Shigellosis are named as *S. dysenteriae*, *S. flexneri*, *S. boydii*, and *S. sonnei*. According to the available reports, at least 54 serotypes or subtypes of Shigella are currently recognized of which *S. dysenteriae* has 16 serotypes, *S. flexneri* has 17 serotypes and subserotypes, *S. boydii* has 20 and *S. sonnei* has a single serotype. The bacteria can survive and transit through the stomach since they are less susceptible to acid than other bacteria. It is observed that a low dose of bacteria as few as 10 to 100 organisms can initiate the disease. Ingested bacteria once they pass into the small intestine start replicating in large numbers and eventually pass into the colon, where they enter the colonic cells. Shigella transmission can occur via contaminated food and water or via direct person-to-person spread. Humans are the only natural reservoir for disease.

Epidemiology

The disease is more prevalent among children causing diarrhoea/dysentery. Children in developing countries of Asia-African regions are more vulnerable to the infection. It is also one of the major cause of foodborne infections in developed countries. During infection, this pathogen secretes a number of virulence factors via the type III secretion system. Recently, Shigella-infected cases have increased considerably in Asia. However the mortality has decreased considerably due to timely nutritional and clinical interventions. Antimicrobials administered in time can reduce the episodes of shigellosis. Due to improper use of antibiotics reports during treatment of diarrhoea, reports on multiple antibiotic resistance in shigellae have increased over the years. Several mechanisms such as plasmids and other mobile genetic elements present in the genome are supposed to be involved in the transmission of resistance in shigellae. Due to poor efficacy of many of the existing Shigella vaccines, none of them has been licensed for use in endemic areas.

Treatment and Prevention

Infection with Shigella is generally self-limited; the average duration of untreated Shigella gastroenteritis is seven days[34]. Antibiotic therapy is not recommended since infection gets cleared spontaneously in most individuals. However when the disease is severe and turns positive with a stool culture, antibiotics are recommended for public health reasons. Antibiotics have been shown to decrease the duration of fever and diarrhea by few days. Shortening the duration of shedding with the administration of antibiotics reduce the risk of person-to-person spread.

E. Clostridium Perfringens

C. perfringens is one of the most common causes of foodborne illness both in developing as well as developed countries. It is estimated that it is capable of causing nearly a million cases of foodborne illness each year. *Clostridium perfringens* produces an enterotoxin that is responsible for food poisoning. In a hospital setting, it also causes wound and surgical infections that may lead to gas gangrene in addition to severe uterine infections. Clostridial hemolysins and extracellular enzymes secreted by the bacteria such as proteases, lipases, collagenase and hyaluronidase mostly contribute to the invasive process.

Pathogenesis

C. perfringens is a ubiquitous, anaerobic, Gram-positive, spore-forming bacillus and a frequent contaminant of meat and poultry. Important symptoms resulting due to *C. perfringens* food poisoning can be characterized by onset of abdominal cramps and diarrhea 8-16 hours after consuming the contaminated meat or poultry items. Using the process of sporulation this organism can survive high temperatures during initial cooking. The spores can germinate during cooling of the food. The vegetative forms of the organism multiply if the food is subsequently held at temperatures of 60-125°F (16-52°C). In case the same food is served without adequate reheating, there is a possibility of ingesting live vegetative forms of *C. perfringens*. The bacteria then secrete the enterotoxin that causes the characteristic symptoms of diarrhoea and abdominal cramping.

Infection and Mode of Transmission

The organism is classified into into 5 types (A-E) on the basis of its ability to produce one or more of the major lethal toxins, alpha, beta, epsilon and iota (α, β, ε, and ι). The type A enterotoxin (CPE)-producing (cpe+) producing *C. perfringens* is reported to be one of the most common food poisoning agents worldwide. The same organism is also implicated as a causal organism in 5%-15% of antibiotic associated diarrhea (AAD) and sporadic diarrhea (SD) cases in humans, as well as diarrhea cases in animals.

Most food poisoning strains studied carry *cpe* gene in their chromosomes; isolates from AAD and SD cases bear *cpe* in a plasmid. Greater heat resistance of the strains with chromosomally located *cpe* is responsible for their survival in cooked food, thus causing instances of food poisonings. The presence of *C. perfringens* strains with chromosomally located *cpe* has been reported in 1.4% of American retail food. This confirms that these strains have an access to the food chain, although sources and routes of contamination are unclear.

An explanation for the strong association between *C. perfringens* strains with plasmid-located *cpe* and cases of AAD and SD disease is thought to be due to in vivo transfer of the *cpe* plasmid to *C. perfringens* strains of the normal intestinal microbiota. Thus, a small amount of ingested *cpe+ C. perfringens* would act as an infectious agent and transfer the *cpe* plasmid to *cpe-C. perfringens* strains of the normal microbiota. This process of conjugative transfer of the *cpe* plasmid has also been demonstrated in vitro, but so far no data exist on horizontal gene transfer of *cpe* in vivo. It is also doubtful whether *cpe+* strains that cause AAD and SD are resident in the gastrointestinal tract or acquired before onset of the disease.

G. Diarrheagenic E. Coli

Diarrhoea is one of the major causes of morbidity and mortality among children below fiveyears of age in developing countries with global mortality rates estimated to be around 1.87 million[35]. In India, around 23% of childhood mortality has been reported due to diarrhea[36]. Diarrheagenic Escherichia coli (DEC) are reported to be a common cause of childhood diarrhea in developing countries with a higher incidence during the first 2 years of life[37].

Pathogenesis and Virulence

The most highly conserved feature of diarrheagenic *E. coli* strains is their ability to colonize the intestinal mucosal surface despite peristalsis and competition for nutrients by the indigenous flora of the gut including other *E. coli* strains. The presence of surface adherence fimbriae is a property of all *E. coli* strains, including nonpathogenic varieties. However, diarrheagenic *E. coli* strains acquire specific fimbrial antigens that enhance their intestinal colonizing ability. It also helps the bacteria to adhere to the small bowel mucosa, a site that mostly remains uncolonized[38,39]. Once colonization is established, the pathogenetic strategies adopted by the diarrheagenic *E. coli* strains exhibit remarkable variety. Three different clinical syndromes result from infection with pathogenic *E. coli* strains. They are categorized as enteric/diarrhoeal disease, urinary tract infection and sepsis/meningitis[40]. Interestingly, as long as these bacteria do not acquire genetic elements encoding virulence factors, they remain benign commensals. Strains that acquire 'foreign' DNA encoding enterotoxins, adhesins or invasion factors become virulent and can cause either a plain, watery diarrhoea or inflammatory dysentery. Three general paradigms have been described by which *E. coli* may cause diarrhea (*i*) Enterotoxigenic *E. coli* (ETEC), (*ii*) enterohemorrhagic *E. coli* (EHEC) and Enteropathogenic *E.coli* (EPEC), (*iii*) enteroinvasive *E. coli* (EIEC). Their clinical and epidemiological features are described in Table 4.2. EHEC is characterized by the presence of two toxin genes, stx1 and stx 2, for toxins that are similar to those of Shigella dysenteriae. The presence of 1 or more heat stable (stI and stII) and labile (ltI and ltII) toxin gene characterizes ETEC strains. In case of EIEC, the invasive property of EIEC is determined by the presence of a 140-MDa virulence plasmid (pInv), which encodes a number of genes for invasion that includes virf, ipah, ipal genes[41].

Table 4.2: Clinical, epidemiological and virulence factors of various E.coli pathotypes[39]

Pathotype	Clinical features	Epidemiological features	Virulence factors
Enteropathogenic	Watery diarrhea and vomiting	Infants in developing countries	Bundle-forming pilus, attaching and effacing
Enterohemorrhagic	Watery diarrhea, hemorrhagic colitis, hemolytic uremic syndrome	Food-borne, water-borne outbreaks in developed countries	Shiga toxins, attaching and effacing
Enterotoxigenic	Watery diarrhea	Childhood diarrhea in developing countries, Traveller's diarrhea	Pili, heat-labile and heat-stable enterotoxins
Enteroaggregative	Diarrhea with mucus	Childhood diarrhea	Pili, cytotoxins
Enteroinvasive	Dysentery, watery diarrhea	Food-borne outbreaks	Cellular invasion, intracellular motility
Uropathogenic	Cystitis, pyelonephritis	Sexually active women	Type I and P fimbriae, hemolysin, pathogenicity islands
Meningitis – associated	Acute meningitis	Neonates	K1 capsule, S fimbriae, cellular invasion

Epidemiology

EPEC is an important category of diarrheagenic *E. coli* which has been linked to infant diarrhea in the developing world. In contrast to the low incidence rate of infection in developed countries, EPEC is a major cause of infant diarrhea in developing countries. Children belonging to 0- to 6-month age group are most vulnerable to EPEC infections. Studies in Brazil[43], Mexico[44] and South Africa[45] have reported that 30 to 40% of infant diarrhea can be attributed to EPEC. In some studies it has been shown that the rate in incidence of EPEC infection exceeds rotavirus infection. EPEC strains remain an important cause of disease in all settings—nosocomial outbreaks, outpatient clinics, patients admitted to hospitals, community-based longitudinal studies, and urban and rural settings. Several studies have shown that breast feeding is protective against diarrhea due to EPEC.

Indian scenario: About 78.26 % atypical EPEC from total EPEC isolates were isolated from the faecal specimens of diarrhea patients in the Kashmir valley[46]. EPEC subtypes were prevalent in children with diarrhea in Delhi and the NCR[47]. Studies on the antibiotic-susceptibility patterns revealed that atypical EPEC are emerging as colonizers of the intestines in children with and without diarrhea, and appropriate usage of antibiotics is required to control the emergence of drug-resistant atypical EPEC strains. Studies also indicated that EPEC do not pose a threat to human health in India.

Epidemiology of ETEC

Studies over the last few years have documented that ETEC is usually a frequent cause of diarrhea in infants younger than two years of age. In Egypt it was found to be the most common cause of diarrhea in infants, accounting for about 70% of the first episodes[48]. The incidence rate was found higher in males than females. In a detailed investigation in children 0 to 5 years in Bangladesh, 90% of cases of ETEC diarrhea reporting to the hospital were children aged from three months to two years[49].

Treatment and Prevention

The treatment of diarrheal disease due to ETEC is the same as that for cholera or any other acute secretory diarrheal disease. The correction and maintenance of hydration is always most important. Antimicrobials are useful only when the diagnosis is correctly done. It is important to distinguish between ETEC and cholera. Provision of adequate nutrition is critical in children in the developing world, where all diarrheal diseases are frequent. The guidelines for therapy of all diarrheas have been widely disseminated by the World Health Organization[50].

Childhood diarrhea is caused not only by ETEC but also by other bacterial and viral agents therefore the clinical presentations are not adequate to differentiate them. Therefore, it has been difficult to study the effect of antimicrobials in children with ETEC disease and therefore antimicrobials are not used routinely in treatment of childhood diarrhea. For adults, a short course of ciprofloxacin, 500 mg every 12 h for 1 day is recommended. This treatment usually stops the illness within 24 h. A new antimicrobial rifaximin has recently been introduced and found to be effective for treatment of traveler's diarrhea in adults. The dose of the drug prescribed is 200 mg two times a day for 3 days. For children, the recommend drug has been azithromycin, 10 mg/kg/day for 2 days, however detail confirmation is still awaited. Although antimicrobials are not used routinely in treatment of children with ETEC diarrhea, the emerging problem of multiple antimicrobial resistance is a cause of concern for the drugs used for traveler's diarrhea. Serious efforts need to be made to improve the awareness of the importance of ETEC, particularly on the health of infants and children living in the developing world.

Based on the great impact of ETEC infections on morbidity and mortality and probably also on nutritional status, particularly in children where the disease is endemic, an effective ETEC vaccine is highly desirable[51]. Such a vaccine is feasible since epidemiologic evidence and results from experimental challenge studies with human volunteers have demonstrated that specific immunity against homologous strains follows ETEC infection. A multitude of *E. coli* pathotypes can be identified causing distinct diseases. Genetic variation is mostly acquired through the horizontal spread of virulence factors present in certain lineages that are inherently more pathogenic. This is perhaps the reason for observed diverse clinical entities.

H. Vibrio Cholerae

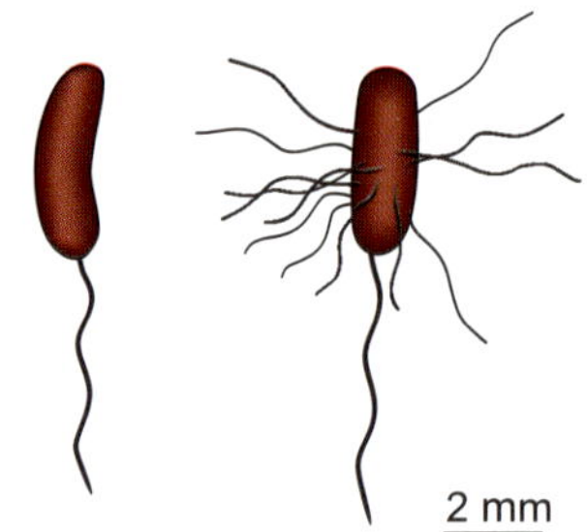

Fig. 4.11: Diagram of the bacterium *Vibrio cholerae*

Cholera is a diarrheal illness caused by infection with the gram-negative bacterium *Vibrio cholerae*. They are comma-shaped rods and possess single polar flagellum. Colonization of the small intestine is mediated by toxin co-regulated pili (Fig. 4.11).

The disease is characterized by severe, watery diarrhea, which can rapidly lead to dehydration and death in case the patient remains untreated. *V.cholera* O1 and O139 serotypes are responsible for epidemic and pandemic cholera in humans. All other serotypes are grouped as "non-O1" strains; they are associated with sporadic cases of gastroenteritis. Many cases of *Vibrio*-associated gastroenteritis are under represented since confirmatory tests using culture based methods are not employed. The routine diagnosis in the clinical laboratories is carried out using the selective medium, thiosulfate-citrate-bile salts- sucrose (TCBS) agar for processing of stool specimens.

Pathogenesis

V cholerae O1 is the primary causative agent of cholera. Infection with this organism manifests in the form of profuse watery diarrhea, vomiting, and muscle cramps. Cholera is a dehydrating diarrheal illness that results in substantial loss of fluid and electrolytes. At some occasion, stool volumes may reach close to 1 L/h. The spectrum of illness in cholera includes asymptomatic infection (75%), mild illness (18%), moderate illness (5%), and severe illness (2%). Severe diarrhea may result in hypovolemic shock when the patient loses blood and fluid from the body at a fast rate due to which death may occur within a few hours if the patient is not treated in an emergency. The incubation period of cholera is usually 2 to 3 days and it may range from 6 hours to 5 days. Severe illness has been associated with high-dose exposure, low gastric acidity, and in the individuals having blood group O. Severe cholera may be characterized by "rice water" stools, sudden loss of 10% or more of body weight, loss of normal skin turgor, dry mucous membranes, sunken eyes, lethargy, anuria, weak pulse, altered consciousness, and eventually may lead to circulatory collapse. Diarrheal fluid loss may result in profound hypokalemia, metabolic acidosis due to loss of bicarbonate and result into renal failure. Severe infections may result in death if the dehydration is not treated aggressively with fluid and electrolyte replacement. In 1992, toxigenic *V cholerae* O139 (the Bengal strain) was recognized as another cause of cholera that was first discovered on the Indian subcontinent[52].

Cholera Toxin: A diarrhoeogenic exo-enterotoxin in cell-free culture filtrates from *Vibrio cholerae* of classical biotypewas demonstrated for the first time by S.N. De in 1959. Among various other microbial toxins studied, research efforts concentrated on cholera toxin has given much insight in understanding the mechanism of action of this toxin. CT is secreted by *V. cholerae* and is transported extracellularly by the so-called type II secretion system. This system comprises more than a dozen interacting proteins and serves to export toxin and other proteins such as extracellular enzymes from the periplasm across the outermembrane[54]. CT is made up of two types of subunits (Fig. 4.12). The larger A subunit (240 amino acids; MW 28 kD) is located centrally, while the five B subunits (103 amino acids;

MW 11 kD each and the aggregate MW ~56 kD) are located peripherally. The upper A1 domain of the wedge-shaped A subunit is held above the plane of the doughnut-shaped pentameric B subunits by the tethering A2 domain. In case of CT, it is an alpha helix for almost its entire length. The beta subunits are arranged in a doughnut fashion, creating an opening for thecarboxy-terminal of the A2 to pass through. The four carboxy-terminal residues of the A2 chain consist of Lys-Asp-Glu-Leu (K-D-E-L).

The intracellular journey of the Cholera toxin begins after binding to the ganglioside GM1 located on the outer leaflet of apical membranes of intestinal epithelial cells. GM1 functions like a vehicle that transports the toxin all the way backwards in the secretory pathway from the plasma membrane to endoplasmic reticulum.

As GM1 forms a cluster by binding to the B-subunit, it gets associated with membrane microdomains that are termed as "lipid rafts". These lipd rafts are composed of gangliosides, cholesterol, GPI-anchored proteins and sphingomyelin. The association of CT with lipid rafts appears to be crucial for toxicity. As the endocytosis progresses, CT or the CTB pentamer travels to the endoplasmic reticulum (ER) via a retrograde transport pathway. Depending on the cell type, this pathway may be either Golgi dependent or independent. The association of the CT-GM1 complex with the actin cytoskeleton takes place via lipid rafts. The actin cytoskeleton plays an important role in CT trafficking from the plasma membrane to the Golgi-ER. After CT gets into the ER, CTA dissociates from CTB. The KDEL sequence of the carboxy-terminal of CTA2 holds a classical eukaryotic signal for retention in the lumen of ER.

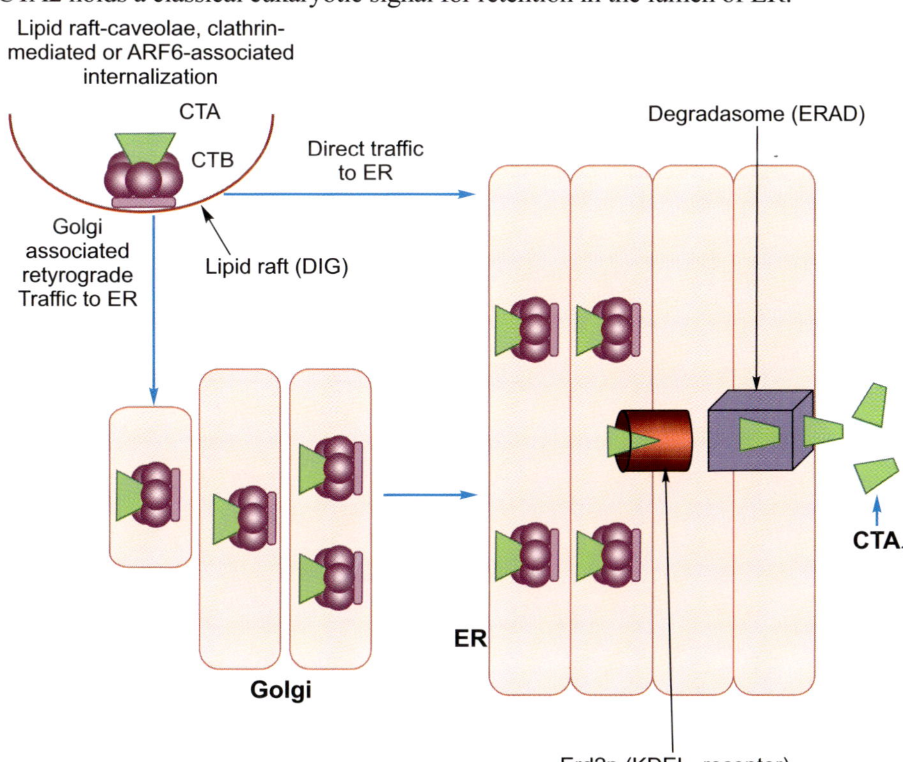

Fig. 4.12: Cholera toxin intracellular trafficking (Adapted from Sánchez J and Holmgren J (2011)[52]

The translocation of CTA1 to the cytosol is mediated by a natural recycling cellular process. This process is referred as ER-associated degradation or degradasome pathway (ERAD). Through this pathway, misfolded proteins are normally retrieved from the ER for their degradation in the cytosol. Although CTA1 is transported by the degradasome, however, CTA1 apparently escapes proteolysis. This is due to its low content of lysine, which is the important target for ubiquitinylation. In order to carery out the process of cell intoxication, the A subunit of CT needs to be transported to the cytosol to induce the activity of adenylate cyclase (AC). After ADP-ribosylation by CT, the AC remains in its GTP-bound state resulting in enhanced AC activity leading to an increased intracellular cAMP concentration. Due to the production of higher levels of cAMP, an imbalance in electrolyte movement in the epithelial cell occurs. This results in a decrease in sodium uptake together with an increase in anion extrusion, mostly chloride with the help of cystic fibrosis trans-membrane conductance regulator (CFTR)[56]. Decreased sodium uptake further reduces water intake by the enterocyte. At the same time, augmented chloride and bicarbonate extrusion gives rise to sodium outflow and thus water secretion follows. The combined effect produces vast fluid loss from the intestine. In extreme cases water loss could be up to 2 liters per hour

Epidemiology

Cholera is a major cause of epidemic diarrhea throughout the developing world. Cholera is identified as an ongoing global pandemic in Asia, Africa and Latin America for the last four decades. In the United States, incidence of cholera is very low and recorded as 0-5 cases per year. If left untreated, 25-50% of typical cholera cases are fatal. Large epidemics are often associated to fecal contamination of water supplies or street vended foods. Due to the migration of large population into urban centers in developing countries where existing water and sanitation infrastructures are not adequate to cater the large population. This increases the disease risk. Epidemics are considered to be a marker for poverty and lack of basic sanitation.

Brackish and marine waters are natural environment for the etiologic agents of cholera including *Vibrio cholerae* O1 or O139. So far no animal hosts have been identified for *Vibrio cholera*. However, the bacteria normally attach themselves easily to the chitin-containing shells of crabs, shrimps, and other shellfish. Therefore, these items become source for human infections when eaten raw or undercooked.

Only toxigenic strains of serogroups O1 and O139 have caused widespread epidemics and are reportable to the World Health Organization (WHO) as "cholera". *V. cholerae* O1 has two biotypes, Classical and El Tor, and each biotype has two distinct serotypes, Inaba and Ogawa.

The symptoms of infection are indistinguishable between the two biotypes. Higher proportion of persons infected with the El Tor biotype remains asymptomatic or suffer from a mild illness. In recent years, infections with the Classical biotype of *Vibrio cholerae* O1 have dropped considerably and now limited to parts of Bangladesh and India.

Recent epidemiological data suggest that an individual is more susceptible to cholera infection and other diarrheal infections if they possess O blood group while those with type AB are more resistant to infection. Between these two extremes are the A and B blood types, where individuals with type A show more resistance to the disease compared to type B group. The dose of bacteria required for establishing the disease in a healthy individual identified as many as 10^{11} organisms.

Increased susceptibility towards the infection may be observed in those with a weakened immune system and individuals with decreased gastric acidity, or those who are malnourished. In areas without clean water or sewage disposal or in the make-shift shelters where displaced populations need to take shelter during natural disaster or during conflict, cholera can spread quickly and have a case fatality rate of as high as 50% in vulnerable groups with limited medical care. The World Health Organization (WHO) reports have confirmed the emergence of new, apparently more virulent, strains of *V. cholera* O1 which now predominate in parts of Africa and Asia. It also indicates the emergence and spread of antibiotic resistant strains. Annual global figures reported by WHO in 2009 included 221,226 cases and 4,946 deaths from 45 countries[57]. The serogroup, *V. cholerae* O139, was discovered as the cause of cholera epidemics in India and Bangladesh in 1992. Since then it has spread to other countries in South East Asia. However, apart from a few imported cases, this serogroup is not known to have occurred outside Asia[58].

Indian Scenario

The first confirmed outbreak of cholera due to *V. cholerae* O139 occurred in October 1992. It began in Madras and spread to Madurai and Vellore. In early 1993, similar epidemics began in Calcutta and southern Bangladesh[59]. Adult cases predominate, which suggests that populations in areas where cholera is endemic are not immune. The organism has spread more quickly than the el tor serogroup. Indigenous cases have been identified in Bangladesh, China, Thailand, Nepal and Malaysia, and imported cases in the United Kingdom, Estonia, Japan, Switzerland, Germany, and the United States. It took the el tor biotype about two years to displace endemic classical cholera in India between 1964 and 1966. *V. cholerae* O139 became the dominant strain in India in less than two months, although the el tor biotype has not been completely displaced[60]. A comparative epidemiological study conducted by Deen et al, 2008 in endemic areas of Asia and Africa revealed that the lowest overall rate was found in Jakarta, where the estimated incidence was 0.5/1000 population/year. The incidence was three times higher in Kolkata (1.6/1000/year) and eight times higher in Beira, Mozambique (4.0/1000/year). In all study sites, the greatest burden was in children under 5 years of age[61].

Treatment and Prevention

During severe disease condition, rapid fluid replacement with a balanced solution of sugar, electrolytes and water termed as oral rehydration saltsis the first line of treatment to be given orally. However, severely dehydrated cases may require intravenous administration. Administration of the antibiotic usually a tetracycline is effective if the organism is sensitive to the antibiotic. This can control the symptoms and also decrease the intestinal excretion of the organism. Patients who are promptly treated normally respond rapidly and recover. For the majority of travellers, advice on food and water hygiene precautions is the most appropriate prevention strategy.

Vaccines are effective for people living where cholera is endemic. They offer significant degrees of protection for up to two years after a single shot. There is one cholera vaccine licensed for use in the UK, Dukoral TM (Crucell), which protects against infection caused by *V. cholerae* serogroup O1. It is inactivated, thiomersal free, and does not contain live organisms, so therefore cannot cause cholera[62]. Each dose contains approximately 1×10^{11} bacteria of *V. cholerae*, combined with recombinant cholera toxin B : 01 Inaba classical biotype (heat inactivated), 01 Inaba El Tor biotype (formalin inactivated), 01 Ogawa classical biotype (heat-inactivated) and 01 Ogawa classical biotype (formalin inactivated).Recombinant cholera toxin B subunit (rCTB) 1 mg (produced in *V. cholerae* 01 Inaba, classical biotype). The vaccine can be considered for travellers visiting areas with cholera of *V. cholerae* serogroup O1 strain. This is also recommended to avoid cholera outbreaks and in the areas with limited access to medical care. This vaccine may also offer some protection against diarrhea caused by *Escherichia coli* producing a heat labile enterotoxin.

4.2.2 Viral Infections of the lower intestinal tract- Pathogenesis, Epidemiology, Treatment and prevention

Viral gastroenteritis commonly referred as stomach flu. Rotavirus and norovirus are common infectious viruses that infect the gastrointestinal tract. Viral gastroenteritis causes millions of cases of diarrhea each year. Viruses are recognized as important causes of childhood diarrhea. Norwalk virus was the first virus identified as a cause of gastroenteritis. Since then the number of viral agents associated with diarrheal disease in humans steadily increased. Rotavirus is the most common cause of severe diarrhea in children under 5 years of age. Other than Rotavirus, Astrovirus, Calicivirus and enteric adenovirus are also reported as etiologic agents of acute gastroenteritis. Few other viruses, such as toroviruses, coronaviruses, picobirnaviruses and pestiviruses are recently identified as causative agents of diarrhea.

4.2.2.1 Rotavirus

Rotavirus is the leading cause of childhood diarrhea worldwide, causing an estimated 600,000 deaths every year. Rotavirus is a double-stranded RNA virus. It belongs to the family *Reoviridae*. The virus is composed of three concentric shells that enclose 11 gene segments. The outermost shell contains two important proteins—VP7, or G-protein, and VP4, or P-protein. VP7 and VP4 define the serotype of the virus and induce neutralizing antibody that is probably involved in immune protection (Fig. 4.13). Rotavirus is very stable and may remain viable in the environment for weeks or months if not disinfected.

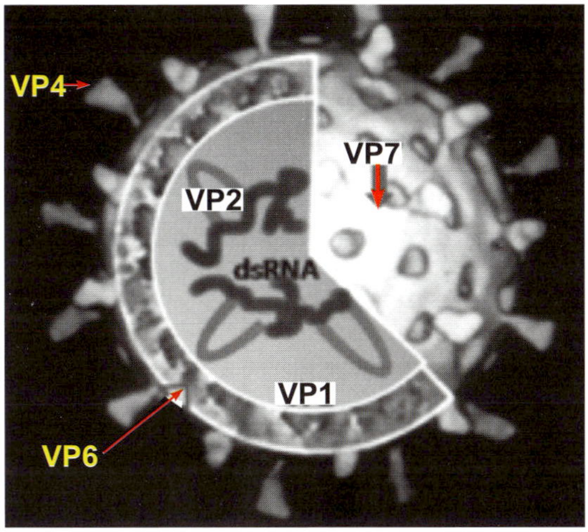

Fig. 4.13: A simplified diagram of the location of rotavirus structural proteins
Source: en.wikipedia.org-475 × 416

Pathogenesis

Rotavirus is recognized as the most common cause of severe gastroenteritis in infants and young children. The virus enters the body through the oral route. Replication of the virus takes place in the villous epithelium of the small intestine. During the viral infection an important change in the intestinal absorption of metabolites occurs. This result in decreased absorption of sodium, glucose, and water and also decreased levels of intestinal lactase, alkaline phosphatase, and sucrase activity that eventually lead to isotonic diarrhea. The incubation period for rotavirus diarrhea is only for 48 hours. The clinical manifestations of infection vary and depend on the age of the child and the number of episodes. The first infection after 3 months of age is generally the most severe. Reinfections are usually less severe. Infection may remain asymptomatic or just may cause self-limited watery diarrhea. In case of severe dehydrating diarrhea, the patient developes fever and vomiting. It has been estimated that around one-third of infected children may develop temperature greater than 102°F (39°C). The gastrointestinal symptoms generally resolve in 3 to 7 days.

Children who are immunocompromised because of congenital immunodeficiency or because of bone marrow or solid organ transplantation normally experience severe or prolonged rotavirus gastroenteritis. In some cases, the patients show evidence of abnormalities in multiple organs, particularly the kidney and liver.

Epidemiology

Rotavirus occurs throughout the world. The prevalence of rotavirus strains varies with geographic area. The reservoir of rotavirus is the gastrointestinal tract and stool of infected humans. Rotaviruses are shed in high concentration in the stool of infected persons. Transmission is by fecal-oral spread, and through close person-to-person contact. Rotaviruses are also probably transmitted by other modes such as locally contaminated food and water. They may also get transmitted through respiratory droplets. Rotavirus is designated as highly communicable as evidenced by the incidence of universal infection in children by the age of 5 years. Infected persons shed large quantities of virus in their stool beginning 2 days before the onset of diarrhea and for up to 10 days after onset of symptoms. Rotavirus may be detected in the stool of immunodeficient persons for more than 30 days after infection. The virus spread at a fast rate within families, institutions, hospitals, and child-care settings .The rapid evolution of rotavirus strains and the emergence of new strains takes place possibly through the transmission of viruses across species or through reassortment between animal and human rotaviruses. Therefore, preventive measures must include intensive strain surveillance. This is as an important component of any vaccine implementation programme. These measures are absolutely necessary in countries such as India where the mortality and economic burden associated with rotavirus is high[63].

Treatment/Vaccination

The Advisory Committee on Immunization Practices (ACIP) recommends routine rotavirus vaccination of all infants without a contraindication. The vaccine should be administered as a series of either two or three oral doses for RV1 and RV5 respectively, beginning at 2 months of age[64]. The vaccination series for both vaccines may be started as early as 6 weeks of age. Subsequent doses in the series should be spaced out from the previous dose by 1 to 2 months. Rotavirus vaccine can be administered at the same visit when other vaccines are given at these ages.

Rotavirus vaccine is contraindicated for infants who are known to have a severe allergic reaction (anaphylactic) to a vaccine component or following a prior dose of the vaccine. Latex rubber is contained in the RV1 oral applicator therefore it is not recommended for infants with a severe allergy to latex however, the RV5 dosing tube is latex free.

4.2.2.2 Enteric Adenoviruses

Enteric adenoviruses, including serotypes 40 and 41, are one of the commonest causes of pediatric viral gastroenteritis, secondary only to rotaviruses. The virus is transmitted via the fecal-oral route. The incidence of enteric adenovirus infection is variable. In industrialized countries, the incidence varies from 1% to 8%, whereas in developing countries it varies from 2% to 31% [65]. Enteric adenovirus infection can be detected by their morphology under electron microscope. They can also be detected by agglutination tests, enzyme-linked immunosorbent assay, or by PCR.

Adenoviruses are resistant to lipid disinfectants. These viruses can be inactivated by formaldehyde and chlorine. Adenoviruses can be inactivated by contact with 1:5 dilution of bleach for 1-2 minutes, and by contact with alcohol-based hand gels. Adenovirus can be inactivated by heating to 56 °C for 30 min, 60 °C for 2 min, and autoclaving can dompletely destroy infectivity. Adenovirus serotype 40 is also sensitive to UV radiation.

4.2.2.3 Norwalk Virus and Other Caliciviruses

Infections caused by these classes of virus also cause severe diarrhea. Various other symptoms associated with this diarrhea are vomiting, abdominal cramps, myalgias, malaise, headache, nausea, and low-grade fever. The transmission of the virus takes place by ingestion of virus particles through oral route. Viruses grow in the small intestine and cause lesions in intestinal mucosa. The viruses are shed along with feces[66].

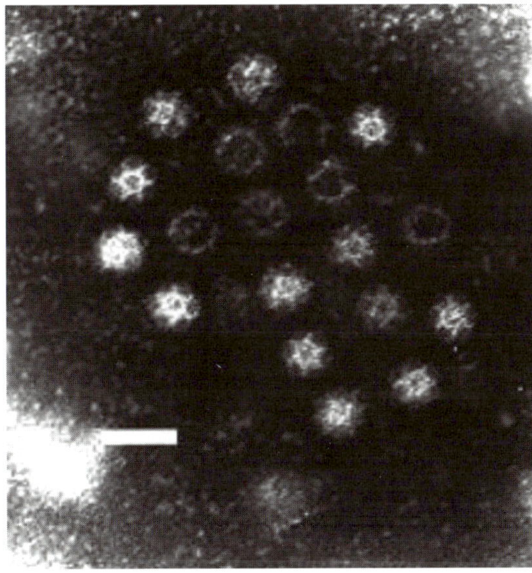

Fig. 4.14: Distinctive star–like structure of Calicivirus
Source: www.virology.net

The caliciviruses are grouped on the basis of morphology, size, protein profile and nucleic acid content. Norwalk virus and other human caliciviruses share considerable genetic homology but are antigenically distinct. The caliciviruses are small, nonenveloped RNA viruses that have characteristic cup-shaped depressions on a spherical capsid surface. Caliciviruses have a distinctive structure (Fig. 4.14). These are also described as a 'star of David', and are also distinctive in their size (27 to 35 nm). They possess only a single major polypeptide and contain only a single-stranded RNA genome.

The disease occurs worldwide and is common in older children and adults. Outbreaks can emerge in camps, schools, nursing homes etc through consumption of contaminated foods, water and uncooked foods.

No specific antiviral therapy is available for Norwalk or other human calicivirus gastroenteritis. Prospects for vaccine development seem poor because of the complex pattern of clinical immunity to Norwalk virus. The development of a vaccine that can offer long-lasting immunity is counterproductive since it can not offer immunity in rechallenged volunteers. Hand washing and careful monitoring of water purification are the most important measures to control the infection.

4.2.3 Fungal Infections-Pathogenesis, Epidemiology, Treatment and Prevention

Fungal infections are becoming more prevalent with the increase in use of immunosuppressive drugs for the treatment of malignant diseases and organ transplant patients. Mycotoxin producing fungi are known to be the aetiological agents. In case they are ingested may lead to severe complications like blood diseases, disorder in nervous system, kidney and liver damage and in some cases it may also be a cause of cancer. Diagnosis is based on detection of the fungi producing mycotoxin from suspected food items.

1. ***Ergot Poisoning:*** *Claviceps purpurea*, a fungus causing smut infections on grain crops like rye, wheat and barley has been identified as one of the major causes of Ergot poisoning. The fungus infected cereal upon ingestion release the toxin that can restrict blood flow in the limbs, with resulting gangrene. It amy also cause hallucinogenic symptoms. The ergot sclerotium contains high concentrations (up to 2% of dry mass) of the alkaloid ergotamine, a complex molecule consisting of a tripeptide-derived cyclol-lactam ring connected via amide linkage to a lysergic acid (ergoline) moiety, and other alkaloids of the ergoline group that are biosynthesized by the fungus[67].

2. ***Aflatoxin Poisoning:*** Aflatoxin is a mycotoxin produced by the fungus *Aspergillus flavus*. Corn, cottonseed and peanuts are the crops most at risk of being contaminated by aflatoxins. Aspergillus infection is also encountered on beans, rice, tree nuts and wheat. The fungus infection is less often seen on other grains and nuts. Aflatoxin poisoning can cause serious damage to livestock when their feed is contaminated with *A. flavus.*

There are three main types of aflatoxin mycotoxins:

(*a*) **Aflatoxins B:** This group consists of two members -aflatoxin B1 and B2. Aflatoxin B1 is quite common and considered to be the most toxic and carcinogenic.

(*b*) **Aflatoxins G:** This group includes aflatoxin G1 and aflatoxin G2

(*c*) **Aflatoxins M:** This group includes aflatoxins M1 and M2. These aflatoxins are metabolic products which are found in the urine and milk produced by animals which have been given feed containing aflatoxins in it.

Developed countries prohibit high levels of aflatoxin mycotoxins in foods. The level of aflatoxins set by United States is under 20 parts per billion in food and specifies that the aflatoxins M must be below 0.5 parts per billion in milk. Once the level cross the permissible limits, it may cause acute toxicity and may even lead to death in birds, fishes and even mammals including humans. Among various other Porgans of the body it mostly affects the liver. High levels of aflatoxin have also been recorded in the lungs, kidneys, brains and hearts of individuals suffering from acute aflatoxicosis. Acute necrosis and cirrhosis of the liver is typical, along with haemorrhaging and oedema. LD 50 (lethal dose) values for animals vary between 0.5 and 10 mg/kg body weight.

4.2.4 Parasitic Infections

Parasitic infections, caused by intestinal helminths and protozoan parasites, are among the most prevalent infections in humans residing in developing countries. In developed countries, the protozoan parasites are the common cause of gastrointestinal infections compared to helminths. Intestinal parasites cause a significant morbidity and mortality in endemic countries[68].

4.2.4.1 Intestinal Helminthes and Nematodes:Pathogenesis, Epidemiology, Treatment and Prevention

Pathogenesis

The intestinal helminthes are worms with multinucleated cells (Fig. 4.15).

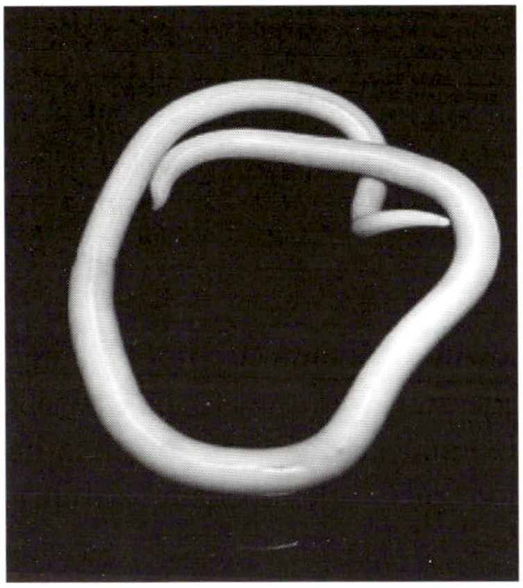

Ascaris lumbricoides (round worm)

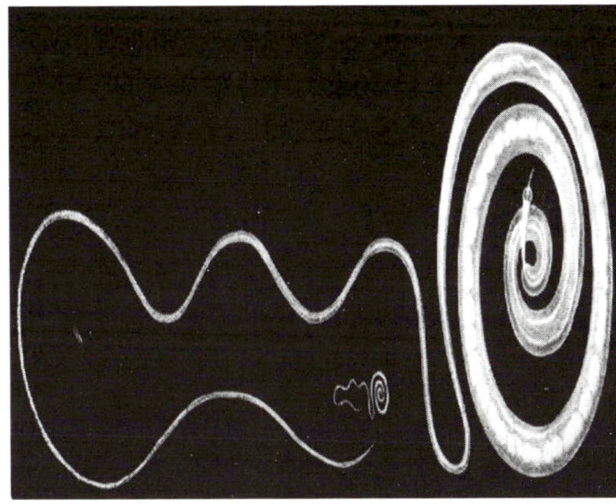

Trichuris trichiura (whip worm)

Strongyloides stercoralis
(thread worm)

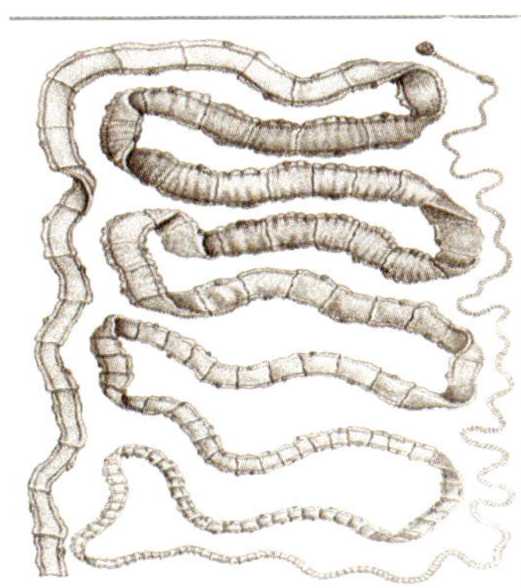

Taenia solium
(tape worm)

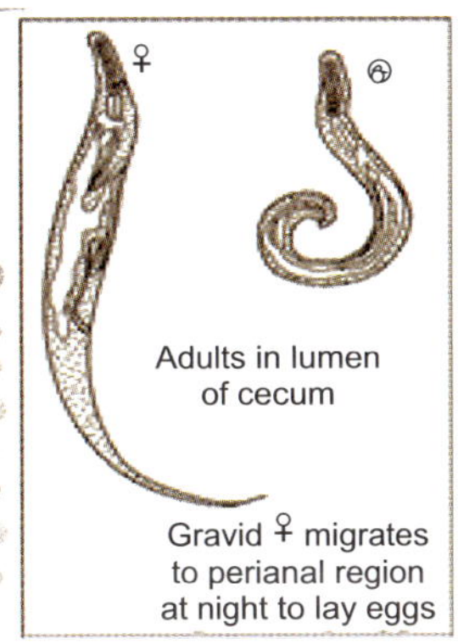

Enterobius vermicularis
(tape worm)

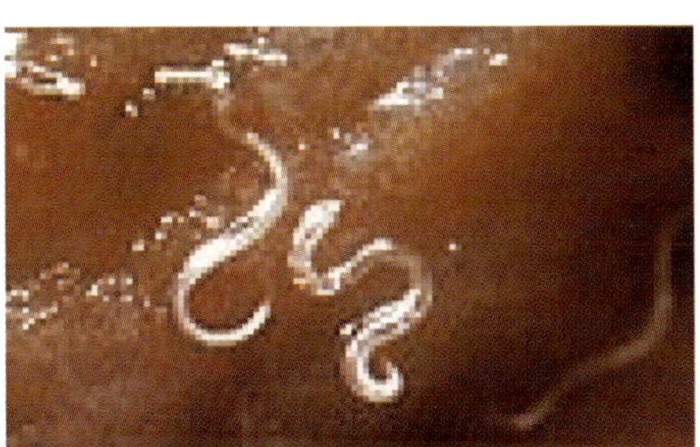

Ancylostoma caninum (hook worm)

Trichinella spiralis (pork worm)

Fig. 4.15: Common worms inhabiting human gut.
Source: CDC Division of Parasitic Diseases)

Nematodes (roundworms), cestodes (tapeworms), and trematodes (flatworms) are among the most common helminths that inhabit the human gut. There are four species of intestinal helminthic parasites known as either geohelminths or soil-transmitted helminthes (STH): *Ascaris lumbricoides* (roundworm), *Trichiuris trichiuria* (whipworm), *Ancylostoma duodenale*, and *Necator americanicus* (hookworms).Infection occurs through accidental ingestion of eggs (*A. lumbricoides and T. trichiura*) or penetration of the skin (by hookworm larvae).

The *A. lumbricoides* is a roundworm that infests the entire small intestine; the adult hookworm of the *Necator* and *Ancylostoma* genera parasite the upper part of the human small intestine; and the adult *T. trichiura* (whipworms) lives in the large intestine, especially in the cecum. The soil-transmitted helminthes (STH) vary greatly in size, and female worms are larger than males. *Trichuris trichiura* has a narrow anterior esophageal end and shorter and thicker posterior anus. These worms are pinkish-white in colour and they appear as threaded structure all through the mucosa lining. They attach to the host through their slender anterior end and feed on secretions releasing out of the tissue. Females are larger (35–50 mm long)in size than males (30–45 mm). STHs do not reproduce within the host. This has an important implication in the epidemiology of these parasites.

Hookworm: Persons infected with hookworms show the symptoms such as iron-deficiency anemia, physical and mental retardation. Sometimes they may show cardiac complication. Diagnosis is made by the finding of ova in stools.

Pinworm (*Enterobius vermicularis*): The most common complaint caused by pinworms is perianal itching. The infected individual may also develop intestinal abscesses and bleeding. Eggs can be observed in the perianal area. For microscopic observations, they can be transferred with a cellophane tape swab to a slide. Worms may also be seen in stools, and sometimes around the anus.

Roundworm (*Ascaris lumbricoides*): Infected individuals may develop severe symptoms-such as intestinal, biliary, pancreatic obstruction or even appendicitis. Eggs of roundworms can easily be diagnosed by a simple smear of stools.

Strongyloidiasis (*Thread worm*): The infectious agent is the nematode Strongyloides stercoralis. Common symptom is diarrhea. Larvae can be easily detected in stools. The infectious larvae penetrate the skin when it contacts soil. The strongyloid's life cycle is quite complex than other nematodes due to its alternation between free-living and parasitic cycles. It also possesses the potential for autoinfection and multiplication within the host. The parasitic cycle is homogenic, while the free-living cycle is heterogonic. The heterogonic life cycle is of advantage to the parasite because during this stage the parasite reproduces for one or more generations in the absence of a host.

Tapeworm (*Taenia solium*): In infected individuals, tapeworms can be found in stools. Eggs can be swabbed from perianal areas and examined.

Trichinosis: Infected individuals often develop severe trichinosis resulting in death within six weeks if not diagnosed and treated in time. Diagnostic procedures are rather difficult since neither eggs nor worms can be detected in stools. Biopsy of muscle samples is the only reliable test that can detect the presence of encysted larvae of *Trichinella spiralis* (pork worm)

Whipworm (*Trichuris trichiura*): Blood-streaked stools, rectal prolapse, iron-deficiency anemia, and malnutrition are the common symptoms of individuals infected with whipworm. Stool examination for eggs and parasites is indicated for an accurate diagnosis.

Epidemiology

These infections are most prevalent in tropical and subtropical regions of the developing world where adequate water and sanitation facilities are lacking[69,70]. Recent estimates suggest that *A. lumbricoides* can infect over a billion, *T. trichiura* 795 million, and hookworms 740 million people[71]. Mostly helminthes have been linked with an increased risk for nutritional anemias, protein-energy malnutrition and may also lead to growth deficits in children. This may also cause low pregnancy weight gain and intrauterine growth retardation followed by low birth weight. About 44 million pregnant women have hookworm infections which cause chronic blood loss from the intestine hence predisposes an individual to the development of iron deficiency anaemia. This causes a major public health problem. Intestinal helminthic infections are observed at a high intensity and quite common among school age children.

Other species of intestinal helminths are not widely prevalent. Though the mortality rate due to intestinal helminths is low however, the burden of disease is related to the chronic effects on health and nutritional status of the host[72]. In addition to their health effects, intestinal helminth infections also impair physical and mental growth of children thus impacts the educational achievement and economic development of a child.

Treatment and Prevention

In accordance with the WHO[73], the recommended drugs for for use in public health interventions to control STH infections are:

- Albendazole (400mg) tablets given in a single dose for adults. However the dose is reduced to 200mg for children between 12 and 24 months;
- Levamisole (40mg) tablets given in a single dose by weight (2.5mg/kg). The drug Levamisole at a dose of 80mg has been successfully used in primary school-age children
- Mebendazole (500mg) tablets given in a single dose;
- Pyrantel pamoate (250mg) tablets given in a single dose by weight (10mg/kg). A combined preparation of pyrantel-oxantel has been proved to be more effective than pyrantel alone in treating *T. trichiura* infection.

Environmental factors such as quality of water supply, sanitation and housing conditions and personal hygiene play a vital role in getting STH infection. Among other factors such as socioeconomic, demographic and health related behavior are also known to influence this infection. Many developing countries are endemic to these heleminthic infections due to two principal factors- 1. Favorable qualities of the soil and 2. Frequent contamination of the environment by human feces. The transmission of these heleminths within the community is predominantly related to their day to day habits related to eating, defecation, personal hygiene and cleanliness.

4.2.4.2. Intestinal Protozoan Parasites

The most common intestinal protozoan parasites are: *Entamoeba histolytica, Giardia intestinalis*, *Cyclospora cayetanenensis*, and *Cryptosporidium* spp. The diseases caused by these intestinal protozoan parasites are known as amoebiasis, giardiasis, cyclosporiasis, and cryptosporidiosis respectively. The common symptom of these diseases is diarrhea.

A. Amoebiasis

The protozoan parasite, *Entamoeba histolytica*, is the causative agent of amoebiasis in humans. According to the available estimates approximately 48 million individuals suffer from amoebiasis throughout the world[74]. In 1984, at least 40,000 deaths were attributed to amoebiasis. Amoebiasis is a major problem in developing countries such as India. This is primarily because of inadequate sanitation and contaminated food and drinking water. Normally present in the large bowel, the ameba usually persist for months or even years as asymptomatic luminal gut infections. However, occasionally the parasite penetrates the intestinal mucosa and induces colitis or disseminates to other organs, most commonly to the liver, where it causes abscess formation. At the present time, the factors that trigger ameba invasion are unknown.

Life Cycle of E. histolytica

The organism exists in two forms—the trophozoite or the dividing form and the cyst which is the dormant form (Fig. 4.16). Human infection usually begins with the ingestion of the cyst which is present in food and/or water contaminated with human fecal material. Cysts normally survive the acidic pH of the stomach and pass into the intestine. In the intestine, the cysts undergo excystment and each cyst gives rise to eight trophozoites. These trophozoites eventually migrate to and multiply in the colon. In asymptomatic individuals the trophozoites in the intestine live as commensals. However, when the disease triggers, the trophozoites attack and invade the intestinal mucosa causing dysentery. In rare cases, the trophozoites progress through the blood vessels to extra-intestinal locations like liver, brain and lungs. In such cases they may form life-threatening abscesses. In the intestine, some of these trophozoites again undergo the process of encystation and produce quadrinucleated cysts. Both trophozoites and cysts are excreted along with the feces. Cysts can survive for prolonged periods outside the host while the trophozoites are unable to survive beyond few hours due to their fragile structure. Cysts are responsible for the transmission of the disease through fecal-oral contamination. Trophozoites play no role in transmission of the disease but are responsible for producing tissue pathology. The reservoir of human infection is the "carrier" or asymptomatic human host who continuously passes cysts[75].

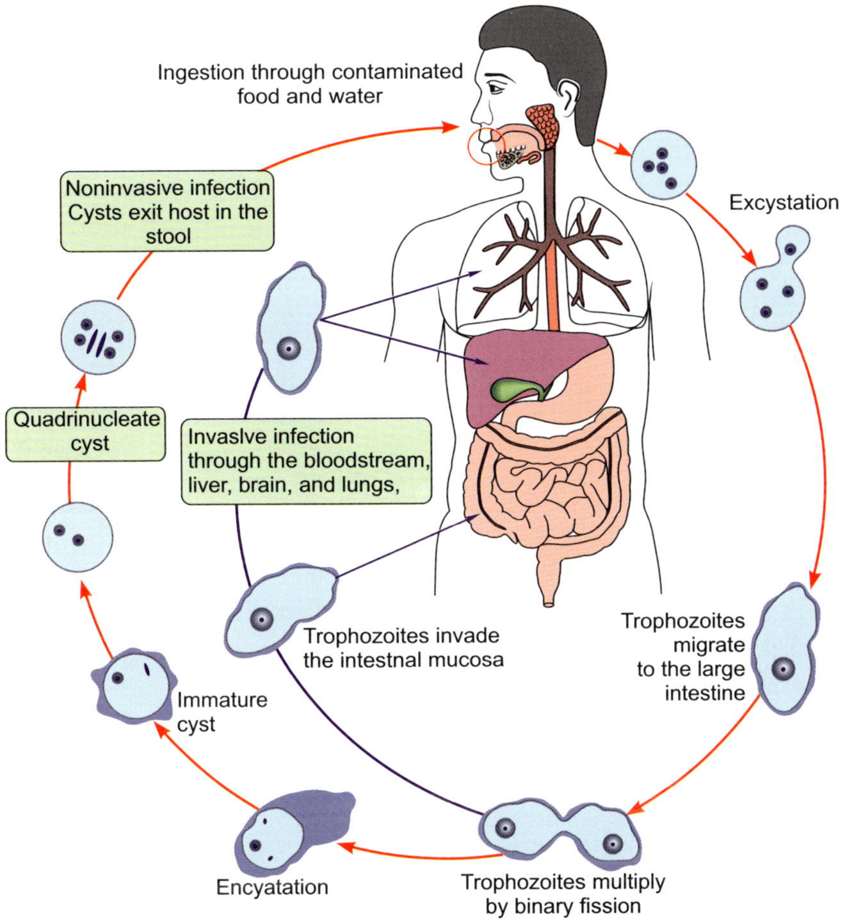

Fig. 4.16: Life cycle of *Entamoeba histolytica*

Pathogenesis

The genus Entamoeba contains many species, some of which (*Entamoeba histolytica, Entamoeba dispar, Entamoeba moshkovskii, Entamoeba polecki, Entamoeba coli, Entamoeba hartmanni*) can reside in the human interstitial lumen. *E histolytica* is, thus far, the only Entamoeba species definitely associated with disease; the others are considered nonpathogenic[76]. Pathogenesis of amoebiasis is believed to be a multistep, multifactorial process. Though a large number of studies have attempted to unravel the factors/molecules responsible for the pathogenesis of amoebiasis, the processes involved in pathogenesis are by no means well understood. The aspects of pathogenesis which have been investigated experimentally can be broadly categorized into different stages (*i*) interactions with the intestinal flora, (*ii*) lysis of target cell by direct adherence, (*i*) lysis of target cell by release of toxins and (*iv*) eventually phagocytosis of target cells.The vast majority of *E. histolytica* infections are asymptomatic in nature. The strains found in asymptomatic individuals (nonpathogenic) were so distant from pathogenic strains that they have now been accorded a separate species status, namely *E. dispar* and the name *E. histolytica* has been retained for the pathogenic strains[77].

Diagnosis

Various diagnostic methods are in use. Routine microscopic examination has low sensitivity and cannot distinguish *E. histolytica* from *E. dispar*. Each of these methods have certain limitations therefore, molecular diagnostics are now been advocated[76].

1. *Microscopy:* Microscopic stool examination for trophozoites from a single stool sample in amebic colitis is only 33-50% sensitive. Examination of 3 stool samples within a span of 10 days can improve the detection rate to 85-95%.

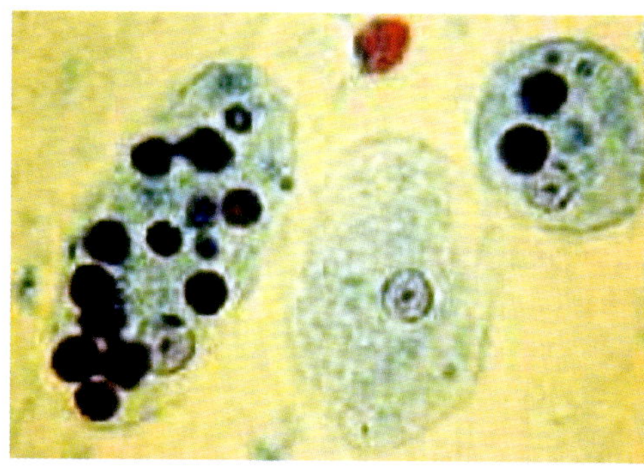

Fig. 4.17: Trichrome stain of *Entamoeba histolytica* trophozoites during amebiasis (http://emedicine.medscape.com/article/212029-workup)

Two diagnostic characteristics can be observed under microscopy (Fig 4.17). Two trophozoites have ingested erythrocytes, and all 3 trophozoites have nuclei with small, centrally located karyosomes. The presence of intracytoplasmic red blood cells in trophozoites is considered to be a gold standard test for *E. histolytica.*

The World Health Organization (WHO) recommends that intestinal amebiasis be diagnosed with an *E. histolytica*-specific test, thus rendering the classic stool ova and parasite examination obsolete.

2. *Culture:* *Xenic cultivation, first introduced in* 1925, is defined as the growth of the parasite in the presence of an undefined flora. This technique is still in use today using modified Locke-egg media.

Axenic cultivation, first achieved in 1961, involves growth of the parasite in the absence of any other metabolizing cells. Only a few strains of *E. dispar* have been reported to be viable in axenic cultures.

Cultures can be performed using fecal or rectal biopsy specimens and liver abscess aspirates. The success rate is low and is technically difficult.

3. *Antigen detection:* Enzyme-linked immunosorbent assay (ELISA) is used to detect antigens from *E histolytica* in stool samples. Antigen-based ELISA kits using monoclonal antibodies against the GAL/GalNAc–specific lectin of *E histolytica* (*E. histolytica* II, TechLab, Blacksburg, VA) yield an overall sensitivity of 71%-100% and specificity of 93%-100%. In patients with amebic liver abscess, serum and liver aspirate antigen detection using the same kit was shown to yield a sensitivity of 96% and 100%, respectively.

Other stool detection kits use monoclonal antibodies against the serine-rich antigen of *E. histolytica* (Optimum S kit, Merlin Diagniostika, Bornheim-Hersel, Germany) or against other specific antigens (Entamoeba CELISA-PATH, Cellabs, Brookvale, Australia; ProSpecT EIA, Remle Inc, Lenexa, KY).

4. ***Serology:*** Multiple serologic assays are available for the diagnosis of amebiasis.

ELISA is the most used assay throughout the world and is used to measure the presence of serum antilectin antibodies (IgG). The sensitivity for detection of antibodies to *E histolytica* in patients with amebic liver abscess is 97.9%, whereas the specificity is 94.8%. False-negative results can occur within the first 7-10 days following infection.

Immunofluorescent assay (IFA) is also rapid, reliable, and reproducible. In the setting of amebic liver abscess, the sensitivity and specificity of IFA was shown to be 93.6% and 96.7%, respectively. Indirect hemagglutination (IHA) is very specific (99.1%) but is less sensitive than ELISA. The seropositivity prevalence is very high in endemic areas, limiting antibody-based testing for diagnosing currently active disease, since antibodies can persist for years after infection.

5. ***Polymerase chain reaction:***

E. histolytica can be identified in various types of clinical specimens, including feces, tissues, and liver abscess aspirates.

A wide variety of polymerase chain reaction (PCR) methods targeting different genes, including a small-subunit rRNA gene (18S rDNA) which is extrachromosomal in nature, 30-kDa antigen gene, serine-rich protein gene, chitinase gene, hemolysin gene have been described for the detection and differentiation of *E. histolytica, E. dispar,* and *E. moshkovskii*. Sensitivities can vary according to sampling and the specific target gene used for detection. When performed on fecal samples, PCR yields a sensitivity that is similar to that of stool antigen-based assay.

PCR-based tests have been strongly endorsed by WHO. Application of PCR-based methods in routine diagnosis is still very limited, as the generation of nonspecific DNA fragments from environmental and clinical samples often leads to false-positive results. Generation of inhibitor free DNA template using commercially available stool kit yielded satisfactory results.

Epidemiology

Amebiasis is second only to malaria in terms of protozoa-associated mortality. The combined prevalence of amebic colitis and amebic liver abscess is estimated at 40-50 million cases annually worldwide, resulting in 40,000-100,000 deaths[78]. Recently, the World Health Organization has recommended reassessment of the epidemiology of *E. histolytica* using test systems, which are able to differentiate between the two ameba species[79]. From the few studies that have been conducted so far, it is evident that *E. dispar* is substantially more prevalent than *E. histolytica,* and that only a small proportion of individuals specifically infected with *E. histolytica* will progress into having amebic disease. Amebiasis primarily affects migrants from and travelers to endemic regions and immunosuppressed or institutionalized individuals. *Entamoeba* species infect approximately 10% of the world's population. The prevalence of Entamoeba infection is as high as 50% in areas of Central and South America, Africa, and Asia[80]. In Japan and Taiwan population, it has been recorded that HIV seropositivity is a risk factor for invasive extraintestinal amebiasis[81]. This has not been observed elsewhere.

Amebic liver abscess is 7-12 times more common in men than in women. It is predominantly observed among men aged 18-50 years. The reason for this sexual disparity is unknown, although hormonal effects may be implicated, as the prevalence of amebic liver abscess is also increased among postmenopausal women. Alcohol may also been an important risk factor. The sexual distribution is equal in children. Very young children seem to be predisposed to fulminant colitis[78].

Treatment and Prevention

Prevention and control measures are similar to other diseases transmitted by the fecal–oral route. The major difference is that humans are the only host for *E. histolytica* and there is no possibility of zoonotic transmission. The disease can be prevented by avoiding the contamination of food or water with fecal material. Health education in context to improving personal hygiene, proper disposal of feces, hand washing and safe drinking water are particularly effective. Although waterborne transmission of *Entamoeba* is lower than other intestinal protozoa, protecting water supplies will lower endemicity and epidemics. *Entamoeba* cysts are resistant to standard chlorine treatment but sensitive to treatment by iodine or boiling. Sedimentation and filtration processes are quite effective in removing *Entamoeba* cysts. Medicines administered for treatment of amoebiasis are presented in Table 4.3

Table 4.3: Treatment of amebiasis

Drugs	Uses
Iodoquinol, paromomycin, ordiloxanidefuroate	Luminal agents to treat asymptomatic cases and as a follow up treatment after a nitroimidazole
Metronidazole or tinidazole	Treatment of non dysenteric colitis, dysentery, and extraintestinal infections
Dehydroemetine or emetine	Treatment of severe disease such as necrotic colitis, perforation of intestinal wall, rupture of liver abscess

Source: http://www.garlandscience.com/res/pdf/9780815365006_ch03.pdf

B. Giardiasis

The causal organism of Giardiasis, *Giardia lamblia* (also known as *G. duodenalis* or *G. intestinalis*) is a flagellated protozoan parasite. *G. lamblia* causes both epidemic and sporadic disease and is an important etiology of waterborne and foodborne diarrhea.

Life cycle of Giardia lamblia

Infection occurs after cysts are ingested through contaminated water or food. This marks the beginning of the life cycle. After ingestion, mature cysts in the small intestine release trophozoites through a process called excystation[82]. Cysts are able to survive exposure to gastric acid. It is thought that gastric acid may actually trigger excystation. The trophozoite stage is responsible for producing clinical disease in humans.

Trophozoites have two distinct nuclei and four pairs of flagellae (Fig. 4.18). They resemble teardrops when viewed from the top and they are spoon shaped from the side. They are 12-15 μm in length. Trophozoites colonize the small intestine, attaching to the mucosa of the bowel using a ventral sucking disks (Ax). Presence of microtubule structure known as median Body (MB) is very conspicuous for this parasite. The trophozoites then multiply by longitudinal binary fission[83]. As the *Giardia* trophozoites move toward the colon, they retreat into the cyst stage known as encystation and the new cysts are excreted in the feces. Bile salts and intestinal mucus boost trophozoite multiplication and encystations.

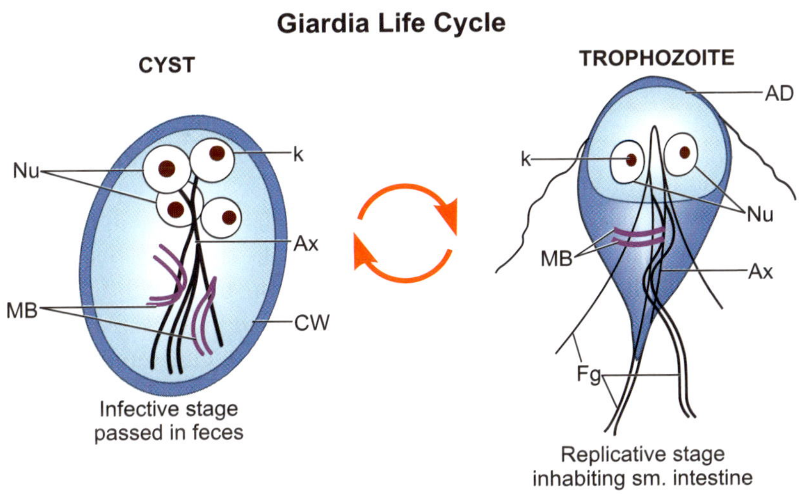

Fig. 4.18: Life cycle of *Giardia*

Symptoms

The major symptom of acute giardiasis is protracted diarrhea seen more commonly in travelers or children. The incubation period for developing the infection is generally 9-15 days. The acute stage usually begins with a feeling of intestinal uneasiness followed by nausea and anorexia. Patients may develop low grade fever and chills. These symptoms are followed by watery, foul-smelling, explosive diarrhea, abdominal pain, passage of foul gas and belching. This stage lasts for 3-4 days.

Pathogenesis

Pathogenesis of giardiasis is not completely understood and several hypotheses exist. There are several proposed mechanisms of disease that involve both mucosal and luminal factors in patients who exhibit abdominal pain, diarrhea, and malabsorption symptoms. Actual invasion of the mucosa by organisms is a rare finding. The intestinal mucosa may be damaged by the trophozoite itself disrupting the epithelial brush border during attachment, or less likely by direct invasion[84]. In addition, release of toxic substances from the organism itself may damage intestinal epithelium. Absorptive activities may also be blocked due to the trophozoites "blanketing" the intestinal mucosa and causing functional mucosal obstruction. Immunologic mechanisms may also play a role since individuals with decreased gamma globulin levels have a higher prevalence of infection and reinfection.

Diagnosis

The traditional diagnosis for giardiasis consists of performing an ova and parasite (O + P) exam of one to three stool specimens on non-consecutive days[85]. Several days of specimen collection are needed to improve sensitivity. Stool specimens may be preserved by formalin or zinc sulfate concentration methods[86].

Stool antigen tests has improved giardiasis diagnostic capabilities. Immunodiagnostic assays are available for the detection of Giardia and are both more sensitive and more specific (>90%) than the traditional ova and parasite (O + P) examinations[87]. *Giardia* specific antigen in stool specimens are detected by several different methods: EIA, indirect and direct immunofluorescent assays using monoclonal antibodies and direct fluorescent assays. They are available from several different commercial sources, including Meridian, Alexon-Trend, Tech-lab, and Biosite.

Epidemiology

The prevalence varies between rural and urban areas depending on economic level and social status of a country. The prevalence of *G.lamblia* exists more among poorer communities, slightly higher in males than in females, with higher prevalence among 2–5-year-old children.

High infection rates for giardiasis have been reported from developing countries, and some recent data are listed. In various Asian countries (Bangladesh, Cambodia, China, India, Indonesia, Laos, Malaysia, Nepal, Philippines, Thailand, Turkey, Saudi Arabia, and Vietnam), North America (Cuba, Mexico, and Nicaragua), South America (Argentina, Brazil, Colombia, and Peru), and Africa including both northern Africa, west Africa, and South Africa. The available data mostly focused on children, and infection rates for *G. duodenalis* fell into the range of 8% to 30% in the majority of these studies

An important aspect of the epidemiology of giardiasis is to take into account the host range of different *Giardia* species, their strains and genotypes, the potential for cross-species transmission, and risk factors involved in the exposure of the pathogen. This is particularly important in determining the zoonotic potential of Giardia infections in domestic animals and in determining the human disease burden attributable to parasites of animal origin. Recently, with the advent of molecular typing tools, the epidemiology of *Giardia* is currently systematically addressed.

Giardia duodenalis is considered to be a species complex of several assemblages, of which assemblage A and B are predominantly associated with human infections. Assemblage type has been associated with risk of occurrence of symptoms and duration of illness. Hence genotyping of giardial isolates help to understand better the epidemiology and transmission ecology of the disease in a particular setting or area[88] (Fig. 4.19). Assemblage B predominates in most studies from south and south east Asia including India. The predominance of assemblage B however, did not extend to all the far eastern countries. Reports from South Korea and Henan province in China based on very few samples, showed predominance of assemblage A.

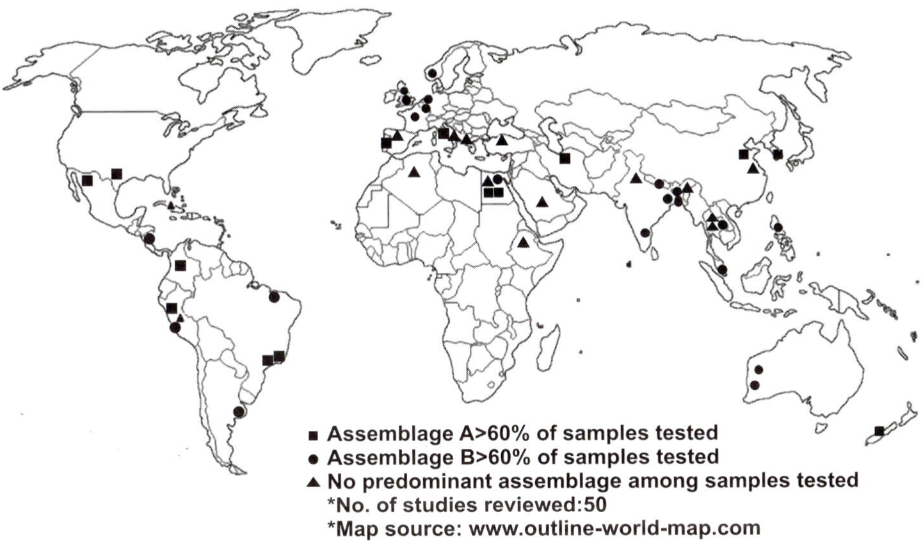

Fig. 4.19: Geographical distribution of giardial assemblages A and B (Adapted from Laishram S *et al*[88])

Any association between assemblage and symptoms has not been proved conclusively so far.

Zoonotic Transmission

The potential for zoonotic transmission of Giardia was recognized in the late 1970s, though it has yet not been conclusively proved. Possible sources of such zoonotic transmission are from dairy cattle, companion animals like cats and dogs or wild life. Giardiasis is widespread among livestock and infected calves that can shed as many as 10^5-10^6 cysts per gram of feces. Transmission of animal associated assemblage AI has been seen among dairy farm workers in West Bengal[89]. A European network of 9 public and veterinary health institutes called ZOOnotic Protozoa NET work (ZOOPNET) (http://www.medvetnet.org/cms/) found that though a majority of the human isolates (n01658) belong to assemblage B (56%) and assemblage A (43%), around 1% of isolates included zoonotic assemblages C, D, E and F[90].

Giardiasis in India

In several epidemiological studies from India, giardiasis has been found to be prevalent throughout the country. In community-based studies from northern India, prevalence rates ranged from 5.5% to 70% with the highest rates reported in a low socioeconomic group in Chandigarh[91]. High rates of asymptomatic cyst passage and diarrhea associated with giardiasis were identified in children[92]. In studies from southern India, the prevalencerates of giardiasis in children with diarrhea and adults with gastrointestinal symptoms (diarrhea, abdominal discomfort, flatulence, weight loss or anorexia) were 8% to 10% and 37.1%, respectively[93]. In Calcutta,Giardia was previously found to be associated with 0.4% to 2.6% of diarrheal episodes in children <5 years; however, more recent studies that used ELISA and PCR-based methods,showed incidence rates of giardial diarrhea of up to 15.8% in this age group[94]. The variations in the prevalence of giardiasis described above may reflect a true variation in infection rates in different areas, or be due to the use of different methodologies used and the number of sequential stool samples tested per patient.

Though Giardia is not considered a classical opportunistic infection, its prevalence rate has been found to be higher among AIDS patients and is associated with diarrhea. In India, the prevalence of giardiasis in HIV-infected patients was found to vary from 1.6% to >30% showing a higher prevalence inpatients with lower CD4 counts[95].

Treatment and Prevention

Metronidazole (Flagyl) is the most commonly used antibiotic for giardia infection. Side effects may develop like nausea and a metallic taste in the mouth. Consumption of alcohol is prohibited during treatment. Both Tinidazole as well as metronidazole has the same side effects. These drugs can be administered in a single dose. Among other drugs, Nitazoxanideis prescribed for children as it is available in liquid form and may be easier for children to swallow. Side effects may include nausea, flatulence, yellow eyes and brightly colored yellow urine. There is no recommended medication for giardiasis during pregnancy.

Since this disease is caused by fecal-oral route, drinking safe water and practicing good hygiene are the best way to prevent the disease. One must avoid swallowing pool water during swimming.

C. Cyclosporiosis

Cyclospora cayetanensis has emerged as an important cause of endemic or epidemic diarrheal illness in children and adults world wide that leads to significant morbidity and mortality[96]. The increasing globalization of the food supply has contributed to the spread of *C. cayetanensis* from developing to developed countries. The parasite has currently emerged as a global public health problem. In recent decades, it has been recognized as having great potential to cause food- and waterborne diseases.

Pathogenesis

Cyclospora cayetanensis is an apicomplexan, cyst-forming coccidian protozoan that causes a self-limiting diarrhea[97]. Morphological features of *C. cayetanensis* represent typical spherical oocysts that vary between 7.5 and 10 micrometers in diameter having a 50 nanometer thick wall with an outer thread like coat (Fig. 4.20). Infection is transmitted through the oral-fecal route, and begins when a person ingests oocysts present in fecally contaminated food or water. The only host *C. cayetanensis* uses are humans.

Life Cycle and Symptoms

Humans ingest sporulated oocysts considered to be the infectious stage of *C cayetanensis*. The oocyst excysts in the jejunum area of the small intestine and subsequently invades the intestinal epithelial cells. This follows the stage of schizogony which begins with the formation of a trophozoite that grows into a mature schizont that contains 8-12 merozoites. Several cycles of type I schizogony follows and finally type II meronts (sexual forms) develop. At this stage, each cell contains four merozoites. As they invade the epithelial cells, few of the merozoites form single macrogametes and rests divide multiple times to form microgametes. Upon release, a microgamete fertilizes a macrogamete and finally develops into a zygote. The zygote then turns into an oocyst by developing a resistanl wall.

The oocyst passes into the environment through fecal material and exist in the environment as nonsporulated and noninfectious oocyst for days to weeks. Hence, direct fecal-oral transmission from relatively fresh stool does not occur. Clinical illness is characterized by persistent diarrhea, bloating, flatulence, abdominal cramps, constipation, and fatigue. The incubation period in the host is typically around a week, and illness can last six weeks. The infection is often self-limiting. However, unless treated, illness may relapse. It has been reported that the more severe forms of the disease can occur in inimmunocompromised patients such as those with AIDS.

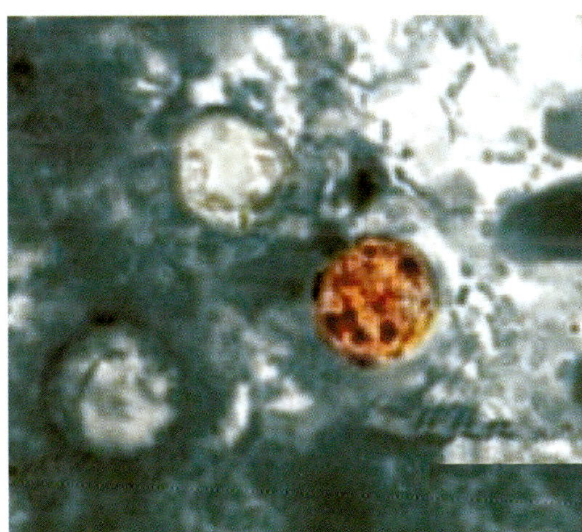

Fig. 4.20: *Cyclospora cayetanensis* oocysts

Epidemiology

Human-to-human spread of the parasite occurs indirectly via the environment through oocysts present in contaminated water, food or soil. In endemic areas, other risk factors associated with the infection include association with animals, type of sanitation and low socioeconomic status. *C. cayetanensis* is an important cause of traveller's diarrhoea and numerous large foodborne outbreaks have been found to be associated with the globalisation of the food supply and importof fruits and vegetables from developing countries. In endemic areas, the most important steps to prevent infection are improving environmental sanitation and health education[98]. In developing countries, risk categories for cyclosporiasis include immunocompromised patients, foreigners and children. It is well known that the infection causes significant morbidity and mortality in immunosupressed individuals, in particular those with HIV infection. The highest risk of infection and diarrhoea occur in the first five years of life[99] .The prevalence of *Cyclospora* markedly increases during the warm and rainy season. Currently there is no confirmed evidence of zoonotic transmission of *C. cayetanensis* and the role of animals as natural reservoir hosts remains to be determined. In India, prevalence rate of cyclosporiasis has been found to be high (~6.6 %) in HIV infected individuals.

Treatment and Prevention

The most important steps to prevent infection are health education including personal hygiene, changing eating habits, proper sanitary infrastructures and safe drinking water. Preventing geophagia in children is important because of the potential soilborne transmission of infection. Foreign residents or visitors from non-endemic regions should also have safe drinking water and avoid fresh produce.It is necessary to implement detection techniques in the laboratories and in the field that would help to control the infection and prevent outbreaks locally.

TMP-SMZ is the drug of choice. Immunocompetent patients become symptom-free within a median of 3 days. Nitazoxanide, a 5-nitrothiazole derivative with broad-spectrum activity against helminths and protozoans, has been shown to be effective against *C cayetanensis*, with an efficacy 87% by the third dose (first, 71%; second 75%)

D. Cryptosporidiosis

Cryptosporidium is a coccidian protozoan parasite that has gained much attention in the last 20 years as a clinically important human pathogen. The discovery of *Cryptosporidium* was made by E.E.Tyzzer in 1907. He described it as a cell-associated organism in the gastric mucosa of mice. *Cryptosporidium* is a protozoan parasite in the phylum Apicomplexa. It affects the intestines and causes typically an acute short-term infection. It spreads through the fecal-oral route, often through contaminated water. The main symptom is self-limiting diarrhea in immunocompetent individuals. However, in immunocompromised individuals, such as AIDS patients, the symptoms are particularly severe and often fatal.

A number of Cryptosporidium species infect different mammals. In humans, the main causes of disease are *C. parvum* and *C. hominis* (previously *C. parvum* genotype 1). *C. canis*, *C. felis, C. meleagridis*, and *C. muris* can also cause disease in humans.

Life Cycle

Cryptosporidium is monoxenous, that is, its life cycle is completed within one host. The parasite moves from one host to the other by fecal-oral route. Mature oocysts are shed in the feces of an infected host. Through contamination of the environment, water or food, the oocysts are ingested by other suitable hosts. In the gastrointestinal tract (usually the ileum or the jejunum), sporozoites excyst from the oocyst and parasitize epithelial cells. The sporozoites and subsequent developmental stages are seen at the lumenal epithelium of the infected host GI tract. The life cycle of the organism is depicted in the Fig. 4.21.

The sporozoite differentiates into a spherical trophozoite and undergoes asexual multiplication occurs, forming two types of meronts (or schizonts). Type I contain 6-8 nuclei, which become incorporated into 6 to 8 merozoites when the meront is mature. Each merozoite is able to infect a new host cell, and then develop either into a new type I meront, or into a type II meront, which contains 4 merozoites when mature.

The merozoites from type II meronts also invade new host cells, but there initiate sexual multiplication by differentiating into either male (microgametocyte) or female (macrogamont) stages. Upon maturation, the microgametocytes contain spermlike microgametes, which fertilize the macrogamonts. The fertilized macrogamont, or zygote, then develops into an oocyst, which undergoes meiosis within the host. When sporogony (meiotic sporozoite formation) is complete each oocyst contains four potentially infective sporozoites. Some oocysts are shed via the feces or perhaps via respiratory secretions. Others release sporozoites within the body, and these may repeat the merogony, gamogony and sporogony cycle. Sporozoites excyst from an oocyst and enter the microvillus of an epithelial cell. Once they enter the epithelial cell, they differentiate into trophozoites. Trophozoites undergo nuclear proliferation to form type I meronts. A type I merozoite leaves the meront to form either a type I or type II meront. A type II merozoite leaves the meront to form microgametes or a macrogamont. The microgamete fertilizes the macrogamont, which then develops into an oocyst. Oocysts sporulate in situ and either release sporozoites for autoinfection or pass from the body in the feces.

"Thick-walled" oocysts, released in the feces, are responsible for long term survival of the parasite in the environment. Each generation of parasites can develop and mature within 12-14 hours. The rapid life cycle plus the auto infective cycles may lead to huge numbers of parasitic cells in the gut, and to secondary infection sites in the duodenum and large intestine. In immunosuppressed individuals, parasites can be detected in the stomach, bile duct, pancreatic duct and the respiratory tract as well. Ingesting small numbers of oocysts is enough to cause severe infections; this may be attributable to two life cycle features that differ from most other mammalian and avian coccidia: the auto infective cycles, and sporogony within the host, so that oocysts are infective when passed into the feces. Attack rates for cryptosporidiosis are on the order of 40% for the population at risk, compared with attack rates of 5-10% for giardiasis, the other waterborne parasite-caused illness of concern to humans.

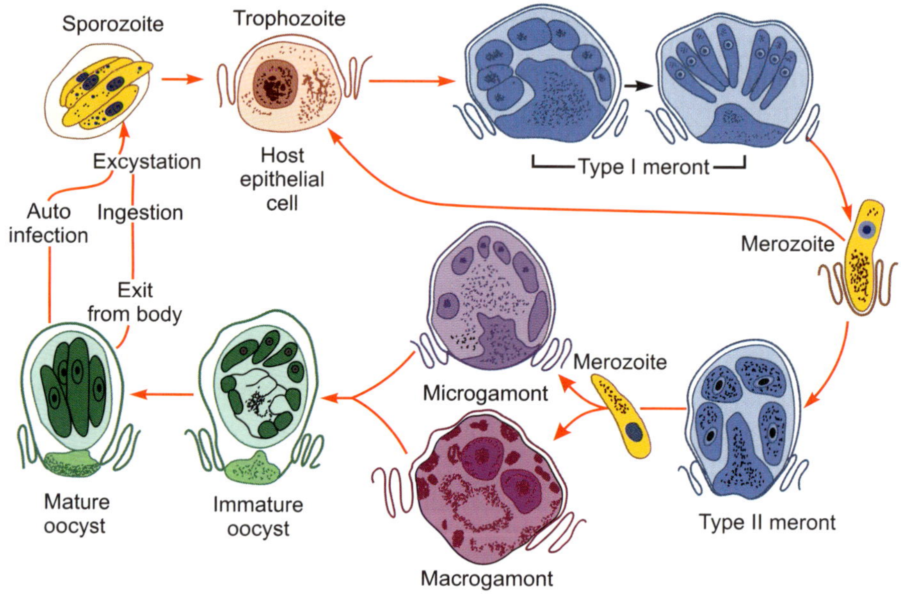

Fig. 4.21: Diagrammatic representation of the life cycle of Cryptosporidium

Pathogenesis

The various symptoms of cryptosporidiosis differ greatly between immunocompetent and immunocompromised individuals. In immunocompetent patients, cryptosporidiosis is an acute, yet self-limiting diarrheal illness and lasts for 1-2 week duration. The symptoms include frequent, watery diarrhea, nausea, vomiting, abdominal cramps associated with low-grade fever[100]. For immunocompromised persons, the illness is much more severe involving debilitating cholera-like diarrhea (up to 20 liters/day), severe abdominal cramps, low-grade fever, weight loss and anorexia. Due to lack of tissue specificity, *C. parvum* infection has also been identified in the biliary tract causing thickening of the gallbladder wall and it may also enter the respiratory system[101].

A sporozoite-specific lectin adherence factor has been identified as the agent of attachment to the intestinal surface. After sporozoite attachment, it has been hypothesized that the epithelial mucosa cells release cytokines that activate resident phagocytes[102]. These activated cells release soluble factors that are responsible for increase in intestinal secretion of water and chloride. These factors also inhibit absorption of nutrients. These soluble factors include histamine, serotonin, adenosine, prostaglandins, leukotrienes, and platelet-activating factor. These factors act on various substrates, including enteric nerves and on the epithelial cells themselves. As a result, epithelial cells are damaged either by cell death as a direct result of parasite invasion, multiplication. Cell damage may also occur through T cell-mediated inflammation, producing villus atrophy and crypt hyperplasia. This leads to distortion of villus architecture, accompanied by nutrient malabsorption and diarrhea.

Detection and Diagnosis

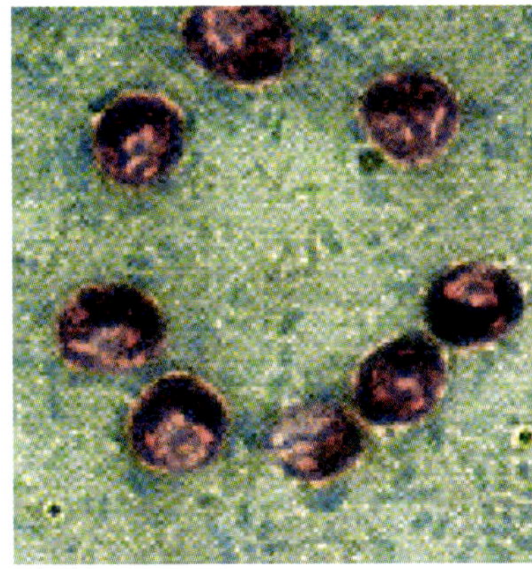

Fig. 4.22: Acid fast stained oocysts of Cryptosporidium (***Source:*** Henricksen et al[103])

Diagnosis of cryptosporidiosis is made by examination of stool samples. The modified acid-fast stain is traditionally used that specifically detect the presence of cryptosporidial oocysts (Fig. 4.22)

Environmental samples containing low number of oocysts are normally processed by EPA method 1623, and *Cryptosporidium* oocysts present are detected by immunofluorescent microscopy or PCR. This process involves the concentration of oocysts by filtration, followed by the isolation of oocysts by immunomagnetic separation (IMS). The recovered oocysts are stained with a fluorescent antibody and 4′,6-diamidino-2-phenylindole dihydrochloride (DAPI). This can then be detected by microscopy and the stained oocysts can also be enumerated[104].PCR-based methods have been used increasingly in reference diagnostic labs for detection of the parasite to species level and also for the detection and analysis of *Cryptosporidium* oocysts in water[105]. Compared to method 1622/1623, the new generation of PCR methods (genotyping techniques) has the ability to differentiate *Cryptosporidium* species that are infective to humans from those that are not infective to humans.

Epidemiology

Surveys in developing countries find a higher prevalence of infection than in industrialised countries[106]. The epidemiology of cryptosporidiosis in humans is not completely understood due to the existence of multiple transmission routes.

Within a large population, there are specific groups that are at greater risk of infection including children, malnourished persons, and a range of immunocompromised individuals including AIDS patients, transplant recipients, patients receiving chemotherapy for cancer, institutionalized patients, and patients with immunosuppressive infectious diseases.

Transmission Routes

Faeces deposited on the ground when subjected to wind and water transport oocysts across or through soil. In some cases humans and animals contribute to the movement of oocysts. To initiate infection, oocysts must be ingested with food, water, or by close personal contact with infected people, animals or contaminated surfaces[107]. Faecal contamination of soil and surface water can ultimately lead to contamination of fresh foods, drinking water and recreational water. Detection of positive oocysts in untreated wastewater, secondarily treated wastewater, activated sludge effluent, combined sewer overfows, groundwater, surface water, and treated drinking water indicate widespread faecal contamination[108]. Foodborne transmission of oocysts have been reported on the surface of raw vegetables from the market place like cilantro leaves and roots, lettuce, radishes, tomatoes, cucumbers, and carrots but not from cabbage. Cool and moist vegetables provide an optimal environment for survival[109]. Various outbreaks highlights that food handlers should thoroughly wash their hands before handling food items and utensils. Raw fruits and vegetables as well as previously cooked items should not be handled with bare hands[110]. Uncooked produce should be thoroughly washed before being placed on kitchen surfaces.

Transmission via recreational water has been reported in various places and affected over 10,000 people. Frequent faecal contamination coupled with oocyst resistance to chlorine facilitates transmission. Recognition of cryptosporidiosis as a major cause of recreational water-borne disease necessitates public health officials, pool operators, and users to collaborate in developing plans to reduce the risk of water-borne transmission.

High-risk Individuals

HIV-positive patients, particularly those with AIDS (as measured by a low CD4 count), are at increased risk of being infected with cryptosporidiosis. This risk is increased when these individuals engage in high-risk sexual behavior. The high prevalence of *Cryptosporidium* spp. in AIDS patients is probably related to an increased risk of contracting infection from infected individuals. AIDS patients are more commonly infected with *Cryptosporidium* spp. other than *C. parvum* than do immunocompetent individuals[111]. *Cryptosporidium parvum* is highly transmissible and infective in the family setting, with transmission rates similar to other highly infectious enteric pathogens such as *Shigella* species. These are cause for added concern because of the rapidly increasing rate of seropositivity for human immunodeficiency virus.

Indian Scenario

In India, *Cryptosporidium* spp. are a leading cause of infectious diarrhea in children. The reported positivity rates vary from 1.1% to 18.9%[112,113]. Molecular epidemiology studies have shown fewer zoonotic species in children in a community setting compared to HIV-infected adults reporting in the hospital. Subgenotypes were also more diverse among the HIV-infected adults than among the children. The predominant subgenotypes recorded in *C. hominis* positive samples were Ib and If in HIV-infected adults. In children infected with *C. hominis* predominantly show Ia genotype. All *C. parvum* samples from children were Ic at the *Cpgp40/15* locus. Interestingly, subgenotypes including IIa, IIb, and Ic, were seen in *C. parvum* samples from HIV-infected adults. These data suggest differences in transmission patterns and susceptibility to disease in these two populations[114].

Treatment and Prevention

Recovery from Cryptosporidium infection depends largely upon the immune status of the host. Immunocompetent patients usually have a spontaneous recovery within a few weeks and normally get rid of the parasite within a few months without requiring any specific therapy.

Administration of Nitaxozanide (500 mg of nitazoxanide (tablet or suspension) or placebo, given twice daily for three days) reduces the load of parasites and may be useful in immunocompetent individuals[115]. Due to the seriousness of the potential outcomes of cryptosporidiosis, the use of nitaxozanide should be considered in immunocompromised patients. The absence of effective therapy highlights the need to ensure that infection is avoided. Following preventive measures are suggested by CDC

(http://www.cdc.gov/parasites/crypto/prevention.html)

1. ***Practicing Good Hygiene:*** Washing hands with soap and water for at least 20 seconds, rubbing hands together vigorously and scrubbing all surfaces:

- Before preparing or eating food
- After using the toilet
- After changing diapers or cleaning up a child who has used the toilet
- Before and after tending to someone who is ill with diarrhea
- After handling an animal or animal waste

2. ***Child care facilities:*** In order to reduce the risk of disease transmission, children with diarrhea should not be admitted to childcare settings until the diarrhea has stopped.

3. ***Recreational water venues (pools, interactive fountains, lakes, ocean):***

- Avoid swimming while experiencing diarrhea. If diagnosed with cryptosporidiosis, avoid swimming for at least 2 weeks after diarrhea stops.
- One must take shower before entering the swimming pool.
- Wash children thoroughly with soap and water after they use the toilet or their diapers are changed and before they enter the water body
- Take children on frequent bathroom breaks and check their diapers often.
- Diapers should be changed in the bathroom, not at the pool side.

4. ***Around animals:***

- Minimize contact with the feces of all animals, particularly young animals.
- One must use disposable gloves while cleaning up animal feces and always wash hands when finished.
- Wash hands after any contact with animals or their living areas.

5. ***Outside:*** Wash hands after gardening, even if wearing gloves

6. ***Immunocompromised persons:*** Must take special precautions to avoid infection with Cryptosporidium

- Avoiding close contact with any person or animal that has cryptosporidiosis as this can become a life threatening disease for immunocompromised persons.
- Avoid handling animal feces because infection can be life threatening for immunocompromised persons.

REFERENCES

1. Topazian RG, Goldberg MH, Hupp JR, eds. Oral Maxillofacial Infections. 4th ed. Philadelphia, Pa: WB Saunders; 2002
2. Mettenleiter TC, Klupp BG, Granzow H (2006). Herpesvirus assembly: a tale of two membranes. Curr. Opin. Microbiol. 9 (4): 423–9.
3. Whitley RJ, Roizman B (2001) Review Herpes simplex virus infections. Lancet. 357(9267) : 1513-8.
4. Subramanian RP, Geraghty RJ (2007). Herpes simplex virus type 1 mediates fusion through a hemifusion intermediate by sequential activity of glycoproteins D, H, L, and B. Proc. Natl. Acad. Sci. U.S.A. 104 (8): 2903–8.
5. Matis J, Kúdelová M (2001). Early shutoff of host protein synthesis in cells infected with herpes simplex viruses. Acta Virol. 45 (5–6): 269–77
6. Granzow H, Klupp BG, Fuchs W, Veits J, Osterrieder N, Mettenleiter TC (2001). Egress of Alphaherpesviruses: Comparative Ultrastructural Study. J. Virol. 75 (8): 3675–84.
7. Whitley, R. J. 2001. Herpes simplex viruses, p. 2461-2509. In D. M. Knipe and P. M. Howley (ed.), Fields virology, 4th ed. Lippincott Williams & Wilkins, Philadelphia, Pa.
8. Halioua B and Malkkin J E. Epidemiology of Gential Herpes -Recent Advances. Eur. J. Dermatol. 1999
9. Corey L and Handsfield H H.(2000) Gential Herpes and Public Health: Addressing a Global Problem. JAMA, 283: 791-794.
10. Megan S. Harrison, Takemasa Sakaguchi, Anthony P. Schmitt (2010) Paramyxovirus assembly and budding: Building particles that transmit infections. Int. J. Biochem. & Cell Biol. 42: 2010, Pages 1416–1429.
11. Sonnenberg A, Everhart JE(1997) Health impact of peptic ulcer in the United States. Am J. Gastroenterol. 92:614-620.
12. Stark RM, Gerwig GJ, Pitman RS et al. (1999). Biofilm formation by Helicobacter pylori. Lett Appl Microbiol28: 121–6.
13. Kusters JG, van Vliet AH, Kuipers EJ (July 2006). Pathogenesis of Helicobacter pylori Infection. Clin Microbiol Rev 19 (3): 449–90
14. Josenhans C, Eaton KA, Thevenot T, Suerbaum S (August 2000). Switching of Flagellar Motility in Helicobacter pylori by Reversible Length Variation of a Short Homopolymeric Sequence Repeat in fliP, a Gene Encoding a Basal Body Protein. Infect Immun 68 (8): 4598–603
15. Parsonnet J, G.D. Friedman, N. Orentreich, H. Vogelman (1997) Risk for gastric cancer in people with CagA positive or CagA negative *Helicobacter pylori* infection Gut, 40 : 297–301.
16. Linz B, Balloux F, Moodley Y, et al. (2007). An African origin for the intimate association between humans and *Helicobacter pylori* Nature445: 915–8.
17. Fischbach L, Evans EL (2007). Meta-analysis: the effect of antibiotic resistance status on the efficacy of triple and quadruple first-line therapies for Helicobacter pylori. Aliment. Pharmacol. Ther.26: 343–57.
18. Yanaka et al.; Fahey, JW; Fukumoto, A; Nakayama, M; Inoue, S; Zhang, S; Tauchi, M; Suzuki, H et al. 2009). Dietary Sulforaphane-Rich Broccoli Sprouts Reduce Colonization and Attenuate Gastritis in Helicobacter pylori-Infected Mice and Humans. Cancer Prevention Research2: 353–360.
19. Khan, M.A.H.A., Bari, A.S.M., Islam, M.R., Das, P.M and Ali, M.Y.(1998). Pullorum disease in seminature chicks and its experimental Pathology. Bang. Vet. J. 32: 124-128.

20. Michael E. Ohl and Samuel I. Miller (2001) Salmonella : A Model for Bacterial Pathogenesis. Annu. Rev. Med. 2001.52:259-274.
21. Misra S, Diaz PS, Rowley AH (1997) Characteristics of typhoid fever in children and adolescents in a major metropolitan area in the United States. Clin Infect Dis. 24(5):998.
22. Crump JA, Luby SP, Mintz ED (2004) The global burden of typhoid fever. Bull World Health Organ. 82(5):346.
23. Fredriksson-Ahomaa M, Stolle A, Korkeala H (2006) Molecular epidemiology of Yersinia enterocolitica infections. FEMS Immunol Med Microbiol. 47:315–29
24. Fàbrega A, Vila J. (2011) Yersinia enterocolitica: Pathogenesis, virulence and antimicrobial resistance. Enferm Infec Microbiol Clin. doi:10.1016/j.eimc.2011.07.017.
25. Cornelis GR, Boland A, Boyd AP, Geuijen C, Iriarte M, Neyt C, et al (1998) The virulence plasmid of Yersinia, an antihost genome Microbiol Mol Biol Rev. 62:1315–52
26. KJ; Ray CG (editors) (2004). Sherris Medical Microbiology (4th ed.). McGraw Hill. ISBN 0-8385-8529-9.
27. Lambrecht E, Baré J, Van Damme I, Bert W, Sabbe K, Houf K.(2013)Behavior of Yersinia enterocolitica in the presence of the bacterivorous Acanthamoeba castellanii. Appl Environ Microbiol.79:6407-13.
28. Beauregard, K. E., K. D. Lee, R. J. Collier, and J. A. Swanson. 1997. pH-dependent perforation of macrophage phagosomes by listeriolysin O from Listeria monocytogenes. J. Exp. Med. 186:1159-1163
29. Tilney LG, Portnoy DA. (1989) Actin filaments and the growth, movement, and spread of the intracellular bacterial parasite, Listeria monocytogenes. J Cell Biol.109:1597–1608.
30. Schmid M. W., Ng E. Y. W., Lampidis R., Emmerth M., Walcher M., Kreft J., Goebel W. et al. (2005). Evolutionary history of the genus Listeria and its virulence genes. Systematic and Applied Microbiology 28: 1–18.
31. Robbins JR, Barth AI, Marquis H, de Hostos EL, Nelson WJ, Theriot JA (1999). *Listeria monocytogenes* Exploits Normal Host Cell Processes to Spread from Cell to Cell. J Cell Biol146: 1333–50.
32. BhujwalaR.A., V. Hingorani, R.K. Chandra. (1973). Genital listeriosis in Delhi: a pilot study. The Indian Journal of Medical Research, 61:1284–1287.
33. Dhanashree B., S.K. Otta, I. Karunasagar, W. Goebel, I. Karunasagar (2003).Incidence of Listeria spp. in clinical and food samples in Mangalore, India. Food Microbiology, 20: 447–453
34. Dupont HL. Shigella species (bacillary dysentery). In: Principles and Practice of Infectious Diseases, 6th Ed, Mandell GL, Bennett JE, Dolin R (Eds), Churchill Livingstone, Philadelphia 2005. p.2655
35. Boschi-PintoC., VelebitL., Shibuya K. (2008) Estimating child mortality due to diarrhoea in developing countries. Bull. World Health Organ., 86: 710–717.
36. Jones G., Schultink W., Babille M. (2006) Child survival in India. Indian J. Pediatr., 73: 479–487.
37. FuleR.P., Kaundinya D.V. (1985) Enteropathogenic E. coli in infantile diarrhea in a rural area. Indian J. Pediatr., 52 (1985), pp. 533–536.
38. Levine M M, Ristaino P, Marley G, Smyth C, Knutton S, Boedeker E, Black R, Young C, Clements M L, Cheney C, Patnaik R (1984) Coli surface antigens 1 and 3 of colonization factor antigen II-positive enterotoxigenic Escherichia coli: morphology, purification, and immune responses in humans. Infect Immun.44:409–420.
39. Vial P A, Robins Browne R, Lior H, Prado V, Kaper J B, Nataro J P, Maneval D, Elsayed A, Levine M M. (1988) Characterization of enteroadherent-aggregative Escherichia coli, a putative agent of diarrheal disease. J Infect Dis.158:70–79.

40. Nataro, J. P. & Kaper, J. B. (1998). Diarrheagenic Escherichia coli. Clin Microbiol Rev 11, 142–201.

41. BerluttiF., CasalinoM., ZagagliaC., FradianiP.A., ViscaP., Nicoletti M. (1998) Expression of the virulence plasmid-carried apyrase gene (apy) of enteroinvasive Escherichia coli and Shigella flexneri is under the control of H-NS and the VirF and VirB regulatory cascade. Infect. Immun. 6: 4957–4964.

42. Donnenberg M.S and Whittam T.S (2001) Pathogenesis and evolution of virulence in enteropathogenic and enterohemorrhagic Escherichia coli. J Clin Invest. 107: 539-548.

43. Gomes, T. A. T., V. Rassi, K. L. Macdonald, S. R. T. S. Ramos, L. R. Trabulsi, M. A. M. Vieira, B. E. C. Guth, J. A. N. Candeias, C. Ivey, M. R. F. Toledo, and P. A. Blake (1991) Enteropathogens associated with acute diarrheal disease in urban infants in Sao Paulo, Brazil. J. Infect. Dis. 164:331–337.

44. Cravioto, A., R. E. Reyes, and R. Ortega (1988) Prospective study of diar- rhoeal disease in a cohort of rural Mexican children: incidence and isolated pathogens during the first two years of life. Epidemiol. Rev. 101:123

45. Robins-Browne, R. M (1987) Traditional enteropathogenic Escherichia coli of infantile diarrhea. Rev. Infect. Dis. 9:28–53.

46. Wani, S. A., Nabi, A., Fayaz, I., Ahmad, I., Nishikawa, Y., Qureshi, K., Khan, M. A. & Choudhary, J (2006) Investigation of diarrhoeic faecal samples for enterotoxigenic Shiga toxin-producing and typical or atypical enteropathogenic Escherichia coli in Kashmir, India. FEMS Microbiol Lett 261, 238–244.

47. P.K Ghosh and Arif Ali (2010) Isolation of atypical enteropathogenic Escherichia coli from children with and without diarrhoea in Delhi and the National Capital Region, India J Med Microbiol. 59:1156-1162.

48. Rao, M. R., R. Abu-Elyazeed, S. J. Savarino, A. B. Naficy, T. F. Wierzba, I. Abdel-Messih, H. Shaheen, R. W. Frenck, Jr., A. M. Svennerholm, and J. D. Clemens (2003) High disease burden of diarrhea due to enterotoxigenicEscherichia coli among rural Egyptian infants and young children. J. Clin. Microbiol. 41:4862–4864.

49. Qadri, F., S. K. Das, A. S. Faruque, G. J. Fuchs, M. J. Albert, R. B. Sack, and A. M. Svennerholm. (2000) Prevalence of toxin types and colonization factors in enterotoxigenic Escherichia coli isolated during a 2-year period from diarrheal patients in Bangladesh. J. Clin. Microbiol. 38:27–31.

50. World Health Organization (1989) The treatment and prevention of acute diarrhea, practical guidelines, 2nd ed. World Health Organization, Geneva, Switzerland.

51. Mata, L (1992) Diarrheal disease as a cause of malnutrition. Am. J. Trop. Med. Hyg. 47:16–27.

52. Ramamurthy T, Garg S, Sharma R, et al (1993) Emergence of novel strain of Vibrio cholerae with epidemic potential in southern and eastern India. Lancet. 341:703-704.

53. De S.N (1959) Enterotoxicity of bacteria-free culture-filtrate of Vibrio cholerae. Nature183(4674):1533-4.

54. Johnson TL, Abendroth J, Hol WG, Sandkvist M (2006) Type II secretion: from structure to function. FEMS Microbiol Lett. 255:175–86.

55. Sánchez J and Holmgren J (2011) Cholera toxin- A foe and a friend. Indian J Med Res.133: 153–163

56. Sonawane ND, Zhao D, Zegarra-Moran O, Galietta LJ, Verkman AS (2007) Lectin conjugates as potent non-absorbable CFTR inhibitors for reducing intestinal fluid secretion in cholera. Gastroenterology132:1234–44.

57. World Health Organization. Cholera, 2009. Wkly Epidemiol Rec. 31:293308, 2010.

58. World Health Organization. Cholera vaccines: Wkly Epidemiol Rec. 85:11728, 2010.

59. Nair GB, Ramamurthy T, Bhattacharya SK, Mukhopadhyay AK, Garg S, Bhattacharya MK, et al (1994) Spread of Vibrio cholerae O139 Bengal in India. J Infect Dis 169: 1029-34.
60. Jesudason MV, Samuel R, John TJ (1994) Reappearance of Vibrio cholerae and concurrent prevalence of O1 and O139 in Vellore, South India. Lancet 344: 335-6.
61. Deen JL, von Seidlein L, Sur D, Agtini M, Lucas MES, et al. (2008) The High Burden of Cholera in Children: Comparison of Incidence from Endemic Areas in Asia and Africa. PLoS Negl Trop Dis 2(2): e173. doi:10.1371 /journal.pntd.0000173).
62. Sinclair D, Abba K, Zaman K, Qadri F, Graves PM (2011).Oral vaccines for preventing cholera.Cochrane Database Syst Rev (3): CD008603. doi:10.1002/14651858. CD008603.pub2. PMID 21412922
63. Ramani S, Kang G. (2007) Burden of disease & molecular epidemiology of group A rotavirus infections in India. Indian J. Med Res.125 :619-32.
64. Glass RI, Parashar UD (2006) The promise of new rotavirus vaccines. N Engl J Med.354:75–7.
65. Wilhelmi I., Roman E., Sanchez-Fauquier A. (2003). Viruses causing gastroenteritis. Clin. Microbiol. Infect.9, 247-262.
66. Blacklow NR, Greenberg HB. Viral gastroenteritis (1991) N Engl J Med. 325:252
67. Tudzynski P, Correia T, Keller U (2001). Biotechnology and genetics of ergot alkaloids. Appl Microbiol Biotechnol. 57 (5–6): 4593–4605.
68. Haque R (2007) Human Intestinal Parasites. J Health Popul Nutr.25: 387–391.
69. Savioli L, Albonico M (2004) Soil-transmitted helminthiasis. Nat Rev Microbiol.2:618–9.
70. Cappello M. Global health impact of soil-transmitted nematodes (2004) Pediatr Infect Dis J. 23:663–4.
71. de Silva NR, Brooker S, Hotez PZ, Montresor A, Engles D, Savioli L (2003) Soil-transmitted helminth infections: updating the global picture. Trends Parasitol.19 :547–51.
72. Stephenson LS, Latham MC, Ottesen EA(2000) Malnutrition and parasitic helminth infections. Parasitology 121: S23–38.
73. World Health Organization (2004) WHO Model Formulary. Based on the 13th Model List of Essential Medicines 2003. Geneva: WHO p. 82-5.
74. Walsh J A (1986) Problems in recognition and diagnosis of amoebiasis: estimation of the global magnitude of morbidity and mortality. Rev. Infect. Dis. 8:228-238.
75. Sehgal D, Bhattacharya A and Bhattacharya S (1996)Pathogenesis of infection by Entamoeba histolytica. J. Biosci.21:423-432.
76. Fotedar R, Stark D, Beebe N, et al (2007) Laboratory diagnostic techniques for Entamoeba species. Clin Microbiol Rev. 20:511-32.
77. Diamond L S and Clark C G (1993) A redescription of E. histolytica Schaudinn, 1903 (Walker, 1911)E. dispar separating it from Brumpt, 1925; J. Euk. Microbiol. 40 340-344.
78. Stanley SL Jr. (2003) Amoebiasis. Lancet 361:1025-34.
79. World Health Organization (1997) Amoebiasis. WHO Wkly Epidemiol Rec 72: 97–100.
80. Caballero-Salcedo A, Viveros-Rogel M, Salvatierra B, et al. (1994) Seroepidemiology of amebiasis in Mexico. Am J Trop Med Hyg.50:412-9.
81. Misra SP, Misra V, Dwivedi M. (2006) Ileocecal masses in patients with amebic liver abscess: etiology and management. World J Gastroenterol.12:1933-6.

82. Jones JE. (1988) Giardiasis.872-882. In A. Balows, WJ Hausler, M. Ohashi, and A. Turano (ed.), Laboratory diagnosis of infectious diseases, vol. 1. Springer-Verlag, New York.

83. Hill DR.(1993) Giardiasis.Issues in diagnosis and management.Infect Dis Clin North Am. 7:503-25.

84. Farthing MJ (1996) Giardiasis. Gastroenterol Clin North Am. 25:493-515.

85. Wolfe MS (1992) Giardiasis.Clin Microbiol Rev. 5:93-100.

86. Ortega YR, Adam RD (1997) Giardia: overview and update. Clin Infect Dis. 25:545-9.

87. Adrabbo K, Peura D (2002) Giardiasis: A Review. Practical Gastroenterology. June; 15-29.

88. Laishram S, Kang G, Ajjampur SS (2012)Giardiasis: a review on assemblage distribution and epidemiology in India.Indian J Gastroenterol. 31:3-12.

89. Khan SM, Debnath C, Pramanik AK, Xiao L, Nozaki T, Ganguly S (2011) Molecular evidence for zoonotic transmission of Giardia duodenalis among dairy farm workers in West Bengal, India. Vet Parasitol.178:342–5.

90. Sprong H, Caccio SM, van der Giessen JW (2009) Identification of zoonotic genotypes of Giardia duodenalis. PLoS Negl Trop Dis.3:e558.

91. Ramesh G, Malla N, Raju G, et al (1991) Epidimiological study of parasitic infestations in lower socio-economic group in Chandigarh (North India). Indian J Med Res.93:47–50.

92. Kaur R, Rawat D, Kakkar M, Uppal B, Sharma V (2002). Intestinal parasites in children with diarrhoea in Delhi, India.Southeast Asian J Trop Med Public Health.33:725–9.

93. Shenoy S, Urs S, Prabhu G, Mathew B, Antony G, Bharati B (1998) Giardiasis in the adult population of Dakshina Kannada District of South India. Trop Doc.28:40–2.

94.Chatterjee BD, Thawani G, Sanyal SN (1989) Etiology of acute childhood diarrhoea in Calcutta. Trop Gastroenterol.10:158–66.

95. autam H, Bhalla P, Saini S, et al (2009) Epidemiology of opportunisticinfections and its correlation with CD4 T-lymphocyte counts and plasma viral load among HIV-positive patients at a tertiary care hospital in India. J Int Assoc Physicians AIDS Care (Chic) 8:333–7.

96. Chacín-Bonilla L (2010) Epidemiology of Cyclospora cayetanensis: a review focusing in endemic areas. Acta Trop.115:181–193.

97. Ortega YR et al. (1993). Cyclospora species: a new protozoan pathogen of humans. New Eng Jr. Med 328:1308–1312.

98. Chacín-Bonilla L (2010) Epidemiology of Cyclospora cayetanensis: A review focusing in endemic areas.Acta Trop.115:181-93.

99. Bern C, Ortega YR, Checkley W, RobertsJM, Lescano AG, Cabrera L,Verastegui M, Black RE, Sterling C, Gilman RH (2002) Epidemiologic differences between cyclosporiasis andcryptosporidiosis in Peruvian children. Emerg. Infect. Dis., 8:581–585.

100. Juranek DD (1995) Cryptosporidiosis: sources of infection and guidelines for prevention." Clin Infect Dis,21: S57-61.

101. Casemore DP, Garder, CA and O'Mahony C(1994)Cryptosporidial infection, with special reference to nosocomial transmission of Cryptosporidium parvum: a reviewFolia Parasitol41:17-21.

102. Goodgame, RW (1996) Understanding intestinal spore-forming protozoa: cryptosporidia, microsporidia, isospora, and cyclospora. Ann Intern Med, 124 : 429-41.

103. Henricksen SA, Pohlenz JFL (1981) Staining of cryptosporidia by a modified Ziehl-Neelsen technique. Acta Vet Scand22:594-96.

104. U.S. Environmental Protection Agency (2001) Method 1623: Cryptosporidium and Giardia in water by filtration/IMS/FA. EPA 821-R-01-025. Office of Water, U.S. Environmental Protection Agency, Washington D.C.

105. Xiao, L, Lal AA, and Jiang J (2004) Detection and differentiation of Cryptosporidium oocysts in water by PCR-RFLP. Methods Mol. Biol. 268:163-176.

106. Ungar BLP (1990) Cryptosporidiosis in humans (Homo sapiens). In: Dubey JP, Speer CA, Fayer R, editors. Cryptosporidiosis of man and animals.Boca Raton, FL: CRC Press, pp. 59-82.

107. Fayer R, Morgan U, Upton SJ (2000) Epidemiology of Cryptosporidium: transmission, detection and identification.Int J Parasitol.30:1305-22.

108. Rose LB, Lisle JT, LeChevallier M (1997) Waterborne cryptosporidiosis incidence, outbreaks, and treatment strategies. In: Fayer R, editor. Cryptosporidium and cryptosporidiosis. Boca Raton, FL: CRC Press, 1997. pp. 93-110.

109. Monge R, Chinchilla M (1996) Presence of Cryptosporidium oocysts in fresh vegetables. J Food Protect 59: 202-3.

110. Rai AK, Chakravorty R and Paul J (2008) Detection of Giardia, Entamoeba, andCryptosporidium in unprocessed food items from northern India. World J Microbiol Biotechnol 24:2879–2887.

111. Paul R. Hunter, and Gordon N (2002) Epidemiology and Clinical Features ofCryptosporidium Infection in Immunocompromised Patients. Clin. Microbiol. Rev. 15 :145-154.

112. Kaur R,Rawat D,Kakkar M, Uppal B, and Sharma VK (2002) Intestinal parasites in children with diarrhea in Delhi, India. Southeast Asian J. Trop. Med. Public Health 33:725-729.

113. Sethi S, Sehgal R, Malla N, and Mahajan RC (1999) Cryptosporidiosis in a tertiary care hospital. Natl. Med. J. India. 12:207-209.

114. Kang G, Samuel R, Vijayakumar TS, Selvi S, Sridharan G, Brown D, and Wanke C (2005) Community prevalence of antibodies to human immunodeficiency virus in rural and urban Vellore, Tamil nadu.Natl. Med. J. India 18:15-17.

115. Abubakar I, Aliyu SH, Arumugam C, Hunter PR, Usman NK (2007) Prevention and treatment of cryptosporidiosis in immunocompromised patients. Cochrane Database Syst Rev.

5

Urinary and Genital Tract Infections

Chapter Outline

5.1 Urinary tract infections

5.2 Genital tract infections

5.1 Urinary Tract Infections

5.1.1 Bladder Infections

Lower urinary tract infections usually develop within a short period of time and may include one or more of the following symptoms:

- Urge for frequent urination
- A painful burning sensation is felt during urination indicating an infection limited to urethra and named as urethritis.
- Discomfort or pressure felt in the lower abdomen.
- Pain in the pelvic area or back.
- The urine often has a strong smell, appears cloudy and often contains blood. This indicates the sign of *pyuria*.
- A high white blood cell count in the urine. This is considered as a reliable indicator of urinary tract infections.
- Occasionally fever develops.

Pathogenesis

After the flu and common cold, urinary tract infections (UTIs) are the most common medical complaint among women during the period of reproduction. UTIs are far more common among women than among men. Most women develop a UTI at some time in their lives, and many will have recurrences. In general, the higher risk in women is mostly due to shorter length of the female urethra, which is 1.5 inches compared to 8 inches in case of male. Due the shorter length of the urethra, bacteria from fecal matter at the anal opening can contaminate the opening of the urethra.

Following bacterial strains are involved that can cause UTIs:

Escherichia (E.) coli, responsible for most cystitis cases in women, especially in younger women. *E. coli* is generally a harmless microorganism originating in the intestine. As and when it reaches the vaginal opening, it may invade and eventually colonize the bladder, causing an infection. The spread of *E. coli* to the vaginal opening most commonly occurs when women wipe themselves from back to front after urinating or after sexual activity.

Staphylococcus saprophyticus accounts for 5 - 15% of UTIs, mostly in younger women.

Other causal organisms of UTI are *Klebsiella, Enterococci* bacteria, and *Proteus mirabilis*. They are generally found in UTIs in older women. Other bacteria include *Ureaplasma urealyticum* and *Mycoplasma hominis*, which are generally harmless organisms.

Epidemiology

UTIs in pregnant women pose serious health risks for both mother and child. UTIs that manifest during pregnancy pose a higher than average risk of developing into kidney infections. Women are significantly more likely to experience UTI than men. By the age of 24 years, nearly 1 in 3 women suffer from at least one episode of UTI requiring antimicrobial therapy. Almost half of all women will experience 1 UTI during their lifetime. Specific subpopulations at increased risk of UTI include infants, pregnant and elderly women, patients with spinal cord injuries and/or catheters, patients with diabetes or multiple sclerosis, patients with acquired immunodeficiency disease syndrome and patients with underlying urologic abnormalities. Catheter-associated UTI is considered to be one of the most common nosocomial infections. More than one million cases per year are reported from hospitals and nursing homes. The risk of UTI increases with duration of catheterization. UTIs are the second most common form of infection in elderly populations, accounting for nearly 25% of all infections.

Treatment and Prevention

Oral antibiotic treatment cures 94% of uncomplicated urinary tract infections, although the rate of recurrence remains high. The following antibiotics are commonly used for uncomplicated UTIs. The standard regimen has traditionally been a 3-day course of trimethoprim-sulfamethoxazole, commonly called TMP-SMX (Bactrim, Cotrim, Septra). TMP-SMX combines an antibiotic with a sulfa drug. A single dose of TMP-SMX is sometimes prescribed in mild cases, but for effective cure the drug is prescribed for three days. Often allergies to sulfa drugs are common and may be serious. In geographic areas that have a high resistance to TMP-SMX, quinolones are being used as the first-line treatment for UTIs. Ciprofloxacin (Cipro) is the quinolone antibiotic most commonly prescribed. Pregnant women should not take fluoroquinolones.

5.1.2 Kidney Infections

The main components of the urinary tract system are kidneys, ureters, bladder, and urethra. Any part of the urinary system may become infected and referred to as urinary tract infection (UTI). When a kidney, located in the upper urinary tract becomes infected, the condition is medically referred as 'pyelonephritis'. Thus, kidney infection is only one of several types of infections encompassed by the term UTI.

Pathogenesis

Pyelonephritis may present as a progressive disease characterized by increasing tissue destruction and associated functional insufficiency. The course of the disease may be variable; in some instances it progresses slowly over many years (pyelonephritis lenta), while in other instances an accelerated form may occur. Factors which may influence the rate of progression of the disease include (*a*) the infecting bacterial species, (*b*) the frequency of reinfection or recurrence of a preexistent infection, (*c*) the nature of predisposing anatomic deformities if any, and (*d*) the presence of certain systemic disorders, such as hypertension, diabetes mellitus and gout. 85 percent of infection is caused by gram negative bacilli from patients own normal flora. Most common gram negative bacteria involved in the pathogenesis of the disease are *E. coli* , Proteus , *Klebsiella* and *Enterobacter*.

Two ways the bacteria can reach the kidney:

1. Through the blood stream - hematogenous infection is however less common. It occurs during septicaemia, bacterial endocarditis and in immune compromised patients.
2. Ascending infection from the lower urinary tract - more common mode of infection

Epidemiology

Incidence is normally highest among young women, followed by infants and the elderly population. The ratio of outpatient to inpatient cases found to be highest among young women (ranging from 5:1 to 6:1). *Escherichia coli* cause 80% of cases of acute pyelonephritis in women and 70% of cases in men and less dominant in older age groups.

Treatment and Prevention

Treatment regimen is same as in other UTI infections. However, relatively few data are available on the antimicrobial resistance patterns of bacterial strains causing acute pyelonephritis. Because resistance to trimethoprim-sulfamethoxazole among *Escherichia coli* strains causing acute cystitis and pyelonephritis is increasing and is associated with treatment failure[1].

5.2 Genital Tract Infections

The infections are of two types:

1. Sexually transmitted infections
    - Chlamydia
    - Gonorrhoea
    - Genital warts
    - Genital herpes
    - Trichomonas
    - Syphilis
    - HIV
2. Other infections
    - Candidiasis
    - Bacterial vaginosis

5.2.1 Sexually Transmitted Infections (STI)

STIs from a huge public health issue in a global scenario. These infections are commonly more active in a community amongst sex workers. Therefore, the epidermiological profile is very distinct and dynamic as well (Fig 5.1)

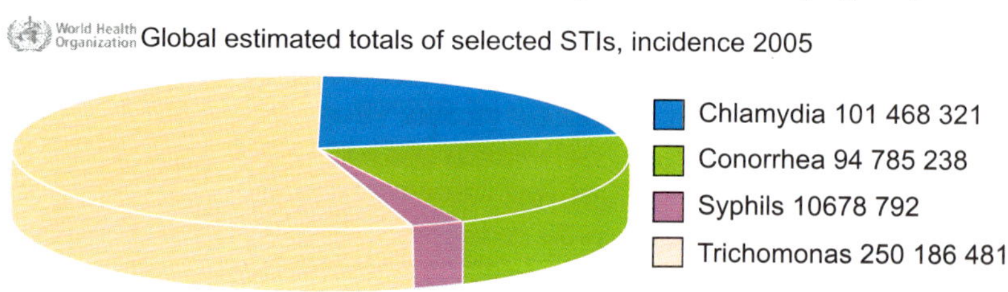

Fig. 5.1

5.2.1.1 Chlamydia

Sexually transmitted *Chlamydia trachomatis* infection is a widespread public health concern because of its prevalence and potentially devastating reproductive consequences, including pelvic inflammatory disease (PID), infertility, and ectopic pregnancy. The bacterium can cause prostatitis and epididymitis in men.

Chlamydia is an etiological agent, with incidences of up to 39.5% reported in patients with prostatitis. Infection of the testis and prostate due to Chlamydia results in the deterioration of sperm quality that may affect fertility. Chlamydia infection may affect male fertility by directly damaging the sperm. Various sperm parameters like proportion of DNA fragmentation, and acrosome reaction capacity are impaired with chlamydial infection.

Pathogenesis

C. trachomatis attaches to sperm and the bacteria reach the fallopian tubes, promoting sterility or ectopic pregnancy, tubal damage can occur with or without symptoms. The symptoms generally appear 7 to 14 days after exposure. In men, the main symptom is a thin, grey-white discharge from the penis, while women most commonly develop an increased vaginal discharge accompanied by painful urination, vaginal bleeding (Fig. 5.2)

C. trachomatis is a spherical, obligate intracellular bacterium. It is a small gram-negative bacterium with unique biologic properties that distinguish it from all other living organisms. It is an obligate intracellular parasite that has a distinct life cycle that consists of two major phases.

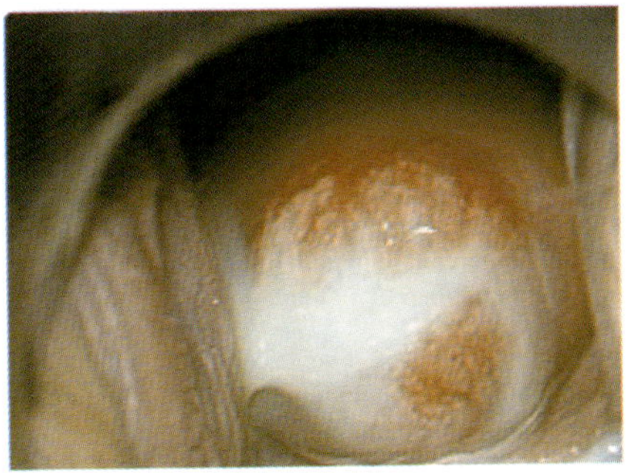

Fig. 5.2: *Chlamydial* cervicitis in a female patient characterized by mucopurulent cervical discharge, erythema, and inflammation.
Adapted from (http://en.wikipedia.org/wiki/Chlamydia_infection)

1. The small elementary bodies attach and penetrate into the host cells. Within a period of six to eight hours, these bodies then transform into the metabolically active form, called the reticulate bodies. These forms create large inclusions within cells.
2. The reticulate bodies then reorganize into small elementary bodies, and within two to three days the cell ruptures, releasing newly formed elementary bodies. Release of the elementary bodies initiates the replicative process, since this is the form which can infect new epithelial cells. The long growth cycle explains why prolonged courses of treatment are necessary.

Diagnosis

The diagnosis of genital chlamydial infections evolved rapidly from the 1990s through 2006. Nucleic acid amplification tests (NAAT), transcription mediated amplification (TMA), and the DNA strand displacement amplification (SDA) now are the common diagnostic methods employed for detection of infection. NAAT for chlamydia may be performed on swab specimens sampled from the cervix (women) or urethra (men), on self-collected vaginal swabs, or on voided urine.[2] NAAT has been estimated to have a sensitivity of approximately 90% and a specificity of approximately 99%, regardless of sampling from a cervical swab or by urine specimen.[3] Women reporting to the STI clinic often show a negative test for the infection by urine test and then they are recommended to undergo a subsequent cervical swab which may turn positive in two percent of cases.

Epidemiology

The generally assumed risk of developing PID after lower genital tract chlamydia infection varies considerably, and is up to 30%. For developing tubal infertility after PID the risks are 10–20%. This implies that the risk of test-positive women of developing tubal infertility would range between 0.1 and 6%. Newborn babies of infected mothers may experience eye infections or pneumonia contracted during passage through the birth canal. Globally, as of 2010, sexually transmitted chlamydia affects approximately 215 million people (3.1% of the population). It is more common in women (3.8%) than men (2.5%).[4]

Treatment and Prevention

The surest way to prevent a chlamydia infection is to abstain from sexual activities. Pregnant women should routinely undergo a screening test for Chlamydia. A course of Tetracyclines or its derivatives Doxycycline is the preferred antibiotic. Since the Tetracyclines are contraindicated in pregnant women, doctors may prescribe erythromycin.

5.2.1.2 Gonorrhoea

It is a contagious disease transmitted most often through sexual contact with an infected person. Gonorrhea may also get transmitted by contact with infected bodily fluids. Even an infected mother could pass on the infection to her newborn during childbirth. Both men and women can get gonorrhea.

Pathogenesis

Gonorrhea is caused by *Neisseria gonorrhoeae*, a bacterium that can grow and multiply easily in mucus membranes of the body. Gonorrhea bacteria normally grow in the warm, moist areas of the reproductive tract, including the cervix, uterus and fallopian tubes in women. It may also grow in the urethra of women and men (Fig. 5.3). The bacteria can also grow in the mouth, throat and anus.

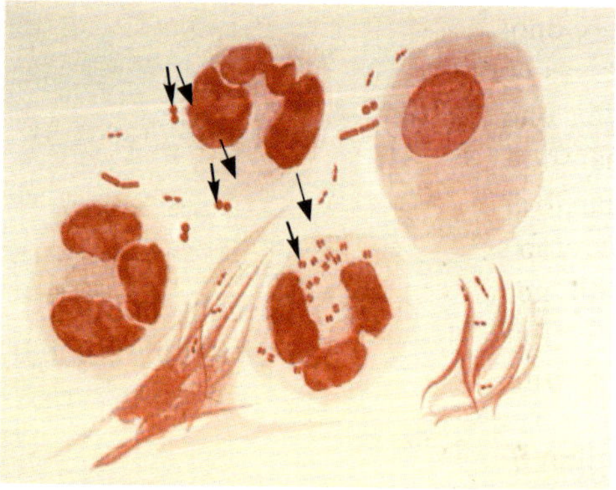

Fig. 5.3: Urethral exudate containing *Neisseria gonorrhoeae* from a patient with gonococcal urethritis. This illustration depicts a Gram-stain of a urethral exudate showing typical intracellular gram-negative diplococci, and pleomorphic extracellular gram-negative organisms, which is diagnostic for gonococcal urethritis. (Adapted from http://commons.wikimedia.org/wiki/File:Neisseria_gonorrhoeae)

The incubation period of gonorrhea is generally short. Symptoms appear two to five days after exposure. In men, it is characterized by inflammation of the urethra with pain during urination and with a thick pus-containing discharge. The pathogenic mechanism involves the attachment of the bacterium to non ciliated epithelial cells via pili (fimbriae) and the production of lipopolysaccharide endotoxin.

Neisseria gonorrhoeae from a patient with gonococcal urethritis. This illustration depicts a Gram-stain of a urethral exudate showing typical intracellular gram-negative diplococci, and pleomorphic extracellular gram-negative organisms, which is diagnostic for gonococcal urethritis. Neisserial LPS is distinguished from enteric LPS by its highly-branched basal oligosaccharide structure and the absence of repeating O-antigen subunits. Therefore, neisserial LPS is referred to as lipooligosaccharide (LOS). The bacterium characteristically releases outer membrane fragments called "blebs" during growth. These blebs contain LOS and participate in the pathogenesis if they are disseminated during the course of an infection. Neisserial LOS has a profound effect on the virulence and pathogenesis of *N. gonorrhoeae*.

Gonorrheal infection is generally limited to superficial mucosal surfaces lined with columnar epithelium. The areas most frequently involved are the urethra, cervix, rectum, pharynx, and conjunctiva.

In the male, the organism may invade the prostate resulting in prostatitis. Sometimes it may also extend to the testicles resulting in orchitis. In the female, it may extend through the uterus to the fallopian tubes resulting in salpingitis, or to the ovaries resulting in ovaritis.

Epidemiology

Gonorrhea is transmitted almost exclusively by sexual contact. Sexually active person can be infected with gonorrhea. The highest rates of infection are reported among sexually active teenagers and young adults. Persons who have multiple sex partners are at higher risk. Rates of gonorrhea are higher in males.

Gonorrhea is usually contracted from a sex partner who is either asymptomatic or has only minimal symptoms. It is estimated that the efficiency of transmission after one exposure is about 35 percent from an infected woman to an uninfected man however, the efficiency of transmission is 50 to 60 percent from an infected man to an uninfected woman. More than 90 percent of men with urethral gonorrhea develop symptoms within 5 days of exposure whereas only 50 percent of women with genital gonorrhea will develop the disease. Women with asymptomatic infections are at higher risk of developing pelvic inflammatory disease and disseminated gonococcal infection.

Indian Scenario

In India, the prevalence of gonorrhoea among STI clinic attendees in different regions varies from 3% to 19%. However, it has been on the decline over the past decade, probably due to the availability of medical facilities at the primary healthcare level, and growing awareness about AIDS. A steady decline in prevalence was observed in Chandigarh, Delhi and Patiala, while a marginal increase was reported from Rohtak and Ahmedabad. In a survey of women attending an STI clinic in Mumbai in 1996, 9.7 percent were positive for gonorrhoea. This was higher than that reported from gynaecological clinics in Amritsar (1995) and Chandigarh (1986), where it was 0.8 percent and 1.8 percent, respectively. The apparent ratio of male-to-female cases is 10:1 with 80–90 percent of men acquiring the infection from commercial sex workers (CSWs)[5].

In India, high co-infection rate (14.4 percent) of gonorrhoea with other sexually transmitted infections from 2003-2008 highlight the considerable burden of the disease. This indicates that appropriate screening measures for STIs must be widely and consistently implemented so as to assure prompt and effective treatment for infected persons and their sexual partners leading to reductions in disease burden[6].

Treatment and Prevention

Despite findings that fluoroquinolones have high cure rates, *N. gonorrhoeae* has become increasingly resistant to fluoroquinolones in some geographical areas. Therefore, the CDC advises against using fluoroquinolones to treat gonorrhea infection in patients who live or may have acquired infection in Asia, the Pacific islands (including Hawaii), and California. A substantial increase in fluoroquinolone-resistant *N. gonorrhoeae* has been noted in many geographical areas therefore, it no longer recommends fluoroquinolones as first-line treatment in these patients.

Ten to 30 percent of patients with gonorrhea will have a concomitant chlamydia infection; therefore, many experts advocate empirically treating chlamydia when treating patients with gonorrhea. The CDC recommends azithromycin (Zithromax) in a single 1-g dose orally, or doxycycline, 100 mg twice a day for seven days, for patients with uncomplicated infections of the cervix, urethra, and rectum and possible concomitant chlamydia infections.

5.2.1.3 Genital Warts

Genital warts are one of the most common types of sexually transmitted infection. An estimated 32 million cases are reported worldwide every year. The warts affect the genital area and cause such symptoms as itching, burning, discomfort, pain, or bleeding with intercourse. Because of the recurrence and the stigma associated with genital warts , it also exerts a psychological burden on the affected individuals.

Pathogenesis

Genital warts are soft growths on the skin and mucus membranes of the genitals (Fig. 5.4). They may be found on the penis, vulva, urethra, vagina, cervix and in the anus. The virus that causes genital warts is called human papilloma virus (HPV). More than 70 different types of HPV exist. In women, HPV can spread to areas inside the walls of the vagina and cervix. They are not easy to see without special procedures. HPV infection spreads from one person to another through sexual contact involving the anus, mouth, or vagina. In males, genital warts may be found on the penis, scrotum, thighs, inside and around anus. Genital warts may also occur on the lips, mouth, tongue and throat. Other symptoms are rare but include increase vaginal discharge, genital itching, vaginal bleeding during and after sex.

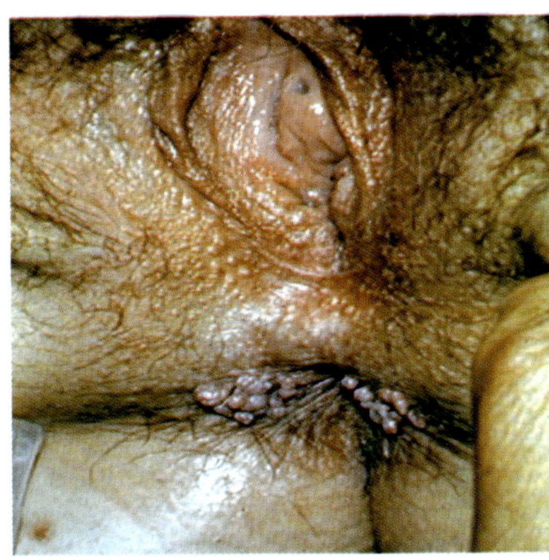

Fig. 5.4: Perianal warts
Source: Seattle STD/HIV Prevention Training Center at the University of Washington/ UW HSCER Slide Bank

The virus that causes genital warts can cause abnormal results on a Pap smear. An HPV DNA test can tell if an individual has a high-risk type of HPV known to cause cervical cancer. This test may be done as a screening test for women over age 30. There is persistent ulceration or bleeding. Visible genital warts are caused by serotypes 6 and 11. These serotypes rarely cause cancer. The serotypes 16 and 18 most likely cause cancer but have a relatively low prevalence rate.

Genital warts may be of following types:

- Condylomata acuminata
  - Cauliflower-like appearance
  - Skin-colored, pink, or hyperpigmented
  - May be keratotic on skin; generally non-keratinized on mucosal surfaces
- Smooth papules
  - Usually dome-shaped and skin-colored
- Flat papules
  - Macular to slightly raised
  - Flesh-colored, with smooth surface
  - More commonly found on internal structures (*i.e.,* cervix), but also occur on external genitalia
- Keratotic warts
  - Thick horny layer that often resemble common warts or seborrheic keratosis

Epidemiology

Genital HPV infection is the commonest viral STI in the developed countries of world. An estimated 30 million new cases are diagnosed annually worldwide[7]. Approximately 15 percent of the general population harbors subclinical infection[8].

In India, the incidence of genital warts has been reported to vary from 2 percent to 25.2 percent among STI clinic attendees[5]. Infection with oncogenic HPV is the leading cause of cervical cancer in India. A study conducted in Delhi during the late 1990s among women STI clinic attendees, HPV was detected in 30 percent that increased to 52 percent and 72 percent among women with precancerous and cancerous cervical lesions, respectively. In Mumbai, using molecular probes of the HPV it was detected in 77% of cervical cancer patients.

Ethnographic variations have also been observed in the epidemiology of HPV infection[9]. Higher rates of prevalence have been reported among African/American women than among whites or Hispanics. This may be due to differences in the probability of encountering an HPV-positive partner or due to genetic predisposition towards a greater susceptibility to the acquisition and persistence of infection, endogenous hormonal factors or differences in sexual behaviour among the ethnic groups.

Treatment and Prevention

Medicines that can be applied include podofilox (Condylox) and trichloroacetic acid (TCA), which kill the wart tissue, or Imiquimod (Aldara), a cream that stimulates the immune system to fight the virus. Among other options, freezing the warts with liquid nitrogen, cauterizing them with heat or a laser, or removing them surgically are quite popular methods adopted.

The best way to prevent genital warts is abstaining from sexual intercourse with an infected individual. Condoms can help lower risk of HPV infection, but they can't fully protect against HPV since there may still be skin to skin contact with an infected area. Condoms are still very useful in protecting against other sexually transmitted infections.

Two vaccines, Gardasil and Cervarix have been approved by the FDA to help reduce the risk of cervical cancer by immunizing young women aged 11-26 against certain strains of HPV which cause 70 percent of cervical cancers. Gardasil also provides protection in men and women against HPV strains that cause 90 percent of genital wart cases. In the late 2010, the vaccine was also approved for the prevention of anal cancer. Ideally, the vaccine should be given prior to becoming sexually active. Both of these vaccines are given in a three-shot series over six months.

5.2.1.4 Genital Herpes

Genital herpes affects the skin or mucous membranes of the genitals. It is caused by the herpes simplex virus (HSV). HSV is spread from one person to another during sexual contact.

- HSV-2 most often causes genital herpes. HSV-2 can spread through secretions from the mouth or genitals.
- HSV-1 usually affects the mouth and lips and causes cold sores or blisters. However, it can spread from the mouth to the genitals during oral sex.

Pathogenesis

Genital infection is caused by inoculation of the virus onto a mucosal surface or through cracks in the skin, usually through close sexual contact. Disease is caused by direct cytopathic effect of the replicating virus, destroying tissue. At primary infection, it evades the immune attack by ascending to the peripheral sensory nerves and becomes established in sensory nerve root ganglia. The disease recurs due to reactivation of the initial strain of the virus from latently infected nerve root ganglia[10].

Epidemiology

HSV is spread by contact, as the virus can be shed through saliva, tears or by genital and other secretions, By far the most common form of infection results from a kiss given to a child or adult from a person shedding the virus. The risk of infection to a non-immune individual in contact with contaminated secretions can be as high as 80%. Genital herpes simplex virus (HSV) infection is the second most prevalent STI worldwide and the commonest cause of genital ulcer disease in the developed world[10]. This infection has important public health implications because: (*i*) undiagnosed cases contribute to the population reservoir and transmission of the virus; (*ii*) perinatal transmission to the neonate may result in disseminated disease, neurological damage resulting in high mortality; and (*iii*) herpetic ulcers also facilitate HIV transmission.

Genital herpes is one of the three most prevalent STIs in the USA (along with chlamydial and human papillomavirus infections), and probably of greatest concern to sexually active people, apart from HIV infection. About 1 in 5 persons with less than 12 years of age (approximately 45 million people) are infected with HSV-2 infection. This leads to 1 million new HSV-2 infections occurring annually.

In India also, there has been a major increase in the proportion of viral STIs, especially HSV infection, with rates varying from 4.1 percent to 27.9 percent among STI clinic attendees in different regions of the country. Using molecular diagnostics, the prevalence rate increased from 11.4 percent to 21 percent on an average.

A cross-sectional study carried out in 1994 of gynaecological clinic attendees in a Delhi hospital revealed that in 20.6 percent of the women, IgA antibodies to HSV were detected. During 1983–86, genital herpes was diagnosed in 19 percent of STI clinic attendees in southern India, with a positive culture in 38.7 percent of the cases with primary infection and 38.2 percent with recurrent disease[11].

Treatment and Prevention

Warts can be treated but not cured, but approximately 90% of cases clear spontaneously within two years. The available methods used for warts are as follows:

1. *Antiviral chemotherapy*: Acyclovir is the drug of choice to treat patients infected with genital herpes. Acyclovir requires the presence of a HSV-encoded thymidine kinase in order to be converted into its active acyclovir triphosphate form. Acyclovir triphosphate is a potent selective inhibitor of HSV DNA polymerase and causes premature chain termination when it competes with guanine triphosphate for incorporation into newly synthesized viral DNA. However, Acyclovir resistant strains of HSV have been reported.
2. *Vaccination*: Vaccines have been used to prevent infection, reduce transmission. Vaccines have been developed using viral surface glycoproteins gB and gD antigens which can completely prevent the acquisition of infection and reduce the frequency of recurrence. A recent small double-blind randomized control trial demonstrated that a gD2 vaccine significantly reduced the number of disease recurrence in the intervention group[12].
3. *Preventive measures*: Educating those with clinical disease to recognize the symptoms of recurrences and avoid unprotected sex and prophylactic use of antiviral agents might reduce the risk of HSV transmission. However, the high prevalence of asymptomatic shedding and the low sensitivity of clinical screening for the detection of genital infection, means that case-finding will not be the basis of an effective population control program.

5.2.1.5 Trichomonas

Trichomoniasis is a sexually transmitted disease (STD) caused by a small organism called *Trichomonas vaginalis*. Women population are more prone to the disease, although men can become carrier and pass the infection to their partners through sexual contact. To diagnose trichomoniasis, a doctor needs to perform a physical examination of the infected individual followed by a laboratory test on a sample of vaginal fluid or urethral fluid to look for the disease-causing parasite. The parasite is difficult to detect in men than in women. Broth culture technique has been the gold standard for *T vaginalis* for the past 40 years.

T. vaginalis is a flagellated protozoan possessing five flagella, four of which are located at its anterior portion whereas the fifth flagellum is incorporated within the undulating membrane of the parasite (Fig. 5.5) which is supported by a slender noncontractile costa. *Trichomonas vaginalis* exists in only morphological stage, a trophozoite, and cannot encyst. The *T.* vaginalis trophozoite is oval or "pear" shaped as seen on a wet-mount. It is slightly larger than a white blood cell measuring 9×7 μm. The nucleus in *T. vaginalis* is located at its anterior portion, and is surrounded by a porous nuclear envelope as seen in many eukaryotes. This organism is considered to be one of the most ancient eukaryotes. *T. vaginalis* has features that are common to anaerobic bacteria as well as higher eukaryotic organisms in terms of its carbohydrate and energy metabolism[13].

Pathogenesis

Trichomoniasis is the most common non-viral sexually transmitted disease however the pathogenicity of *T. vaginalis* is not yet thoroughly understood. Trichomonads after getting an entry in the host, participate in a host-parasite relationship. They adhere to epithelial cells. The ability of trichomonads to adhere is affected by time, temperature, and pH level. *T. vaginalis* grows best in an anaerobic environment with a pH > 6. Binding of *T. vaginalis* to vaginal epithelial cells for colonization and infection is dependent upon specific parasite surface proteins. When the patients are treated with tinidazole, metronidazole or other nitroimidazoles, the parasites lose their ability to adhere, making them ineffective disease agents. Hemolysis, the destruction of red blood cells such that hemoglobin is released, is also correlated with virulence. Increase in severity of Trichomoniasis has been reported during or slightly after menstruation[14].

Fig. 5.5: An electron micrograph depicts the *Trichomonas vaginalis* parasite adhering to vaginal epithelial cells collected from vaginal swabs. A non-adhered parasite (right) is pear-shaped, whereas the attached parasite is flat and amoeboid.

Credit: Image courtesy of: Antonio Pereira-Neves and Marlene Benchimol, Santa Ursula University, Rio de Janeiro, Brazil (http://www.niaid.nih.gov/topics/vaginitis/pages/tvaginalisparasite_photo.aspx)

The vaginal epithelium is the most important site for the initial contact of the parasite infection in humans. *T. vaginalis* infects the squamous epithelium in the genital tract and the clinical manifestation could be:

Acute:
- Diffuse vulvitis
- Yellow or green discharge
- Hemorragic spots in the vaginal mucosa

Chronic:
- Mild symptoms with pruritus
- Vaginal discharge mixed with mucus

The infection can result in varying degrees of vaginitis, and also infertility, premature labour, premature rupture of the placental membranes, and low birth weight infants. There also exist a relative risk of developing invasive cervical cancer and six fold higher probability of infection by human immuno deficiency virus (HIV).

Genetic Diversity

T. vaginalis is a genetically diverse parasite with a unique population structure consisting of two types present in equal proportions world-wide. Parasites belonging to the two types-type 1 and type 2 differ significantly in the rate at which they harbor the *T. vaginalis* virus (TVV). It is a dsRNA virus implicated in parasite pathogenesis. The two types differ in sensitivity to the widely-used drug, metronidazole. Evidences of genetic exchange indicate that a sexual life-cycle of the parasite occurs despite an absence of morphologically-distinct sexual stages[15].

Epidemiology

Trichomoniasis occurs worldwide. There is a higher prevalence among people with multiple sexual partners and/or other venereal diseases. An estimated 170 million cases of trichomoniasis occur worldwide every year. Trichomoniasis is one of the most common sexually transmitted infections found in the United States. According to the CDC estimates 5 million new cases occur annually. It is found in 3-5 percent of females in North America. In urban centers, Trichomonas vaginalis has a high prevalence among African-Americans. According to the CDC, *T. vaginalis* infection is an important cofactor in HIV transmission and may increase HIV shedding (risk of contraction from an infected individual). Vertical transmission of *T vaginalis* during birth is possible and may persist up to 1 year. Around 2-17 percent of female offspring of infected women acquire infection.

Infection is often overlooked in studies due its mild clinical presentation. The disease does not have adverse effects on fertility and so far no evidence has been reported of poor birth outcomes. It is estimated that about 50 percent of cases are under diagnosed. However, *T. vaginalis* may persist in a host for more than two years if untreated.

In India, in a community setting, the burden of *T. vaginalis* infection is around 8.5 percent and is relatively high among young reproductive aged women. Because this infection increases the risk of HIV transmission and is often associated with adverse pregnancy outcomes, there is a need for increased screening and treatment of this easily curable but sexually transmissible infection in India[16].

Treatment and Prevention

Oral metronidazole (Flagyl) remains the treatment of choice for trichomoniasis. In severe cases, members of other nitroimidazoles or high doses of metronidazole may be used. Drug resistance is rare, despite the prevalent use of nitroimidazole drugs in the treatment of trichomoniasis. Topical metronidazole and other antimicrobials are not effective in eradicating the disease. Education concerning STI treatment and prevention is vital. Because *T vaginalis* infection is strongly associated with the presence of other STIs (gonorrhea, chlamydia, and sexually transmitted viruses such as HIV), service providers should provide appropriate counseling, testing, and required treatment specially to sex workers.

5.2.1.6 Syphilis

Syphilis is a sexually transmitted disease (STD) and requires immediate attention or it may lead to serious health problems and increase the risk for getting human immunodeficiency virus (HIV). Syphilis also can be passed from a mother to the child during pregnancy and may cause serious complications.

Pathogenesis

Syphilis is a chronic infection caused by the bacterium *Treponema pallidum. T. pallidum* is a member of the order Spirochaetales, family Spirochaetaceae, and genus *Treponema*, which includes four human pathogens and at least six human nonpathogens[17]. The manifestations of disease are liable to change, occurring in any one individual in different stages over time.

Primary Syphilis

The symptoms of the first stage of the infection known as primary syphilis appears withitin 10 days to 3 months after sexual contact with an infected person. A painless red sore called a chancre may appear on the genitals in the area where the infection occurred. Lymph nodes get enlarged in the infected area. Depending on the type of sexual contact, a chancre can develop either on the mouth or in the rectal area. Chancres are defined as a firm, painless, non-itchy skin ulceration. These are the primary way that syphilis is transmitted between people, but often are unrecognized. In chancroid, a swollen painful ulcer is formed on the genitals that involves infection of the adjacent lymph nodes. The causative agent is *Haemophilus ducreti*, a small gram negative rod that can be isolated from the exudates of lesion. The recommended antibiotics at this stage are erythtomycin and ceftriaxone. Even without treatment, chancres will heal after 3 to 6 weeks but if the infection is not treated in time, the disease may progress to the second stage.

Secondary Syphilis

The secondary stage usually begins within weeks to months after the chancre sore appears. Syphilis bacteria enter the blood and spread through the body and various symptoms are noticed including rash (small red spots), fever, headache, loss of appetite, weight loss, sore throat, muscle aches, joint pain. The patient experiences generally ill feeling with enlarged lymph nodes. The rash of secondary syphilis can develop anywhere on the body including the palms and the soles of the feet. Gray or white wart-like patches of skin called condylomata can appear on the moist areas around the mouth, anus, and vagina. These lesions are full of bacteria and contagious. But during this stage, syphilis can also affect other organs like liver, kidneys, and eyes. It may also cause meningitis. The symptoms of secondary syphilis eventually fade away however without treatment, the infection can advance to the third stage. This can also happen in infected person who did not show symptoms of primary or secondary syphilis.

Late (Tertiary) Syphilis

At the end of the secondary stage, untreated individuals progress to the latent stage, where they do not show any symptoms but are still infected. Symptoms of late syphilis can appear many years later and can damage many vital organs of the body like the eyes, large blood vessels, heart, bones, and central nervous system

A wide range of opportunistic infections (OI) and tumors develop additionally, HIV directly damages some organs. The patterns of opportunistic diseases (OD) are different in different parts of the world, depending on the local prevalence of latent and acquired infections and on the survival of HIV infected patients (Table 5.1).

Table 5.1: Main opportunistic diseases associated with HIV infection, seen in AIDS[30]

Viral infections	**Bacterial infections**
CMV Herpes simplex JC virus causing progressive multifocal leukoencephalopathy (PML) Molluscum contagiosum Herpes zoster Measles Human papilloma virus (HPV) Human herpes virus 8 (HV8) Epstein-Barr virus (EBV)	*Recurrent bacterial pneumonia* (commonly *Streptococcus pneumoniae)* *Mycobacterium tuberculosis* Non-tuberculosis mycobacteriosis (particularly *M. avium*-intracellulare complex) Systemic non-typhoid Salmonella infections (notably S. *enteritidis* and *S. typhimurium)* *Pseudomonas spp. septicaemia and `vasculitis'* *Bartonella spp.* (causing bacillary angiomatosis) *Rhodococcus equi* *Nocardia spp*
Fungal infections	**Protozoal infections**
Severe infection of candida *Pneumocystis carinii* *Cryptococcus neoformans* *Histoplasma capsulatum* *Coccidioides immitis* *Aspergillus spp.* *Penicillium marneffei*	*Toxoplasma gondii* *Cryptosporidium parvum* *Isospora belli* *Leishmania spp.* *Microsporidia spp. (commonly Encephalitozoon intestinalis, Enterocytozoon bieneusi)* *Acanthamoeba spp.* *Trypanosoma cruzi*
Tumours	**Other conditions**
Kaposi's sarcoma Primary cerebral lymphoma High-grade non-Hodgkin lymphoma Carcinoma (invasive) of the cervix Carcinoma of the conjunctiva Carcinoma of the anus T-cell lymphoma Hodgkin's disease Lymphoproliferative disease, pre-lymphomatous	HIV-wasting syndrome (fever, weight loss, diarrhoea) HIV-associated dementia Various dermatitis patterns (e.g. pruritic rash, eosinophilic folliculitis) Skeletal myopathy Peripheral and autonomic neuropathy Cardiomyopathy Pulmonary hypertension Vasculitis HIV-associated nephropathy (HIVAN) Haemolytic uraemic syndrome (HUS) and thrombotic thrombocytopaenic purpura (TTP) Oral and oesophageal ulcers

Paediatric HIV Disease

While disease in adults varies globally, the clinical pathology of children infected with HIV congenitally is remarkably uniform in the absence of medical intervention. Infants suffer from pneumocystosis and bacterial pulmonary and systemic infections. Measles is more common in case the child is not given the required vaccination. Infants often develop Tuberculosis also[31].

Epidemiology

Current Global Snapshot: According to the latest estimates from UNAIDS[32],

- There were 34.0 million people living with HIV in 2011. The figure was 29.4 million in 2001. The reasons for the increase in trend are attributed to as a result of continuing new infections, people living longer with HIV, and general population growth.
- The global prevalence rate (the percent of people in the age group of 15–49 who are infected) has leveled since 2001 and was 0.8 percent in 2011.
- 1.7 million died of AIDS in 2011, a 24 percent decrease was observed since 2005. Deaths have declined due to the intervention by antiretroviral treatment (ART) in a big way. HIV is considered to be the leading cause of death worldwide and the number one cause of death in Africa.

Asia subcontinent: More than 5 million people are living with HIV across South/South-East Asia and East Asia. While most national epidemics appear to have stabilized, HIV incidence has increased by more than 25 percent since 2001 in Bangladesh, Indonesia, the Philippines, and Sri Lanka. The region also includes two highly populated subcontinents like China and India however the prevalence rate of AIDS is under control in these countries[32].

Treatment and Prevention

Treatment as prevention (TASP) is a term increasingly used to describe HIV prevention methods that use antiretroviral treatment to decrease the chance of HIV transmission. People living with HIV often take treatment for their own health; ARV treatment help to decrease the amount of the virus in infected person's body fluids, known as the 'viral load', thereby reducing the likelihood of AIDS-related illnesses[33].

An HIV-positive person's viral load is the single biggest risk factor in the transmission of HIV. From 1996, combination therapies are in use, comprising highly active anti-retroviral therapy (HAART)[34]. The four main classes of ARV are: the nucleoside and non-nucleoside analogues; protease inhibitors; and a ribonucleotide reductase inhibitor. They affect the pathology of HIV in two ways: by their toxicity and through immune restoration[30]. Highly active anti-retroviral chemotherapy prevents many of the common opportunistic infections, but introduces a new range of side effects causing pathological damage. Longer survival of an infected individual may lead to development of new HIV-related diseases. Therefore the pathology of HIV/AIDS is not well defined and may alter depending on the immune system of an individual.

Risk Factors

Risk factors for possible exposure to HIV, includes the following:

- Unprotected sexual intercourse
- Number of sexual partners
- Previous or current sexually transmitted diseases (STDs)
- Sharing of intravenous (IV) drug paraphernalia
- Mucosal contact with infected blood or needle-stick injuries
- Maternal HIV infection

5.2.2 Other Infections

- Candidiasis
- Bacterial vaginosis

5.2.2.1 Candidiasis

In developing countries, Vulvovaginal candidiasis (VVC) is a substantial cause of lower genital infections in women. This causes significant loss of work hours and significant financial burden on the already weak economy. Nearly 75 percent of all adult women have had at least one genital "yeast infection" in their lifetime. The most prevalent species associated with vulvovaginal candidiasis is *C. albicans*. VVC occurs more frequently and more severely in immunocompromised people. There are some other conditions that may put a woman at risk for genital candidiasis like Pregnancy, Diabetes mellitus, use of corticosteroid medications or use of broad spectrum medications.

Pathogenesis

To be invasive, Candida must follow a three- stage mechanism.

(*a*) *Adhesion*: Adhesion of *Candida albicans* is crucial for the survival of blastospore. Blastopore adherence seems to be based on a mannoprotein in fungal membrane, previously known as adhesin. This protein is able to anchor itself on the receptor on surface cells in vaginal epithelium. Adherence is enhanced by estrogen impregnation in vaginal epithelium. Estrogen increases surface exposure of glycoprotein complexes acting as receptors for blastopore adherence. Thus, candidal vaginitis is uncommon during hypoestrogenic periods, such as premenarche and postmenopause. On the other hand, pregnancy — as well as use of oral contraceptives containing a high estrogen dose — causes an increased blastopore adhesion and, so, a higher symptomatic infection rate. It is therefore stated that symptomatic vaginal candidosis is a hormone-dependent (and more specifically, estrogen-dependent) disorder. Another factor affecting candidal adhesion is competition with vaginal lactobacilli. Thus, a reduction in normal lactobacilli flora can cause an imbalance and provoke a candidal vaginitis. In fact, a low number of lactic bacilli has been reported in vaginal exudate samples from patients with CVV[35].

(*b*) *Blastospore germination*: Blastospore forms are not able to penetrate into epithelial tissues. Exogenous factors promoting spore germination can trigger candidal vaginitis in women. Again, estrogen promotes mycelium development in candida[36].

(*c*) *Epithelial invasion*: Once its mycelium has been developed, candida is able to penetrate and invade vaginal surface cells. Epithelium invasion also causes a release of several substances, such as prostaglandins and bradykinin, inducing inflammatory changes in tissues. This process can lead to edema, erythema, exudate increase, and cell shedding. Another factor involved in epithelial invasion is blastopore load. An increase in the number of spores in vaginal environment triggers invasion and candidal colpitis.

Epidemiology

Candida albicans was the most prevalent cause in 87.9 percent of an epidemiologic study of VVC cases. An estimated two thirds of women will suffer at least one episode of VVC in their lives and 40-50 percent may experience multiple episodes. In 20 percent of healthy asymptomatic women, Candida species can be found in the lower genital tract flora. Vulvovaginal candidiasis is common in adults: 70-75 percent of women are affected by this infection at least once in their lives. Nearly 50 percent of adult women may experience a second episode and 5-8 percent of women will report four or more episodes[37]. Among clinical populations of postmenopausal women not using estrogen therapy, vulvovaginal candidiasis is less common than in younger women within menopausal period[38]. Although greater than 50 percent of women in the age group of 25 years or more develop vulvovaginal candidiasis at some time in their life time, however, less than 5 percent of these women experience recurrences. It is essential to evaluate the recurrent episodes of this disease.

Treatment and Prevention

On physical examination, the patient with vulvovaginal candidiasis usually has vulvar erythema and a thick, white to yellow discharge in the vaginal vault. Though Potassium hydroxide (KOH) testing and setting up fungal cultures have some limitations but they are still useful for identifying the infecting species and guiding treatment. Microscopic examination of vaginal secretions in a 10 percent KOH preparation may demonstrate hyphae.

Based on the study findings, prescribed drugs are ketoconazole (Nizoral) administered orally once a day, clotrimazole (Gyne-Lotrimin) administered intravaginally twice weekly, terconazole administered intravaginally once a week, and fluconazole (Diflucan) or itraconazole (Sporanox) administered orally once a month. They have been relatively effective in reducing the recurrence rate for vulvovaginal candidiasis[39]. Since Candida often develop very fast resistance to anti-fungal drugs therefore recurrent infections cannot be successfully treated with a drug that used to work in an earlier infection.

A study assessed the association between the daily ingestion of yogurt containing *Lactobacillus acidophilus* and the prevention of recurrent vulvovaginal candidiasis[40]. The study concluded that daily ingestion of 8 oz of yogurt containing *L. acidophilus* decreased the rate of candidal infection.

5.2.2.2 Bacterial Vaginosis

This is a disease of vagina caused by bacteria. Bacterial vaginosis (BV) is the most common cause of vaginal discharge in women during childbearing age. It accounts for 40 to 50 percent of cases. The absence of inflammation is the basis of the term "vaginosis" rather than "vaginitis."

Pathogenesis

BV represents a complex change in the vaginal flora characterized by a reduction in concentration of the normally dominant hydrogen-peroxide producing lactobacilli and an increase in concentration of other organisms, especially anaerobic gram negative rods[41]. This competitive change allows other bacteria to proliferate, producing amines that contribute to a further rise in pH. The major bacteria detected are *Gardnerella vaginalis*, *Prevotella sp.*, *Porphyromonas sp.*, *Bacteroides sp.*, *Peptostreptococcus sp.*, *Mycoplasma hominis*, *Ureaplasma urealyticum, and Mobiluncus sp*[42]. The most common symptom of BV is an abnormal homogeneous cream color vaginal discharge released specially after vaginal intercourse.

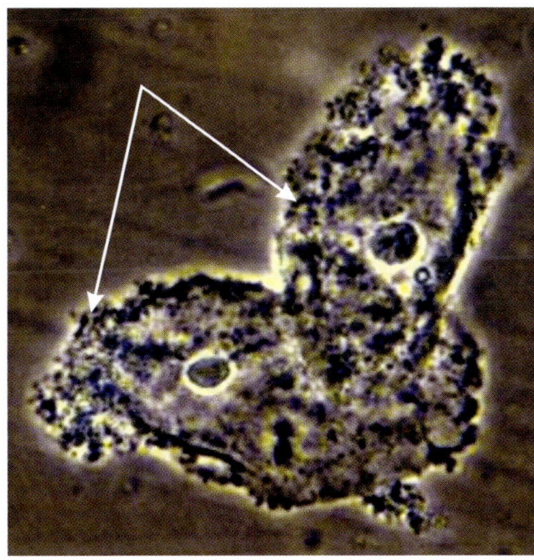

Fig. 5.8: Clue cells (shown by white arrows) *Gardnerella vaginalis* bacteria coat the surface of vaginal epithelial cells (light microscopy)
Source: medind.nic.in[43]

Common diagnostic symptoms are: Homogeneous, thin, greyish-white discharge that smoothly coats the vaginal walls. During the discharge, the vaginal pH increases more than 4.5 and shows a Positive whiff-amine test. The presence of a fishy odor is a characteristic feature of the discharge fluid. A wet mount of Clue cells on saline can be prepared. Clue cells are vaginal epithelial cells studded with adherent coccobacilli that are best observed at the edge of the cell under microscope (Fig. 5.8). In women with BV, at least 20 percent of the epithelial cells on wet mount should be clue cells.

Epidemiology

Prevalence studies indicate that there is a potentially large reservoir of BV infection in the population. Due to high proportion of asymptomatic cases, the prevalence of BV is under-estimated by most studies.

BV renders women particularly vulnerable to the acquisition of *Trichomonas vaginalis*, *Neisseria gonorrhoea*, *Chlamydia trachomatis,* HSV-2 and HIV-1. Moreover, it has been documented that bacterial vaginosis also propagates viral replication and vaginal shedding of the HIV-1 and HSV-2 viruses, thereby further enhancing the spread of these viruses[44].

Bacterial vaginosis (BV) has been linked to sexual behaviour, and the epidemiological profile of BV relates to that of established sexually transmitted infections (STIs). It remains debatable whether BV pathogenesis does actually involve sexual transmission of pathogenic micro-organisms from men to women.

Treatment and Prevention

The recommended antibiotics for treating bacterial vaginosis are metronidazole and clindamycin. Both of these medications are available as oral medication and as vaginal cream or gel. Prophylactic antibiotic treatment often helps to prevent the formation of pelvic inflammatory disease resulting from any bacteria that gains an entry into the pelvic region during or after surgery. Pregnant women with bacterial vaginosis also face the risk of premature delivery or giving birth to low birth weight babies if they remain untreated. BV is not completely understood by scientists, and therefore the best ways to prevent the disease is unknown. However, it is known that BV is associated with having a new sex partner or having multiple sex partners.

REFERENCES

1. Gupta K, Sahm D, Mayfield D, Stamm WE (2001) Antimicrobial resistance among uropathogens that cause community-acquired urinary tract infections in women: a nationwide analysis. Clin Infect Dis. 33:89-94
2. Gaydos CA et al (2004). Comparison of three nucleic acid amplification tests for detection of Chlamydia trachomatis in urine specimens. J Clin Microbiol 42:3041-3045.
3. Haugland S, Thune T, Fosse B, Wentzel-Larsen T, Hjelmevoll S, Myrmel H (2010) Comparing urine samples and cervical swabs for Chlamydia testing in a female population by means of Strand Displacement Assay (SDA) BMC Women's Health 10: 9
4. Vos T et al (2012) Years lived with disability (YLDs) for 1160 sequelae of 289 diseases and injuries 1990-2010: a systematic analysis for the Global Burden of Disease Study 2010. Lancet 380 (9859): 2163–96
5. Sharma VK and Khandpur S (2004) Changing patterns of sexually transmitted infections in India. Natl. Med Jr Ind. 17: 310-319.
6. Bala M, Mullick JB, Muralidhar S, Kumar J, Ramesh V (2011) Gonorrhoea & its co-infection with other ulcerative, non-ulcerative sexually transmitted & HIV infection in a Regional STD Centre. Indian J Med Res 133:346-349.
7. Khanna J, van Look PFA, Griffin PD (1992) Reproductive health: A key to a brighter future. Biennial report 1990–91. Geneva:World Health Organization.
8. Koutsky L. Epidemiology of genital human papillomavirus infection (1997). Am J Med 02:3–8.

9. Tortolero-Luna G (1999) Epidemiology of genital human papillomavirus. Haematol Oncol Clin North Am 13:245–57.
10. Brugha R, Keersmaekers K, Renton A, Meheus A (1997) Genital herpes infection: a review. Int J Epidemiol. 26:698-709.
11. Jacob M, Rao PS, Sridharan G, John TJ (1989) Epidemiology and clinical profile of genital herpes. Indian J Med Res. 89:4–11.
12. Edwards S, White C (1994) Genital herpes simplex virus type 1 in women. Genitourin Med.70: 426.
13. Petrin D, Delagly K, Bhatt R, Garber G (1998). Clinical and microbiological aspect of *Trichomonas vaginalis*. Clinical Microb. (11): 300 – 317.
14. Graves A, Gardner WA Jr. (1993) Pathogenicity of *Trichomonas vaginalis*. Obstet Gynecol. Clin Obstet Gynecol.36:145-52.
15. Conrad MD, Gorman AW, Schillinger JA, Fiori PL, Arroyo R, Malla N, Dubey ML, Gonzalez J, Blank S, Secor WE, Carlton JM (2012) Extensive genetic diversity, unique population structure and evidence of genetic exchange in the sexually transmitted parasite *Trichomonas vaginalis*. PLoS Negl Trop Dis. 6(3): e1573.
16. Madhivanan P, Bartman MT, Pasutti L, Krupp K, Arun A, Reingold AL, Klausner JD (2009) Prevalence of *Trichomonas vaginalis* infection among young reproductive age women in India: implications for treatment and prevention. Sex Health. 6:339-44.
17. Minkoff H L, McCalla S, Delke I, Stevens R, Salwen M, Feldman J (1990) The relationship of cocaine use to syphilis and human immunodeficiency. Am J Obstet Gynecol. 163:521–526.
18. Creighton E T. Darkfield microscopy for the detection and identification of *Treponema pallidum*. In: Larsen S A, Hunter E F, Kraus S J, editors. A manual of tests for syphilis. Washington, D.C: American Public Health Association; 1990. pp. 49–62.
19. Norris S J, Larsen S A. Treponema and other host-associated spirochetes. In: Murray P R, Baron E J, Pfaller M A, Tenover F C, Yolken R H, editors. Manual of clinical microbiology. 6th ed. Washington, D.C: ASM Press; 1995. pp. 636–651.
20. Willcox R R, Guthe T. *Treponema pallidum* A bibliographical review of the morphology, culture and survival of *T. pallidum* and associated organisms. Bull WHO.35:1–169.
21. Larsen S A, Steiner B M, Rudolph A H (1995) Laboratory diagnosis and interpretation of tests for syphilis.Clin Microbiol Rev.8:1–21.
22. WHO Office of HIV/AIDS and STDs. An overview of selected curable STDs. Syphilis estimates, Geneva, Switzerland: World Health Organization; (1995).
23. Centers for Disease Control and Prevention. Trends in reportable sexually transmitted diseases in the United States, 2005. Available from: [http://www.cdc.gov/std/stats/trends2005.htm]
24. Ray K, Bala M , Gupta SM , Khunger N , Puri P , Muralidhar S et al. (2006) Changing trends in sexually transmitted infections at a Regional STD Centre in north India. Indian J Med Res.124:559-68.
25. Fiumara NJ. (1980) Treatment of primary and secondary syphilis. Serological response. JAMA. 243:2500-2
26. Workowski KA, Berman S. (2010) Sexually transmitted diseases treatment guidelines, 2010. MMWR Recomm Rep.59:1-110.
27. Parran T. Shadow on the land. New York, N.Y: Reynal and Hitchcock; 1937
28. WHO case definitions of HIV for surveillance and revised clinical staging and immunological classification of HIV-related disease in adults and children. Geneva: World Health Organization. 2007. pp. 6–16. ISBN 978-92-4-159562-9.
29. Mandell, Bennett, and Dolan (2010). Principles and Practice of Infectious Diseases: Chapter 118. Churchill Livingstone ISBN: 978-0-443-06839-3.

30. Lucas, S (2002) The pathology of HIV infection. Lepr Rev. 73, 64-71.
31. Lucas SB, Peacock CS, Hounnou A et al.(1996) Disease in children infected with HIV in Abidjan, CoÃte d'Ivoire. Br Med J, 312: 335-338.
32. UNAIDS. Report on the Global AIDS Epidemic; 2012
33. Williams B. et al (2010) Treatment as prevention: preparing the way. Journal of the International AIDS Society, 14
34. Paxton LA, Janssen RS (2000) The epidemiology of HIV infection in the era of HAART. J HIV Ther, 5: 2-4.
35. Ferrer J (2000) Vaginal candidosis: epidemiological and etiological factors. Int. Jr. Gynecol & Obstet. 71:21–27.
36. Powell BL, Drutz DI (1983) Estrogen receptor in Candida albicans. A possible explanation for hormonal influences in vaginal candidiasis. Twenty third Interscience Conference on Antimicrobial Agents and Chemotherapy Abstract, 751 p. 222.
37. Sobel, JD (2007) Vulvovaginal candidosis. Lancet 369: 1961-1971.
38. Sobel, JD (2005) Genital Candidiasis. Medicine 33: 62-65.
39. Ringdahl EN (2000) Treatment of Recurrent Vulvovaginal Candidiasis Am Fam Physician. 61:3306-3312.
40. Hilton E, Isenberg HD, Alperstein P, France K, Borenstein MT(1992) Ingestion of yogurt containing Lactobacillus acidophilus as prophylaxis for candidal vaginitis. Ann Intern Med 116:353–7.
41. Hill GB (1993) The microbiology of bacterial vaginosis. Am J Obstet Gynecol. 169:450.
42. Ling Z, Kong J, Liu F, et al (2010) Molecular analysis of the diversity of vaginal microbiota associated with bacterial vaginosis. BMC Genomics 11:488.
43. Khan KJ, Shah R, Gautam M, Patil S (2007) clue cell. Indian J Sex transm Dis 28:108-9
44. Verstraelen H, Verhelst R, Vaneechoutte M, Temmerman M (2010) The epidemiology of bacterial vaginosis in relation to sexual behaviour.BMC Infect Dis.10:81.

6

Blood and Lymphatic Infections

Chapter Outline

6.1 Bacterial infections of the Blood Vascular system

6.2 Viral diseases of the lymphoid system

6.3 Protozoal Infections

The blood and lymphatic systems lie deep within the body and are considered to be sterile. The cardiovascular system consists of the heart, blood and blood vessels. The lymphatic system is composed of the lymph, lymph vessels, lymph nodes and the lymphoid organs such as the tonsils, appendix, spleen and thymus. Fluids in both systems circulate throughout the body where they intimately come in contact with vital tissues and organs. However both the systems at some point of time can function as a vehicle for spreading infection. Infection of the blood and lymphatic system can have devastating effects because infectious agents become 'systemic' since they can be carried to all parts of the body and may lead to disease in one or more vital organs.

6.1 Bacterial Infections of the Blood Vascular System

6.1.1 Bacterial Endocarditis

Infections involving heart valves and linings of the hearts are called endocarditis. The inner layer, called the 'endocardium', lines the heart muscle and covers the valves. An inflammation of the endocardium leads to endocarditis.

Acute Bacterial Endocarditis

1. This is caused when virulent bacteria enter the bloodstream from an infection elsewhere in the body such as in the teeth or tonsils. Microorganisms often released by tooth extractions or tonsillectomies may enter the blood and find their way to the heart. The other source of contamination is through blood stream when contaminated needles are used for injections.
2. *Staphylococcus aureus* and *Streptococcus pneumoniae* are the predominating organisms for this infection.
3. In case of acute infection by virulent forms of the above bacteria, rapid destruction of the heart valves may occur within few weeks which may be fatal if untreated.
4. Members of Streptococci can also cause pericarditis leading to inflammation of the sac around the heart, the pericardium.

Subacute Bacterial Endocarditis

This is usually caused by less virulent forms of Streptococcus and Staphylococcus.

1. Commonly caused by oral *Streptococcus viridans* and *Staphylococcus epidermidis*
2. Normal body flora gain entrance to bloodstream through dental procedures or through other trauma
3. In abnormal or damaged hearts, due to the formation of a thin clot, traps circulating bacteria resulting in the formation of a biofilm
   (*a*) A thin slime layer is formed by the aggregation of bacteria
   (*b*) The biofilm makes the bacteria inaccessible to phagocytes
4. In addition, the blood vessels become weak and increases the chance of stroke and aneurysm
5. Other symptoms include: fever, loss of energy and may also lead to aorta valve stenosis where the opening of the aortic valve narrows down due to infection.

6.1.2 Septicemia or "Blood Poisoning"

An acute illness that is associated with the presence and persistence of pathogenic microorganisms or their toxins in the blood is termed as septicemia.

1. The disease is mainly caused by gram-negative bacteria that are commonly associated with nosocomial illness. Individuals suffering from some chronic diseases like cancer or diabetes often get infected with these organisms during hospitalization.
2. Endotoxins typical of many Gram-negative bacteria are probable cause of this infection. These endotoxins are released upon lysis of the cell.
3. These endotoxins can cause severe drop in blood pressure. Less than one millionth of a milligram of endotoxin is capable of causing severe symptoms.
4. In response to the release of large quantities of endotoxin, the body also releases massive amounts of inflammatory mediators like cytokines
5. Symptoms include: violent shaking, fast heart rates, chills and fever, high counts of white blood cells and often accompanied by anxiety. Shock is common and despite treatment only about half of all people afflicted would survive.

6. Cultures of blood from these patients usually reveal enterobacteria such as *Escherichia coli, Enterobacter aerogenes, Serratia marcescens and Proteous mirabilis*. Other than this, an organism commonly found associated with scepticemia is *Pseudomonas aeruginosa*, commonly found in the natural environment. This organism has such extraordinary biochemical capabilities that it can metabolize a variety of organic compounds, resist antibiotics and disinfectants.
7. Many gram positive bacteria have been identified now causing sepsis. Both Streptococci and Staphylococci produce potent exotoxins. The frequent use of invasive procedures in hospitals allows gram-positive bacteria to enter the blood stream.
8. There is also an emerging evidence of Group B Streptococci (GBS) responsible for life-threatening neonatal sepsis.

Pathogenesis of Scepticemia

Infection results from a disturbance of the normal balance between host defense mechanisms and determinants of microbial pathogenicity. Quite often the infected bacteria can lyse the red blood cells and thereby release large amount of iron-containing hemoglobin that accelerates the growth of bacteria. The disease appears like visible red streaks on the body due to inflamed lymph vessels. If the disease is not controlled quickly by the defense mechanism of the body it results in the progression of the infection and becomes fatal. At one end of this spectrum are the septicaemias caused by organisms of high virulence such as *Streptococcus pneumoniae*, Group A β-haemolytic streptococci, *Salmonella typhi*, *Neisseria meningitidis, Haemophilus influenzae* type b and *Escherichia coli* type Kl. These bacteria can invade the bloodstream from mucosal surfaces with no requirement of extravascular foci of infection. In contrast are the septicaemias involving *Staphylococcus aureus*, coagulase negative staphylococci or many species of Gram negative bacilli. With these organisms of lower virulence septicaemia often involves the presence of a foreign body, severe immunosuppression or trauma, or the presence of a clear focus of infection preceding the septicaemic episode. Once the bacteria get an entry into the circulatory system, it proliferates and during this process generates lot of toxins.

Gram-positive bacteria contain the lipid membrane of cell wall peptidoglycan acid and forms a complex with the damaged tissue thereby activate a battery of cytokines like TNF, IL-l, IL-6, IL-8, INFγ which trigger the body's response to the invasion, known as the systemic inflammatory response syndrome (SIRS) or severe sepsis. This is considered as a debilitating clinical condition that substantially alters the lives of those afflicted. The systemic inflammatory response syndrome (SIRS) is defined by the presence of two or more of the criteria listed in Table 6.1.

Table 6.1: Criteria for the systemic inflammatory response syndrome (SIRS)[1]

Criterion Value	
Temperature	>38°C or <36°C
Heart rate	>90 beats per minute
Respiratory rate	>20 or $PaCO_2$ <32 mm Hg
White blood cell count	>12 K or <4 K mm^{-3}, or >10% bands

The septic response accelerates due to continued activation of neutrophils and macrophages / monocytes. Upregulation of lymphocyte costimulatory molecules and rapid lymphocyte apoptosis, delayed apoptosis of neutrophils, and enhanced necrosis of cells/tissues also contribute to the pathogenesis of sepsis[2].

Therapeutic Interventions

Lists the sepsis trials registered with **http://www.clinicaltrials.gov** as of February 2010. This addresses the need for timely and active supportive care of the septic patient **(Table 6.2).** The overall goals of these guidelines are to maintain adequate organ perfusion, control infection, limit barotrauma due to mechanical ventilation and control hyperglycemia. Other specific interventions, such as the use of low-dose steroids or activated protein C, are suggested if required. Multiple centers demonstrated a decrease in mortality associated with increased compliance with the guidelines.

Table 6.2: Guidelines set for Sepsis management[3]

To be completed within 6 h of identification of severe sepsis
1. Measure serum lactate.
2. Obtain blood cultures prior to antibiotic administration..
3. Administer broad-spectrum antibiotic..
4. If patient is hypotensive and/or serum lactate is >4 mmol liter−1: (*a*) Deliver an initial minimum 20 ml kg−1 of crystalloid or an equivalent. (*b*) Administer vasopressors for hypotension not responding to initial fluid resuscitation to maintain mean arterial pressure>65 mm Hg.
5. In the event of persistent hypotension despite fluid resucitation (septic shock) and/or lactate >4 mmol per liter: (*a*) Achieve a central venous pressure of >8 mmHg. (*b*) Achieve a central venous oxygen saturation of >70% or mixed venous oxygen saturation of >65%.
To be completed within 24 h of identification of severe sepsis
1. Administer low-dose steroids for septic shock.
2. Administer drotrecogin alfa (activated protein C) in accordance with a standardized intensive care unit policy.
3. Maintain adequate glycemic control (usually between 70 and 150 mg dl−1).
4. Maintain a median inspiratory plateau pressure <30 cm H2O for mechanically ventilated patients. Recent international guidelines for management of severe sepsis and septic shock, 2013 are available at the following web link http://www.survivingsepsis.org/SiteCollectionDocuments/Implement-PocketGuide.pdf

Prevention

Treating the infections in time can prevent septicemia. The *Haemophilus influenza* B (HIB) vaccine and *S. pneumoniae* vaccine have already reduced the number of septicemia cases in children. Both are recommended in childhood immunization program. In such cases, where a probable contact with a scepticemia patient has occurred preventive antibiotics may be prescribed.

6.1.3 Bacterial Diseases Involving Lymph Nodes and Spleen

The enlargement of the lymph nodes and spleen is a prominent feature of certain infectious diseases that involve mononuclear phagocyte system. The cells are primarily monocytes and macrophages, and they accumulate in lymph nodes and the spleen. Three examples of these diseases are: **tularemia, brucellosis and plague.** They all represent a constant threat because they exist as zoonoses (diseases of animals transmissible to humans)

6.1.3.1 Tularemia

Tularemia is an example of a zoonotic disease, a disease transmitted by contact with infected animals. It is also known as Pahvant Valley plague, rabbit fever, deer fly fever, and Ohara's fever[4]. This is a serious infectious disease caused by the bacterium *Francisella tularensis*, a Gram-negative, nonmotile coccobacillus, the bacterium has several subspecies with varying degrees of virulence[5].

Pathogenesis

The causative organism enters through unnoticed scratches or by penetration of mucous membrane. The disease is normally acquired from the bites of flies and ticks or even by inhalation of causative organism. The incubation period for tularemia is one to 14 days; most human infections become apparent after three to five days. The clinical symptoms include fever, lethargy, anorexia, signs of septicemia, and often cause death.

Fancisella tularensis is a facultative intracellular pathogen that multiplies predominantly within macrophages. The organisms initially enter macrophages through phagocytosis. It adopts a novel process of engulfment within asymmetric pseudopod loops and then finally disrupt the phagosomal membrane to gain direct access to the cytoplasm. *Francisella tularensis* is a gram-negative coccobacillus. The disease usually have at least one of the six major clinical forms of tularemia[6]. They are:

- Ulceroglandular tularemia
- Glandular tularemia
- Oculoglandular tularemia
- Pharyngeal (oropharyngeal) tularemia
- Typhoidal tularemia
- Pneumonic tularemia

In the ulceroglandular form, the organisms proliferate locally and cause a papule to develop at the site of inoculation within 3 to 5 days after initial exposure[7] (Fig.6.1). Lymphatic vessels draining the area carry the organisms to the regional lymph nodes. These nodes then become direct inoculation of the eye resulting in oculoglandular tularemia. The patient usually presents with unilateral conjunctivitis, prominent swelling of the eyelids, photophobia, and a purulent secretion. Another form known as oropharyngeal tularemia is the result of ingestion of contaminated water or food. Typhoidal tularemia is used to describe tularemia patients with severe systemic symptoms but without any obvious signs of the port of entry.

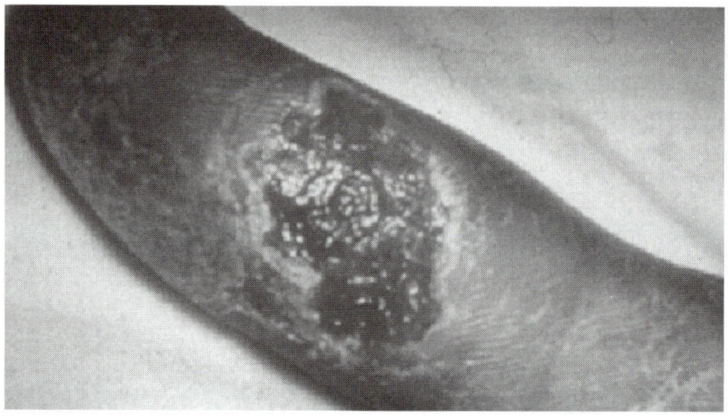

Fig. 6.1: Finger of patient infected with tularemia
Source: CDC file

Epidemiology

The majority of infections in humans and animals are caused by *F. tularensis* subspecies tularensis (virulent species) and *F. tularensis* subspecies holarctica. *F. tularensis* has been described in the United States, Canada, and Mexico. In the United States, the majority of cases have been reported from Arkansas, Missouri, South Dakota, and Oklahoma. Other countries where tularemia is considered to be endemic include most European countries, the former Soviet Union, Tunisia, Turkey, Israel, Iran, China, and Japan. However,Tularemia is relatively uncommon in African countries, Australia, England, and South American countries.

F. tularensis subspecies tularensis may be relatively restricted in their locations, although subspecies *holarctica* strains are more geographically dispersed. Within a single region, animal and human outbreaks may involve multiple *F. tularensis* types[8]. The high incidence of tularemia is observed among males and among children aged less than[10] years. This might be associated with increased opportunity for exposure to infected ticks or animals and not practicing any personal protective measures against tick bites and lack of facilities for diagnosis or reporting bias. The high incidence among American Indians/Alaska Natives might be associated with their increased risk for exposure. *Francisella tularensis* is highly infectious. A small number of bacteria (10-50 organisms) can cause disease. Outbreaks of pneumonic tularemia, particularly in low incidence areas has prompted considering this organism as a bioterrorism weapon[9].

Treatment and Prevention

The goal of treatment is to cure the infection with antibiotics. Streptomycin and tetracycline are commonly used to treat this infection. Tetracycline and chloramphenicol can be used alone, but they have a high relapse rate and are not considered a first-line treatment. The antibiotics are given by injection into a muscle or vein. Health officials must test the bacteria in the early stages of the response to determine which antibiotics will be most effective.

Possible complications are (1) Bone infection (osteomyelitis) (2) Infection of the sac around the heart (pericarditis) (3) Meningitis and (4) Pneumonia

No licensed vaccine against tularemia currently is available. Person-to-person transmission of tularemia has not been documented.

Standard Precautions to avoid infection include the practices related to direct patient care: (1) Hand washing; (2) Using gloves; (3) Masks for eye protection and to prevent aerosol contamination (4) Wearing gown to protect skin and to prevent soiling of clothing; (5) Proper handling of Patient-care equipments.

Autopsy personnel should wear N-95 respirators while carrying out autopsies, regardless of suspected or known pathogens. Powered air-purifying respirators (PAPRs) equipped with N-95 or high-efficiency particulate air (HEPA) filters should be considered. (http://www.cidrap.umn.edu/cidrap/content/bt/tularemia/biofacts)

6.1.3.2 Brucellosis

Brucellosis is an infectious disease caused by bacteria belonging to the genus *Brucella*. The disease is considered as world's most common bacterial zoonosis. It is an infection that affects mainly animals, including goats, sheep, camels, pigs, elk, deer, cattle, and dogs. Humans develop brucellosis when they come in contact with contaminated animals or animal products. The symptoms of brucellosis often resemble a flu-like illness. This disease has had several names throughout its history, including Mediterranean fever, Malta fever, Crimean fever, Bang's disease, and undulant fever (because of the relapsing nature of the fever associated with the disease. In the mid-20th century, the *Brucella* bacteria was also developed for use as a biological weapon by the United States but its use as biological weapon was banned in 1969.

Six species of Brucella are currently known, of which *Brucella melitensis, Brucella suis* and *Brucella abortus* have public health implications. *Brucella melitensis* occurs more frequently than the other types in the general population, predominantly among Hispanic members of the population and it is the most pathogenic and invasive species of *Brucella* followed by *Brucella suis* and *Brucella abortus*.

Pathogenesis

Human brucellosis is considered as a life-threatening debilitating disease characterized by weakness, fever, malaise, arthritis, osteomyelitis, endocarditis or meningoencephalitis[10]. In domestic animals, the disease occurs as a chronic infection that results in placentitis and abortion in pregnant females[11]. Brucellosis induces constant fevers, sweating, weakness, anaemia, headaches, depression, and muscular pain. The incubation period is usually 1 to 3 weeks but might be much longer. Brucella bacteria are small, aerobic gram-negative coccoid rods. They can also become airborne therefore, appropriate precautions are necessary while handling in the laboratory.

An important aspect of *Brucella* infection is its ability to persist and replicate within phagocytic cells of the reticuloendothelial system as well as in non-phagocytic cells such as trophoblasts. *B. melitensis* is considered the most virulent Brucella for humans with a few organisms (10 to 100) being sufficient to cause a debilitating chronic infection.

(*a*) *Source of infection:* In most cases, human infections occur through ingestion of contaminated milk and unpasteurized dairy products. Camel milk is also considered to be the important source of the infection in Middle East countries and Mongolia. However, occupational exposure of mucosa or skin abrasions to fluids and tissues from aborted fetuses of infected animals or carcass is also an important source of infection[12].

(*b*) *Virulence factors:* Virulence factors involved in Brucella replication and Brucella's strategies to circumvent the immune response are quite interesting. Brucella does not have classic virulence factors. VirB proteins that form the type IV secretion system and that are involved in intracellular replication are considered as one of Brucella's virulence factors. In addition to this secretion system, bacterial outer membrane components also help the bacteria to survive in the host. This bacterium possesses an unconventional non-endotoxic lipopolysaccharide that confers resistance to anti-microbial attacks and also participates in modulating the host immune response. These properties make lipopolysaccharide an important virulence factor for Brucella survival and replication in the host.

Epidemiology

Human brucellosis is prevalent worldwide. It has an annual occurrence rate of more than 500,000 cases. Brucellosis is more prevalent in areas lacking in infrastructures for appropriate animal- disease control programs and public-health programs. High-risk areas include the Mediterranean Basin (Portugal, Spain, Southern France, Italy, Greece, Turkey, and North Africa), South and Central America, Eastern Europe, Africa, Asia, the Caribbean, and the Middle East. In Mediterranean and Middle East countries the annual incidence of brucellosis in human varies from 1 up to 78 cases per 100 000. Nearly, 550 cases have been reported from confined endemic areas where no animal control measures are followed. Increasing trend of brucellosis is also reported from urban setting of China indicating that the strategy of vaccination and effective quarantine for infected animals has failed in those areas[13].

Indian Scenario

Brucellosis is a significant and increasing veterinary and public health problem in India. In India 80% of the population living in rural areas are vulnerable due to their close contact with domestic/wild animal population owing to their occupation. Hence, human population stands at a greater risk of acquiring zoonotic diseases including brucellosis. Species of main concern in India are *B. melitensis* and

B. abortus. *B. melitensis* is the most virulent strain. This is also considered to be the common strain infecting human and causing severe and prolonged disease. The disease also has a risk of disability. *B. abortus* is the predominant species infecting cattle. Chahota et al (2003)[14] have reported a severe outbreak of brucellosis in an organized dairy farm leading to abortions, retained placenta and still birth in cows. The diagnosis was made by serology employing rose Bengal plate agglutination test (RBPT) and standard tube agglutination test (SAT) and confirmed by the isolation of *B. abortus* biotype1. The prevalence of this disease in different parts of India has been reviewed by Mantur and Amarnath (2008)[15].

Treatment and Prevention

Elimination of the disease in domestic animals is the best prevention for brucellosis. Brucellosis control programs involve either vaccination or test-and-slaughter of infected animals. These programs have greatly reduced the incidence of the disease in animals as a result of which a substantial reduction in the number of human cases recorded. Human infections can be reduced considerably by avoiding consumption of unpasteurized milk and dairy products and limiting exposure to infected domestic animals. Proper heat treatment of milk or milk products is important for effective prevention of brucellosis in humans.

The gold standard treatment for adults is daily intramuscular injections of streptomycin 1 g for 14 days and oral doxycycline 100 mg twice daily for 45 days (concurrently). Gentamicin 5 mg/kg by intramuscular injection once daily for seven days is an acceptable substitute when streptomycin is not available or difficult to obtain[16]. Another widely used drug is a combination of doxycycline plus rifampin twice daily for at least six weeks. This has an advantage since the drug can be administered orally. A combination of three drugs, doxycycline with rifampin and cotrimoxazole has been used successfully to treat neurobrucellosis[17].

Possible drug resistance Antimicrobial resistance in few strains of Brucella are reported, however, their clinical implications are not yet fully understood. Some of the commonly-used antimicrobials for brucellosis treatment like Rifampicin and Streptomycin are also first line drugs for the treatment of tuberculosis. The present worldwide occurrence of multi-drug resistant strains of pathogenic *Mycobacterium tuberculosis* poses the urgent need for developing an alternative treatment for brucellosis using antimicrobial agents that are not employed for tuberculosis.

6.1.3.3 Plague

A highly fatal infectious disease is caused by the Enterobacterium *Yersinia* (syn. *Pasturella*) *pestis*. The organism gets transmitted primarily by the bite of a rat flea. The disease manifests in three different forms, bubonic, pneumonic, and septicemic forms. The organism is named after the French-Swiss bacteriologist Alexandre Yersin. The disease is notorious throughout history, due to high scale of death and devastation it brought. About ten years back, plague was considered as one of the three epidemic diseases specifically reported to the World Health Organization. The two other ones were cholera and yellow fever.

The great Plague of London, in 1665, killed up to 100,000 people (*adapted* from: http://en.wikipedia.org/wiki/Black_Death)

Infection and Mode of Transmission of Bubonic Plague

Yersinia pestis circulates through animal reservoirs, particularly rodents. This organism has been reported from all continents except Australia. The natural foci of plague are situated in a broad belt in the tropical and sub–tropical latitudes and the warmer parts of the temperate latitudes around the globe, between the parallels 55 degrees North and 40 degrees South[18]. The symptoms of plague depend on the concentrated areas of infection in each person. For example, bubonic plague manifests in lymph nodes, septicemic plague in blood vessels and pneumonic plague in lungs. Plague is treatable if detected early. Plague is still endemic in some parts of the world.

When a flea (Fig. 6.2) bites a human and contaminates the wound with regurgitated blood, the plague carrying bacteria are passed into the host tissue. *Y.* pestis can multiply even if they enter the phagycytic cells. Once in the body, the bacteria are capable of entering the lymphatic system, which drains interstitial fluid. Plague bacteria secrete several toxins, one of which is highly potent and known to cause dangerous beta-adrenergic blockade.

The other symptom of bubonic plague is an infection of the lymph glands known as lymphadenitis due to which the glands become swollen and painful and are known as buboes. As and when the organism gets transmitted via the bite of an infected flea, *Y. pestis* bacteria become localized in the inflamed lymph node where they begin to colonize and start replicating. Buboes associated with the bubonic plague commonly manifest in the armpits, upper femoral, groin and neck region. Acral gangrene that includes the fingers, toes, lips and nose is another common symptom of bubonic plague (Fig. 6.3). The bacterial infection stimulates severe haemorrhagic inflammation that causes the lymph nodes to expand. The expansion of lymph nodes is the cause of the characteristic "bubo" associated with this disease.

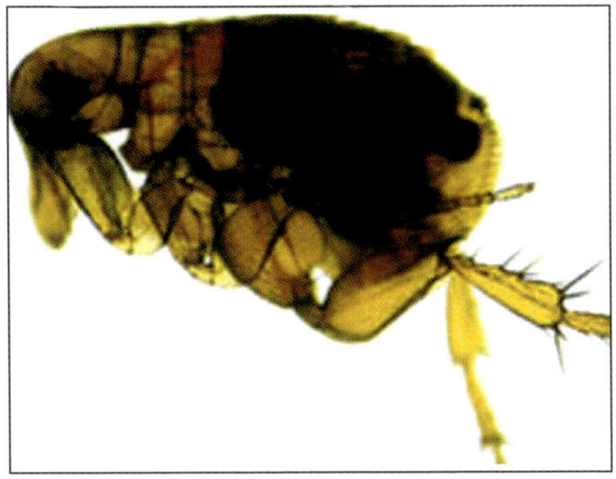

Fig. 6.2: A flea (*Xenopsylla cheopis*) infected with the *Yersinia pestis* bacterium which appears as a dark mass in the gut. The foregut of this flea is blocked by a *Y. pestis* biofilm; when the flea attempts to feed on an uninfected host Y. pestis from the foregut is regurgitated into the wound, causing infection

Symptoms

Following symptoms are characteristic feature of this disease

- Chills
- General ill feeling (malaise)
- High fever (39° Celsius; 102° Fahrenheit)
- Muscle Cramps
- Seizures
- Smooth, painful lymph gland swelling called a buboe, commonly found in the groin, but may occur in the armpits or neck,
- Most often buboe is seen at the site of the initial infection because of the bite or scratch
- Pain may occur in the area before the swelling appears
- Skin color changes to a pink in some very extreme cases

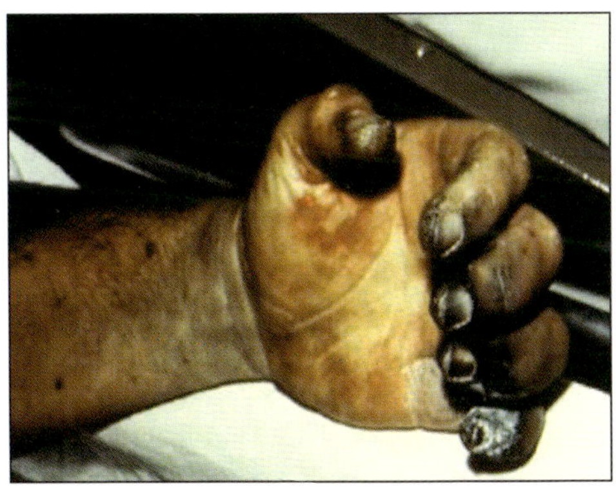

Fig. 6.3: Bubonic-Plague/hand-gangrene

Other symptoms include heavy breathing, continuous vomiting of blood referred as hematemesis, aching limbs, coughing, and extreme pain. The pain is usually caused by the decay or decomposition of the skin while the person is still alive. Additional symptoms may develop that include extreme fatigue, gastrointestinal problems, lenticulae (black dots scattered throughout the body), delirium and coma.

Virulence and Adaptation

The virulence of this bacterium results from the 32 genes of *Y pestis* located chromosomally and two genes which are located specifically on plasmids. The two plasmid located genes contribute in the formation of new genetic material acquired since its evolution from its predecessor.[19] It is assumed that these acquired genetic changes have allowed the pathogen to colonize fleas and to use them as vectors for transmission.[20]

The capacity of the plague bacillus to form permanent foci under highly diverse ecological conditions attests to its extraordinary adaptability. During its emergence in Central Asia, *Y. pestis* had accumulated copies of insertion sequences rendering its genome highly plastic[21]. The capacity to undergo genomic rearrangements may thus be an efficient means for the plague bacillus to adapt to new ecological niches. *Yersinia pestis* has also acquired antibiotic resistance plasmids under natural conditions, probably during its transit in the flea midgut[22].

Epidemiology

A few natural plague foci are located in the western United States. From the states of Arizona, California, Colorado, New Mexico, and Utah, 49 cases of plague and 3 attributed deaths were reported from 1994-1999[23]. In 2006, 13 plague cases were reported among residents of New Mexico, Colorado, California, and Texas, two of which resulted in death. From 1977-1998, 5 of 17 bubonic plague cases resulted in death.

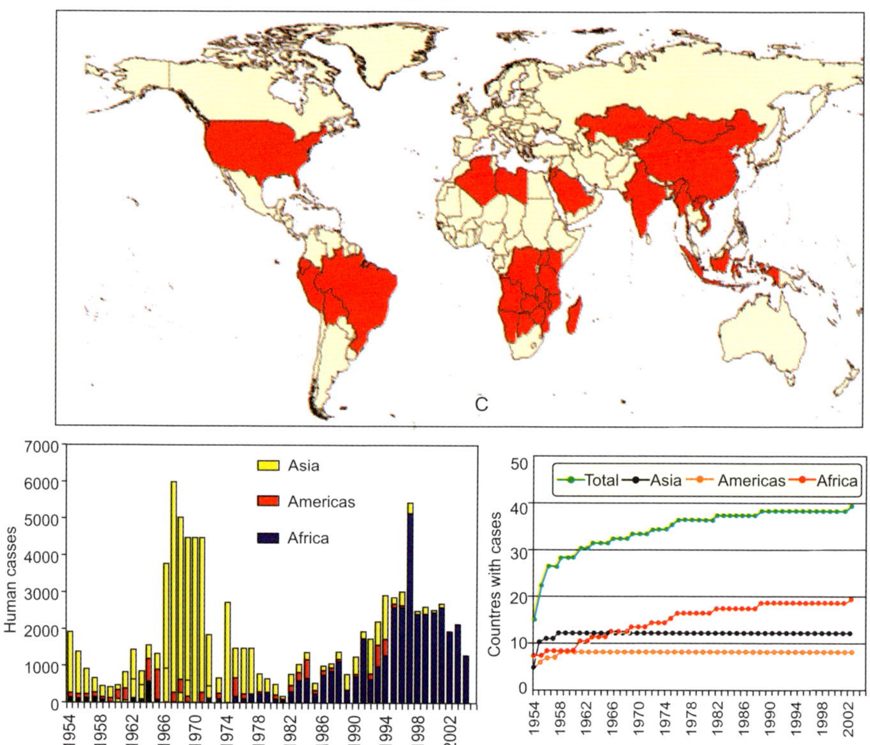

Fig. 6.4: The Global Distribution of Plague (A) Map showing countries with known presence of plague in wild reservoir species (red). (B) Annual number of human plague cases over different continents, reported to WHO in the period 1954–2005. (C) Cumulative number of countries that reported plague to WHO since 1954 [adapted from Stenseth et al, (2008) with permission][24]

Most cases of plague reported outside the United States are from developing countries of Africa and Asia subcontinent. During 1990-1995, a total of 12,998 cases of plague were reported to the World Health Organization (WHO), particularly from countries such as India, Zaire, Peru, Malawi, and Mozambique (Fig. 6.4).

The following countries reported more than 100 cases of plague: China, Congo, India, Madagascar, Mozambique, Myanmar, Peru, Tanzania, Uganda, Vietnam, and Zimbabwe. Several foci are located in the semi-arid regions of northeastern Brazil. Several outbreaks have also been reported from Malawi and Zambia. Australia is the only continent that is considered free of plague[25].

Indian Scenario

In the early twentieth century, plague epidemics accounted for about 10 million deaths in India. Since August 26, 1994, outbreaks of bubonic and pneumonic plague have been reported from areas in south-central, southwestern, and northern India. In 1994, there was a pneumonic plague epidemic in Surat, India that resulted in 52 deaths and resulted in a high rate migration of about 300,000 residents, who fled fearing quarantine. A combination of heavy monsoon rain and clogged sewers led to massive flooding which resulted in unhygienic conditions and a number of uncleared animal carcasses helped in the precipitation of the epidemic. There was widespread fear that the flood of refugees might spread the epidemic to other parts of India and the world, but that scenario was averted as a result of effective public health response mounted by the Indian health authorities[26].

Treatment and Prevention

A patient detected with bubonic plague needs to be admitted to the hospital and must be admitted in an isolation ward. The most effective antibiotics recommended are streptomycin and gentamicin. Antibiotics are administered either intravenously or intramuscularly. The antibiotics are given for 7 to 10 days. People who have been exposed to a person with the bubonic plague, even if they have no symptoms, still need to be treated with antibiotics as a preventive measure. There is no vaccine for the plague. One must avoid travelling to an area known to have had recent plague outbreaks.

Plague cannot be eradicated, since it is widespread in wildlife rodent reservoirs. Hence, there is a critical need to understand the epidemiology of this disease and critically analyze the dynamics of these wildlife reservoirs to understand the risk factors associated with the diseases in humans. One can prevent the plague by avoiding contact with sick or dead animals or their house should be treated with rodent proofing, attempt to keep pets flea free and take precautions such as using insect repellent and watching children and pets closely while outdoors.

6.2 Viral Diseases of the Lymphoid System-Burkitt's Lymphoma, Mononucleosis, HIV Infection, Chikungunya Virus, Dengue Virus, Ebola Virus

6.2.1 Burkitt's Lymphoma

It is named after Denis Parsons Burkitt, a surgeon who first described the disease in 1958 while working in equatorial Africa. Burkitt's lymphoma (BL) is a rare monoclonal proliferation of B-lymphocytes and is classified as a poorly differentiated lymphocytic lymphoma[27a]. This tumor was first noted in Africans. The cause of this tumor is debatable, but strong evidence implicates Epstein-Barr virus contributes in its development. This tumor predominantly affects children and is probably the fastest growing tumor in humans, with a fast rate of proliferation.

Clinical Variants: Burkitt's lymphoma can be divided into three main clinical variants: the endemic, the sporadic, and the immunodeficiency-associated variants.

1. The **endemic variant** occurs in equatorial Africa. It is the most common malignancy among children in this area (Fig. 6.5) Children affected with the disease often also found to suffer from chronic malaria, which is believed to have reduced resistance to Epstein–Barr virus (EBV). The affected areas of the body involve the jaw or other facial bone, distal ileum, cecum, ovaries, kidney or the breast.
2. The **sporadic type** of Burkitt lymphoma is also known as "non-African" type. This is another form of non-Hodgkin lymphoma found outside of Africa. The tumor cells resemble in appearance to the cancer cells of classical African or endemic Burkitt lymphoma. Sporadic lymphomas are rarely associated with the Epstein–Barr virus[27b]. Non-Hodgkin lymphoma, which includes Burkitt's, accounts for 30–50% of childhood lymphoma. The jaw is less commonly involved, compared to the endemic variant whereas the ileocecal region is the common site of involvement.

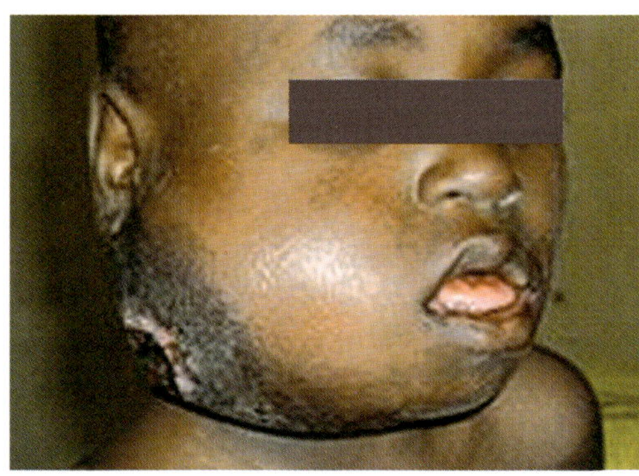

Fig. 6.5: A child with Burkitt's lymphoma showing ulcerated tumor

3. **Immunodeficiency-associated** Burkitt lymphoma is usually co-associated with HIV infection[28] or occurs in the setting of post-transplant patients who are taking immunosuppressive drugs. Burkitt lymphoma can be one of the diseases associated with the initial manifestation of AIDS.

Indian Scenario

Clinical presentation of Burkitt's lymphoma in India does not resemble either endemic or non-endemic areas. Jaw involvement was observed in 27 of 52 (51 percent) cases with Burkitt's lymphoma on retrospective analysis of published cases along with 14 recently reported cases. Higher incidence of jaw involvement is unusual as compared to the pattern of Burkitt's lymphoma of non-endemic areas of the world. Abdominal involvement was seen in 30 of 52 patients (58 percent) in India. In Indian scenario, during clinical presentation of Burkitt's lymphoma cases, involvement of jaw and abdomen is found to be almost equal in number[29]. In a case control study, it has been reported as a very rare malignancy accounting for only 0.76 percent of solid malignant tumors among Indian children[30].

6.2.2 Mononucleosis

Infectious mononucleosis, sometimes called "mono" or "the kissing disease," due to its oral transmission pattern. This is an infection usually caused by the Epstein-Barr virus (EBV). EBV is very common, and many people are exposed to the virus at some time in childhood. The disease was first described in 1889 and was referred to as "Drüsenfieber," or glandular fever. The term infectious mononucleosis was first used in 1920 when an increased number of lymphocytes were found in the blood of a group of college students who had fever and symptoms of the condition.

Symptoms and Mode of Transmission

The characteristic symptoms of infection with EBV include fever, fatigue, malaise, and sore throat accompanied by swollen lymph nodes. The designation "mononucleosis" refers to an increase in a particular type of white blood cells (lymphocytes) in the bloodstream relative to the other white blood cells as a result of the EBV infection. Lymphocytes with unusual lobed nuclei proliferate in the blood during the acute infection (Fig. 6.6).

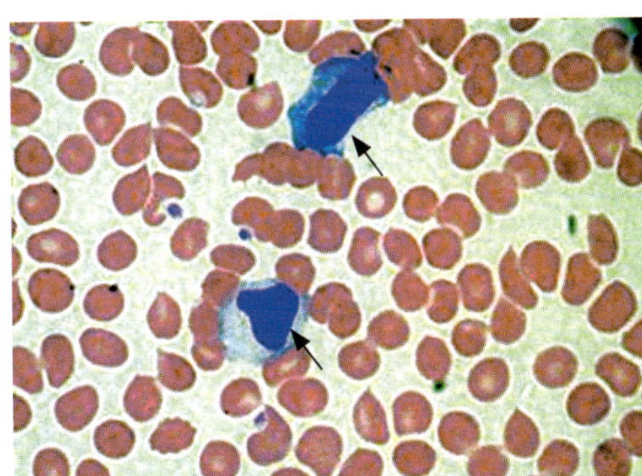

Fig. 6.6: Infectious mononucleosis, peripheral smear, showing reactive lymphocytes with lobed structure (adapted from http://en.wikipedia.org/wiki/Infectious_mononucleosis)

It is presumed that the resting B cells located in lymphoid tissue are the primary site of replication and persistence. The infection can be spread by saliva, and the incubation period is four to eight weeks and the communicability period commonly lasts up to three months[31]. Clinical syndromes encountered are fever, pharyngitis, and adenopathy.

The diagnosis of mononucleosis is confirmed by blood tests. However, a fluorescent-antibody test that detects IgM antibodies against EBV is the most specific diagnostic method. The virus can cause liver inflammation (hepatitis) and enlargement of the spleen. Infected persons continue to shed virus particles in their saliva during reactivations of the viral infection throughout their lifetime. The Epstein Barr virus also is associated with a variety of neoplasms, particularly those of the lymph system such as Burkitt's lymphoma, nasopharyngeal tumors, some leukemias, and lymphomas. In fact, many neoplasms also cause an infectious mononucleosis-like syndrome. One can observe elevated lymphocyte levels or presence of Downey cells (damaged B cells).

Epidemiology

The incidence of IM currently ranges from 50 to 100 cases per 100,000 population, and peaks at ages 15 to 19 years. Infants become susceptible to EBV as soon as maternal antibody protection which is present at birth starts to disappear. Children if infected with EBV usually remain asymptomatic and hence are indistinguishable from the other mild, brief illnesses of childhood. However, when the infection occurs during adolescence or young adulthood, it causes acute infectious mononucleosis.

Treatment and Prevention

Avoid activities involving the transfer of body fluids (commonly saliva) with someone who is currently or recently infected with the disease. At present, there is no vaccination available to prevent infectious mononucleosis.

6.2.3 HIV Infection

Human immunodeficiency virus (HIV) kills or damages the body's immune system cells. AIDS stands for acquired immunodeficiency syndrome. It is the most advanced stage of infection with HIV. Destruction of T-lymphocytes of the immune system was associated with these infections. By 1984, the responsible virus had been identified, and in 1986, it was given the name human immunodeficiency virus (HIV). **Described in the section 5.2.1.7 of Chapter 5.**

6.2.4 Chikungunya Fever

The disease occurs in Africa, Asia and the Indian subcontinent. In recent decades mosquito vectors of chikungunya have spread to Europe and the Americas. The disease shares some clinical signs with dengue, and can be misdiagnosed in areas where dengue is common. Since 2004, chikungunya fever has reached epidemic proportions, with considerable morbidity and suffering.

Chikungunya virus (CHIKV) is a member of the family *Togaviridae*, genus Alphavirus comprising of enveloped, positive single-stranded-RNA viruses. CHIKV replicates in the skin, in fibroblasts, and disseminates through the blood stream to the liver, muscle, joints, lymphoid tissue (lymph nodes and spleen) and brain.

Signs and Symptoms

Chikungunya is characterized by an abrupt onset of fever frequently accompanied by joint pain. Other common signs and symptoms include muscle pain, headache, nausea, fatigue and rash and even massive blisters. The joint pain with high fever is often very debilitating, but usually lasts for a few days however sometime it may prolong to weeks. The joint pain may occur in the wrists, fingers and ankles.

Most patients recover fully, but in some cases joint pain may persist for several months, or even years. Some patients develop complication of eye, neurological and heart complication. Some patients also develop gastrointestinal complaints. Serious complications are not common, but in older people, the disease can contribute to the cause of death. Often symptoms in infected individuals are mild and the infection may remain unrecognized, or at times misdiagnosed in areas where dengue occurs. The death rate is very low.

Transmission

This is a vector borne disease. The vector is the *Aedes aegypti* (yellow fever mosquito) (Fig.6.7) which spreads the disease widely among continents. These mosquitoes can be found biting throughout daylight hours, though there may be peaks of activity in the early morning and late afternoon. After the bite of an infected mosquito, onset of illness occurs usually between four and eight days but may extend from two to 12 days. *Aedes albopictus* (the Asian tiger mosquito) has also played a role in human transmission in Asia, Africa, and Europe. (Fig. 6.8)

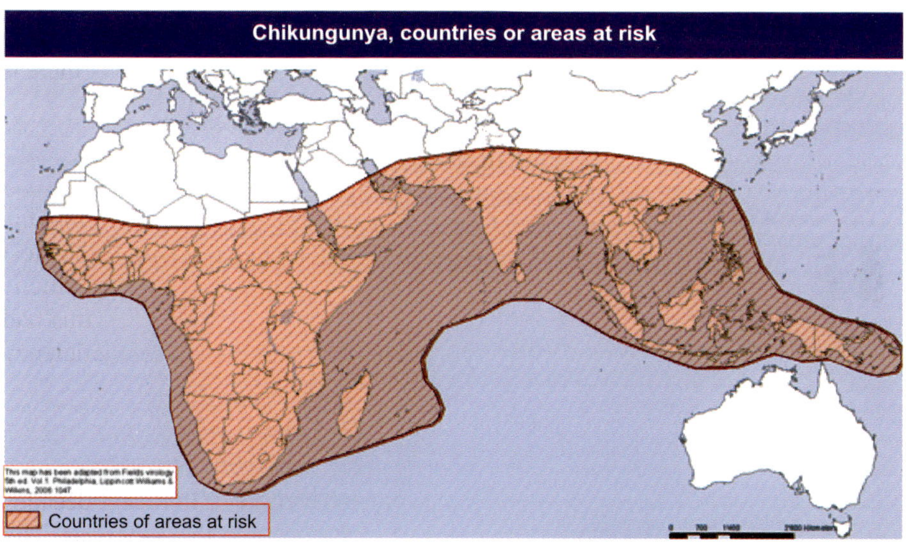

Reproduced with Permission from WHO http://www.who.int/mediacentre/factsheets/fs327/en/

Fig. 6.7: Geographic Distribution of Chikungunya virus (reproduced with permission from CDC)

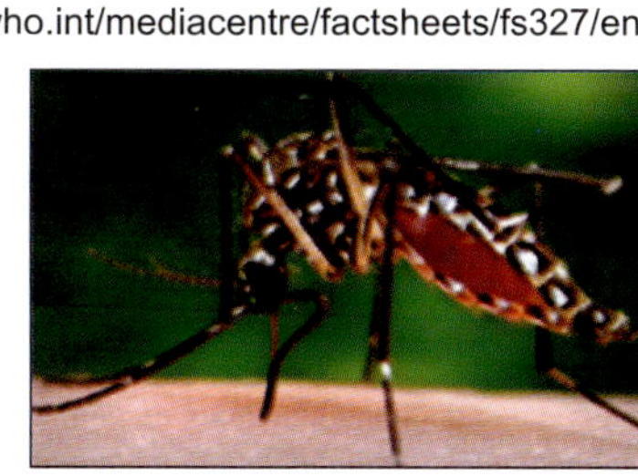

The vector: *Aedes aegypti*

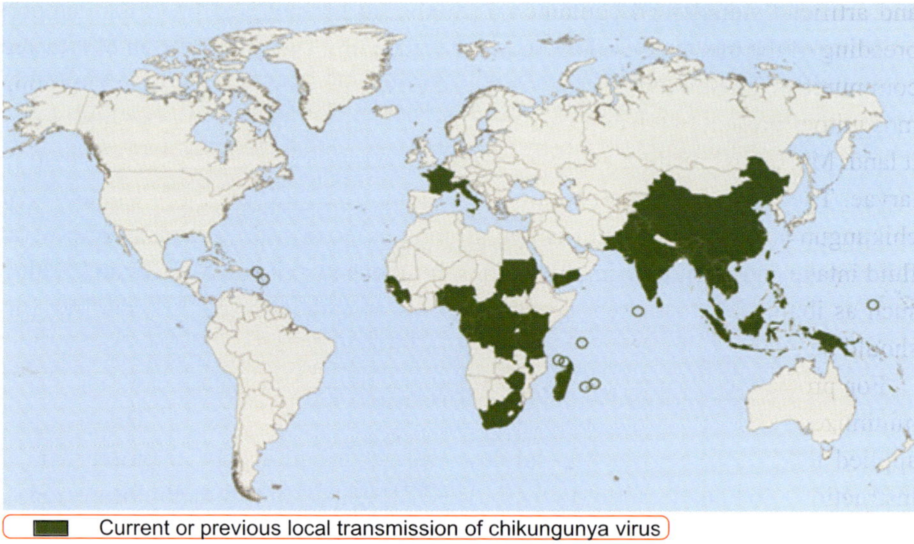

Fig. 6.8: Geographic Distribution of Chikungunya virus (reproduced with permission from CDC)

Diagnosis

Serological tests, such as enzyme-linked immunosorbent assays (ELISA), may confirm the presence of IgM and IgG anti-chikungunya antibodies. IgM antibody levels reach a maximum value within three to five weeks after the onset of illness and persist for about two months. Samples collected during the first week after the onset of symptoms should be tested by both serological and virological methods.

The virus may be isolated from the blood during the first few days of infection. Various reverse transcriptase–polymerase chain reaction (RT–PCR) methods are available but the sensitivity is variable. Some are suited for clinical diagnosis. RT–PCR products from clinical samples may also be used for genotyping of the virus, allowing comparisons of virus samples from various geographical sources.

Treatment

There is no specific antiviral drug treatment for Chikungunya. Treatment with anti-pyretics, optimal analgesics and fluids can control some of the associated symptoms like joint pain etc. There is no commercial chikungunya vaccine available.

Prevention and Control

The proximity of mosquito vector breeding sites to human habitation is a significant risk factor for chikungunya as well as for other diseases that these species transmit. Prevention and control measures are directed to reduce the number of natural and artificial water-filled containers available in the neighborhood that support breeding of the mosquitoes. This requires counseling and mobilization of affected communities. In the event of an outbreak, insecticides may be sprayed to kill flying mosquitoes present in the open water containers and applied on its surfaces where it land. Mosquito breeding grounds should be heavily sprayed to kill the immature larvae. There is no vaccine or specific antiviral treatment currently available for chikungunya fever. Treatment is symptomatic that include bed rest, increase in fluid intake and administering medicines to relieve symptoms of fever and aching such as ibuprofen, naproxen, acetaminophen, or paracetamol. However, Aspirin should be avoided.

For protection during outbreaks of chikungunya, appropriate clothing which minimizes skin exposure to the day-biting vectors is advised. Repellents can be applied to exposed skin or to clothing in strict accordance with product label instructions. Repellents should contain DEET (N, N-diethyl-3-methylbenzamide), IR3535 (3-[N-acetyl-N-butyl]-aminopropionic acid ethyl ester) or icaridin (1-piperidinecarboxylic acid, 2-(2-hydroxyethyl)-1-methylpropylester).

Indian Scenario

In India, three major epidemics of chikungunya fever have been reported during the last century- 1963 in Kolkata, 1965 in Pondicherry and Chennai in Tamil Nadu, Rajahmundry, Vishakapatnam, and Kakinada in Andhra Pradesh, Sagar in Madhya Pradesh, and Nagpur in Maharashtra, and in 1973 from Barsi in Maharashtra[32]. The disease re-emerged in year 2006 and became epidemic affecting 17 states of the country, starting from the state of Andhra Pradesh, including Delhi, infecting more than 1.39 million people before the end of the year[33]. An outbreak of fever with severe joint pain started in the Palam area of Delhi in August 2010. An entomological and epidemiological investigation of this outbreak was conducted to ascertain the nature and cause of the outbreak. Out of the 130 blood samples tested, 97 (70.8%) were positive for the IgM antibodies against chikungunya virus. House-to-house survey was conducted in the affected areas to find out the vector-breeding sites[34].

6.2.5 Dengue Virus

Dengue is an acute viral infection with potential fatal complications. Dengue viruses (DV) belong to family *Flaviviridae* and there are four serotypes of the virus referred to as DV-1, DV-2, DV-3 protein genes, which encode the nucleocapsid or core (C) protein, a membrane-associated (M) protein and DV-4. DV is a positive-stranded encapsulated RNA virus and is composed of three structural, an enveloped (E) glycoprotein and seven non-structural (NS) proteins. It is transmitted mainly by two species of mosquito, *Aedes aegypti* and also by *Ae.*

albopictus. The virus consists of four serotypes that spread by the bite of the Aedes mosquito that can cause the full spectrum of disease from a subclinical infection to a mild self limiting disease, the dengue fever (DF) and a severe disease that may be fatal referred as the dengue haemorrhagic fever/dengue shock syndrome (DHF/DSS)[35a]. The WHO 2009 classification divides dengue fever into two groups: uncomplicated and severe, though the 1997 WHO classification is still widely used[35b].

Symptoms and Pathogenesis

The time between exposure and onset of symptoms ranges from 3–14 days, but most often it is 4–7 days. The characteristic symptoms of dengue are sudden-onset fever, headache (typically located behind the eyes), muscle and joint pains, and a rash. The alternative name for dengue, "breakbone fever", comes from the associated muscle and joint pains. The course of infection is divided into three phases: febrile, critical, and recovery. Four main characteristic manifestations of dengue illness are (*i*) continuous high fever lasting 2-7 days; (*ii*) haemorrhagic tendency as shown by a positive tourniquet test, petechiae or epistaxis; (*iii*) thrombocytopoenia (platelet count less than 100 × 109/l); and (*iv*) evidence of plasma leakage manifested by haemoconcentration (an increase in haematocrit 20 percent above average for specified age, sex and population), pleural effusion and ascites[35a].

Diagnosis

The diagnosis of dengue fever may be confirmed by microbiological laboratory testing. This can be done by virus isolation in cell cultures, nucleic acid detection by PCR, viral antigen detection (such as for NS1) or detection of specific antibodies (serology).Virus isolation and nucleic acid detection are more accurate than antigen detection. However, these tests are not widely practiced due to high cost. Detection of NS1 during the febrile phase of a primary infection may be greater than 90 percent however it reduces in subsequent infection to only 60–80 percent[36a]. Tests for dengue virus-specific antibodies, types IgG and IgM, can be useful in confirming a diagnosis in the later stages of the infection. Both IgG and IgM levels can be quantified after 5–7 days. The highest titres of IgM are detected following a primary infection, but IgM is also produced during reinfection. IgM becomes undetectable 30–90 days after a primary infection, but can be detected again following re-infections. IgG, by contrast, remains detectable for over 60 years and therefore, can be used as an useful indicator to detect past infections. After a primary infection IgG reaches peak levels in the blood after 14–21 days.

In subsequent re-infections, levels peak earlier and the titres are usually higher. Both IgG and IgM provide protective immunity to the infecting serotype of the virus[37a,b].

Epidemiology

Dengue is considered to be one of the most common viral diseases transmitted by arthropods. The disease burden estimated to be 1600 disability-adjusted life years per million of population. World Health Organization counts dengue as one of sixteen neglected tropical diseases.

Dengue virus is maintained in nature in a cyclic fashion that revolves between blood-sucking vectors and vertebrate hosts[38]. In the forests of Southeast Asia and Africa, the virus is maintained by the process of transmission from female Aedes mosquitoes to her offspring and also to lower primates. In urban areas, the virus is primarily transmitted by *A. aegypti* which is a domesticated species. In rural settings the virus is transmitted to humans by *A. aegypti* and other species of Aedes such as *A. albopictus*[38]. Both these species have expanded its ranges in the second half of the 20th century[36]. In all settings the infected lower primates or humans greatly increase the number of circulating dengue viruses in a process called amplification. Consequently, it can affect countries at every level of the socio-economic spectrum.

According to WHO report 2009, thirty fold increase in the rates of dengue has been observed between 1960 and 2010. Various factors implicated for this change are- urbanization, population growth, increased international travel and global warming[39]. The geographical distribution of the disease pattern shows highest concentration of infected population (70 percent) living around the countries near equator. These countries are the endemic areas from Asia and the Pacific. It is also predicted that dengue fever may spread outside the current tropical zones due to potential extension in the habitat of its vector. An exhaustive assembly of known records of dengue occurrence worldwide, and a formal modeling framework to map the global distribution of dengue risk is being carried out[40] (Fig. 6.9). It was estimated that in 2010, there were 96 million dengue infections globally out of which around 67 million infections were recorded from Asia alone. The high burden of disease is characterized by large swathes of densely populated regions coinciding with very high suitability for disease transmission. India alone contributed 34 percent (33 million infections) of the global total[41,42]. Available information on the disease occurrence is less compared to those of prevalence. As more and more data accumulate, future targets would focus on assessing relationships between seroprevalence and clinical incidence as a means of assessing risk[43].

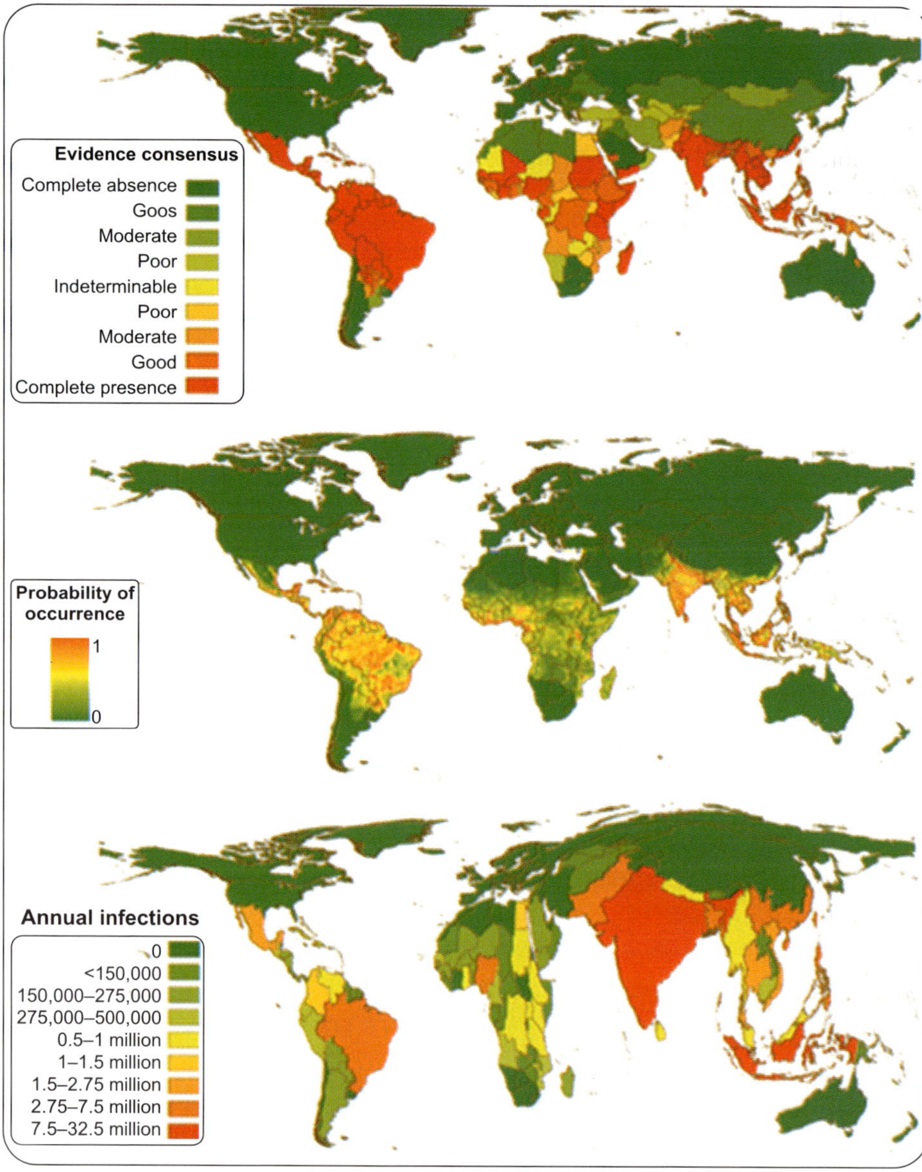

Fig. 6.9: Global evidence consensus (a) risk (b) and burden (c) of dengue in 2010 (Bhatt et al[40]) (reproduced with permission)

Prevention and Management

There are no approved vaccines for the dengue virus. Prevention thus depends on control of and protection from the bites of the mosquito that transmits it. The World Health Organization recommends an Integrated Vector Control program consisting of five elements[44].

1. Advocacy, social mobilization and legislation to ensure that public health bodies and communities more proactive;

2. Collaboration between the health department between public and private sectors;
3. An integrated approach to disease control to maximize use of resources;
4. Evidence-based decision making to ensure any interventions are targeted appropriately; and
5. Capacity-building to ensure an adequate response to the local situation.

The primary method of controlling *A. aegypti* is by eliminating its habitats. This is done by destroying the breeding grounds of mosquitoes such as emptying containers of water or by adding insecticides or biological control agents to these areas.

There are no specific antiviral drugs for dengue, however maintenance of proper fluid balance in infected individuals is important[45a]. Treatment depends on the symptoms, varying from oral rehydration therapy at home with close monitoring to hospital admission with administration of intravenous fluids and if required may be given blood transfusion as advised by the attending physician. One of the studies suggests that bleeding occurs more often in patients with severe thrombocytopenia[45b]. Patients having platelet count less than 20,000/cumm are considered to be at high risk and require urgent platelet transfusion. Patients with platelet count 21-40,000/cumm are in moderate risk and require platelet transfusion only if they have any haemorrhagic manifestations and other superadded conditions.

Distinction between Chikungunya and Dengue Fever (DF)

Dengue and chikungunya are co-circulating vector-borne diseases with substantial overlap in clinical presentations. It is important to differentiate between them during first presentation as their management, especially if diagnosed for dengue hemorrhagic fever (DHF), is different.

Almost 89 percent of cases can be correctly differentiated between DF and Chikungunya on the basis of Platelets count. If the count is less than 118 × 109/L, it is classified as DF. For DHF versus chikungunya using platelets less than 100 × 109/L and the presence of bleeding, the overall correct classification was 98%. The time course analysis supported platelet count as the key distinguishing variable[46].

6.2.6 Ebola Virus

Ebola virus disease (EVD), formerly known as Ebola haemorrhagic fever, is a severe, often causes fatal illness in humans. The fatality rate is as high as 90%. Ebola first appeared in 1976 in 2 simultaneous outbreaks, in Nzara, Sudan, and in Yambuku, Democratic Republic of Congo. The latter was in a village situated near the Ebola River, from which the disease takes its name.

Genus Ebolavirus is a member of the *Filoviridae* family (filovirus) which includes two other genera -genus Marburg virus and genus Cuevavirus Ebola virus is a nonsegmented, negative-sense, single-stranded RNA virus. Genus

Ebolavirus comprises 5 distinct species: Bundibugyo ebolavirus (BDBV); Zaire ebolavirus (EBOV); Reston ebolavirus (RESTV); Sudan ebolavirus (SUDV); Taï Forest ebolavirus (TAFV). BDBV, EBOV, and SUDV have been associated with large EVD outbreaks in Africa. The RESTV species, found in Philippines and the People's Republic of China, can infect humans, however no illness or death in humans from this species has been reported to date.

Transmission

Because the natural reservoir host of Ebola viruses has not yet been identified, the way in which the virus first appears in a human at the start of an outbreak is unknown. However, scientists believe that the first patient becomes infected through contact with the blood, secretions, organs or other bodily fluids of infected animals, such as a fruit bat or primate (apes and monkeys). This event is referred as a spillover event.

In Africa, infection has been documented through the handling of infected chimpanzees, gorillas, fruit bats, monkeys, forest antelope and porcupines found ill or dead. Ebola spreads in the community through human-to-human transmission, with infection resulting from direct contact either through damaged skin or mucous membranes. The infection can also spread through the blood, secretions, organs or other body fluids of infected people. It can also spread through indirect contact with environments contaminated with such fluids. Men who have recovered from the disease can still transmit the virus through their semen for up to 7 weeks after recovery from illness.

Recent Outbreaks

The Zaire species of Ebola virus is the causative agent of the 2014-2015 epidemic in West Africa, in which the case fatality rate has been reported to be as high as 70 percent[47]. Earlier outbreaks of this disease were reported mainly from Central Africa however an epidemic was recorded in later part of 2013 from Guinea located in West Africa. This report was confirmed by WHO[48]. The first victim was identified as a two year child who complained of fever, vomiting and black stools. Further outbreaks were reported from Liberia, Sierra Leone, Nigeria, Senegal, and Mali. On tracking the infection pattern by sequence analysis of the viral genome, it was confirmed that the epidemic transmitted through person-to-person contact[49,50,51]. There was no intervention from animal reservoirs.

As per the report from WHO, (August 9, 2015) the cumulative number of probable, suspected, and laboratory-confirmed cases attributed to Ebola virus is 27,965, including 11,298 deaths[52]. These cases include over 870 infected healthcare workers, of whom approximately 60 percent have died[52]. The magnitude of the outbreak, especially in Liberia and Sierra Leone, has probably been underestimated, According to WHO reports, cases of Ebola virus disease have also occurred in residents and healthcare workers who were exposed to the virus in West Africa, and were then treated in hospitals in the United States and Europe.

Signs and Symptoms

EVD is a severe acute viral illness when the patient develops sudden onset of fever, intense weakness, muscle pain, headache and sore throat. This is followed by other symptoms like vomiting, diarrhea, rash formation. This is followed by impairment of kidney and liver functions, and in severe cases, the patient develops both internal and external bleeding. The walls of the blood vessels are damaged. Then the virus interferes with coagulation and therefore continues to leak into surrounding tissues. Laboratory findings include low white blood cell and platelet counts accompanied by elevated levels of liver enzymes. People remain infectious as long as their blood and secretions contain the virus. The incubation period from infection with the virus to onset of symptom, is 2 to 21 days. During the pathogenesis of the disease, defects in blood coagulation are noticed in Ebola virus disease. It appears to be induced indirectly, through the host inflammatory response. Virus-infected macrophages synthesize cell-surface tissue factor (TF), triggering the extrinsic coagulation pathway; proinflammatory cytokines also induce macrophages to produce TF[53]. The simultaneous occurrence of these two stimuli helps to explain the rapid development and severity of the coagulopathy in Ebola virus infection.

Diagnosis

Ebola virus infections can be diagnosed definitively in a laboratory through several types of tests:

1. antibody-capture enzyme-linked immunosorbent assay (ELISA)
2. antigen detection tests
3. serum neutralization test
4. reverse transcriptase polymerase chain reaction (RT-PCR) assay
5. electron microscopy
6. virus isolation by cell culture.

Samples from patients are an extreme biohazard risk and therefore testing should be conducted under maximum biological containment conditions.

Vaccine and Treatment

No licensed vaccine is available for the treatment of EVD. Patients are frequently dehydrated and require oral rehydration and intensive supportive care.

Prevention and Control

Health-care workers caring for patients with suspected or confirmed Ebola virus should apply standard precautions and control measures to avoid any exposure to the patient's blood and body fluids and direct unprotected contact with the possibly contaminated environment. When in close contact (within 1 metre) of patients with EBV, health-care workers should wear face protection (a face shield or a medical mask and goggles), a clean, non-sterile long-sleeved gown, and gloves (sterile gloves for some procedures).

During EVD out breaks following measures are suggested:

1. Reducing the risk of wildlife-to-human transmission from contact with infected fruit bats or monkeys/apes and must stop the consumption of their raw meat.
2. Reducing the risk of human-to-human transmission in the community arising from direct or close contact with infected patients, particularly with their bodily fluids.
3. Communities affected by Ebola should inform the population residing in neighborhood areas about the nature of the disease.
4. Must follow outbreak containment measures stringently, including burial of the dead individuals. People who have died from Ebola should be promptly and safely buried.

WHO provides expertise and disease documentation to support investigation and control. (http://www.who.int/mediacentre/factsheets/fs103/en/)

6.3 Protozoal Diseases of the Blood and Lymphatic System

6.3.1 Chagas Disease

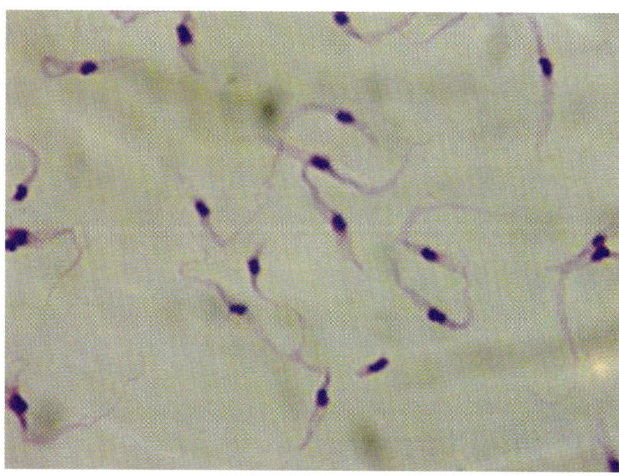

Fig. 6.10: Flagellated *Trypanosoma cruzi* crithidia stained with Giemsa (adapted from http://en.wikipedia.org/wiki/Trypanosoma_cruzi)

This is also known as American Trypanosomiasis caused by *Trypanosoma cruzi,* a flagellated protozoan (Fig. 6.10). This was discovered by a Brazilian scientist Carlos Chagas in 1910. *T. cruzi* is commonly transmitted to humans and other mammals by an insect vector, the blood-sucking "kissing bugs" of the subfamily Triatominae (family Reduviidae), most commonly from species belonging to the *Triatoma, Rhodnius,* (Fig. 6.11) and *Panstrongylus* genera. Chagas disease is common in South America, Central America and Mexico considered to be the primary home of the triatomine bug.

Fig. 6.11: (A) *Rhodnius prolixus* is the principal vector in Colombia, Venezuela, Guatemala,Honduras, and some parts of Nicaragua and El Salvador. (B) *Triatoma infestans,* widespread in the South America: Bolivia, Argentina, Uruguay, Paraguay, Chile, Brazil and Peru. (***Source:*** en.wikipedia.org)

Transmission

The arthropod vector is the reduviid bug, also known as 'kissing bug' because it often bites a person near the lips. The parasite Trypanosome grows in the gut of bug and gets transmitted to the host if the bug defecates while feeding. The person can become infected with the parasite when *T. cruzi* parasites present in the bug feces enter the body through mucous membranes or through any damage in the skin. The disease may also spread from human to human through other means like during blood transfusion or while undergoing organ transplantation, or even by ingestion of food contaminated with parasites. The parasite may also get transmitted from a mother to her fetus.

Pathogenesis

The parasite has two forms, a trypomastigote found in human blood and an amastigote found in tissues. The *Trypanosoma cruzi* life cycle starts in an animal reservoir, usually mammals, wild or domestic, including humans (Fig. 6.12). A triatomine bug serves as the vector. During a blood meal, it ingests *T. cruzi.* In the triatomine bug (*Triatoma infestans*) the parasite remains at the epimastigote stage which is a reproductive stage and non infective. After reproducing through binary fission, epimastigotes transit onto the rectal cell wall, where they become infectious. Infectious *T. cruzi* are then referred as metacyclic trypomastigotes. When the triatomine bug subsequently takes a blood meal from a human, it normally defecates. The trypomastigotes present in the fecal material are motile and capable of swimming into the host's cells using flagella. The trypomastigotes enter the human host through a bite, wound or by crossing mucous membranes. The host cells contain macromolecules such as laminin, thrombospondin, heparin sulphate, and fibronectin that are responsible to protect their surface. These macromolecules are responsible in the process of adhesion between parasite and host and also contribute in the process of host invasion by the parasite. The trypomastigotes further cross a network of proteins that line the exterior of the host cells in order to make contact and invade the host cells. As they enter a human cell, they get converted to amastigotes. Then they multiply through binary fission. The amastigotes then turn back into trypomastigotes, and the cell bursts. The trypomastigotes swim along to infect other cells or get sucked up by other reduviid bugs.

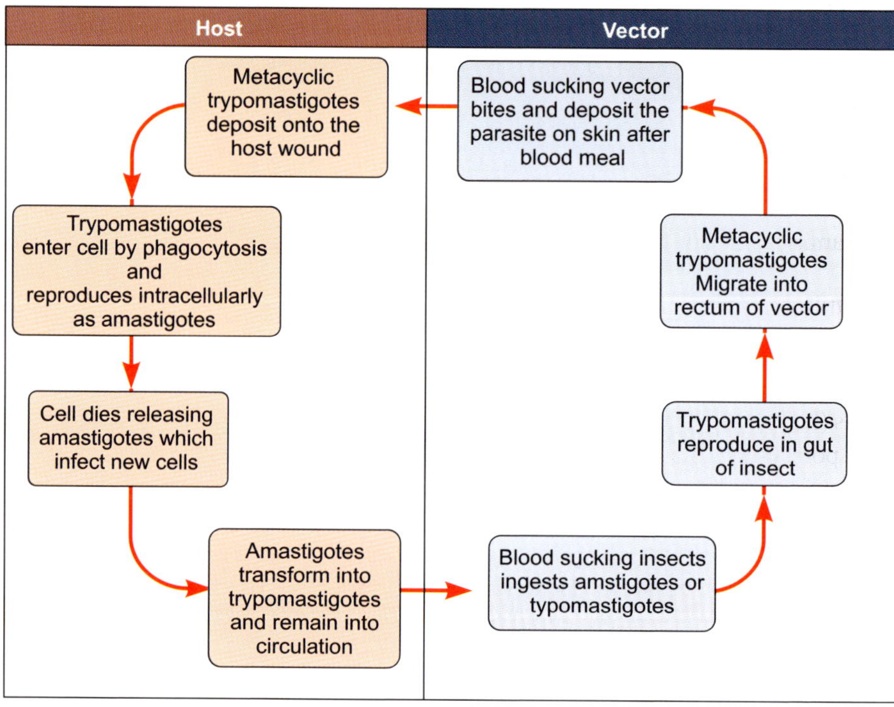

Fig. 6.12: Life Cycle of Trypanosoma cruzi

The infection progresses in stages. In the early, acute stage, symptoms are mild and usually produce no more than local swelling at the site of infection. The acute stage is characterized by fever and swollen glands lasting for few weeks. The initial acute phase is responsive to antiparasitic treatments, with 60–90 percent cure rates. If the infection continues for 4–8 weeks, individuals enter the chronic phase of Chagas disease, which is asymptomatic for 60–80 percent of chronically infected individuals through their lifetime. However, 20–40 percent of chronically infected individuals will still eventually develop life-threatening heart and digestive system disorders. End stage of Chagas disease is achieved by a devastating invasion of the autonomic nervous system, especially the parasympathetic influence of the vagus nerve. Gradual amputation of parasympathetic vagal influence begins in the upper fields of vagal innervation resulting in grossly distended cardiac ventricles, nonischemic cardiomyopathy, and/or megaesophagus. The parasite can progress distally, causing megacolon. Most deaths are caused by damage to the heart, which occurs in about 40 percent of chronic cases.

Diagnosis

The most recognized marker of acute Chagas disease is called Romaña's sign, which includes swelling of the eyelids on the side of the face near the bite wound or where the bug feces were deposited or accidentally rubbed into the eye. It can be detected by microscopic examination of fresh anticoagulated blood, or its buffy coat where motile parasites can be detected. For direct visualization of parasites microscopically, thin blood smears stained with Giemsa can be prepared. *T. cruzi* can be confused with *Trypanosoma rangeli,* which is not known to be pathogenic in humans. Isolation of *T. cruzi* can occur by inoculation into mice, by culturing into specialized media (for example, NNN, LIT); and by xenodiagnosis where uninfected Reduviidae bugs are fed on the patient's blood, and their gut contents are examined for the parasite[54].

Epidemiology

Chagas disease is endemic throughout Mexico, Central America, and South America where an estimated 8 million people are infected. According to CDC reports (http://www.cdc.gov/parasites/chagas/epi.html) , the triatomine bug thrives under poor housing conditions for example, mud walls, thatched roofs, therefore in endemic countries, people living in rural areas are at greatest risk for acquiring infection. Public health efforts aimed at preventing transmission have successfully decreased the number of newly infected people and completely halted vector borne transmission in some areas. However, infection acquired from blood products, organ transplantation, or congenital transmission continues to pose a threat.

By applying published seroprevalence figures to immigrant populations, CDC estimates that more than 300,000 persons with *Trypanosoma cruzi* infection live in the United States. Most people with Chagas disease in the United States acquired their infections from endemic countries. Although there are triatomine bugs in the U.S., only rare vector borne cases of Chagas disease have been documented. A study carried out in a sample of inhabitants from a region considered to be at high risk of natural transmission of Chagas disease in Colombia revealed seroprevalence of antibodies against *T. cruzi* was 16.91 percent, considering an estimated population of 44,355 aged between 15 and 89 years. The factors significantly associated with the infection were: 1-Housing materials like vegetable material, adobe or unfinished brick walls; 2- having a previous history of Chagas disease[48].

Treatment and Prevention

Chagas disease is curable if treatment is initiated soon after infection. There is no vaccine for this disease. Vector control is the most effective method of preventing Chagas disease. Blood banks in Central and South America screen blood donors for exposure to the parasite.

WHO recommends the following approaches to prevention and control:

- insecticide spraying in houses and surrounding areas;
- Improving the housing condition to prevent vector infestation;
- personal preventive measures such as use of bednets;
- good hygiene practices in food preparation, transportation, storage and consumption;
- screening of blood donors;
- testing for the parasite in organ, tissue or cell donors and receivers;
- screening newborns and other children of infected mothers to provide early diagnosis and treatment.

Two drugs are used to treat this infection: benznidazole and nifurtimox. Both the drugs often have side effects. The side effects may be worse in older people.

Side effects may include:

- Headaches and dizziness
- Neuropathy
- Skin rashes
- Loss of appetite and weight loss
- Sleeping problems

6.3.2 Toxoplasmosis

Toxoplasmosis is considered one of the Neglected Parasitic Infections. Toxoplasmosis is a disease caused by a single-celled parasite called *Toxoplasma gondii* (*T. gondii*). Toxoplasmosis is one of the most common parasitic diseases and has been found in nearly all warm-blooded animals, including pets and humans. Charles Nicolle and Louis Manceaux first described the organism in 1908, after they observed the parasites for the first time in the blood, spleen, and liver of a North African rodent, *Ctenodactylus gundi.*

Life Cycle and Transmission

The organism has a complicated life cycle (Fig. 6.13) including two stages that are significant to the spread of the parasite: the cyst stage and the oocyst (egg) stage. Cats are known as the definitive hosts for the parasite since it is unable to complete its entire life cycle without these animals. Other animals like pigs, sheep, rats and birds are considered as intermediate hosts as they play a role as carriers of *T. gondii.* Newly infected cats, especially kittens, often shed millions of oocysts in their feces through a period of 3-10 days. Once they are released in the environment, oocysts start to sporulate and become infectious to a new host. Infected animals can develop cysts in their muscles and organs of the body (brain, liver, heart), but do not develop or shed oocysts as cats do. Infection occurs when animals ingest sporulated oocysts present in soil, feed, water, cat feces or any other materials or cysts contained in the tissues of rodents, wildlife or other animals.

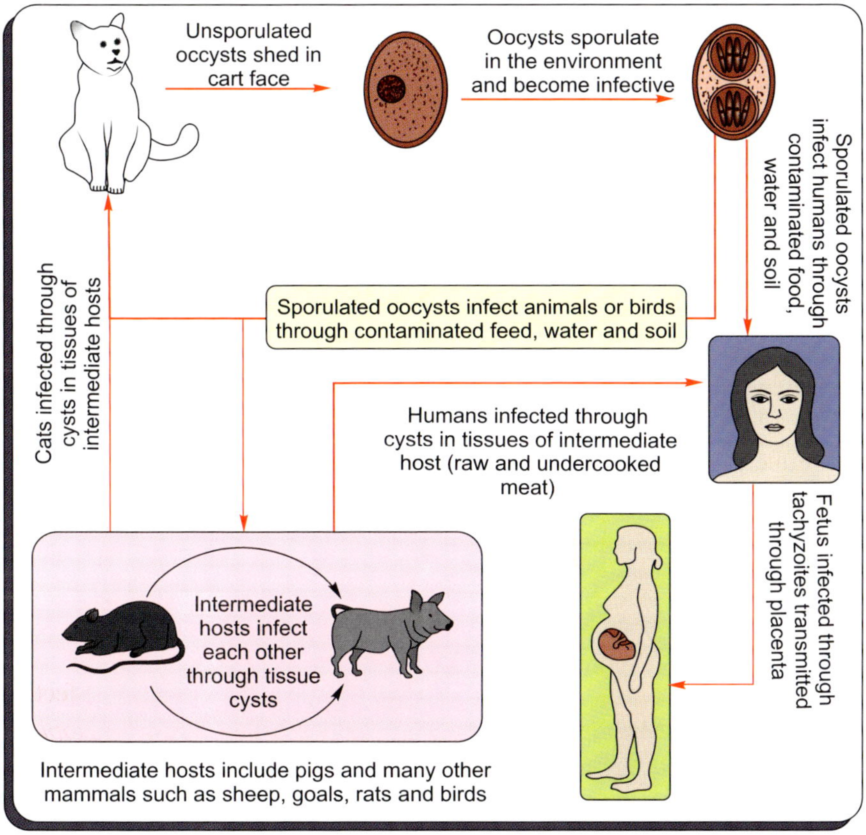

Fig. 6.13: Life cycle of *Toxoplasma gondii*

The entire life cycle of the parasite was clearly understood only in the late 1960s, with the discovery of the central role of the cat as a definitive host harboring the sexual parasitic cycle and spreading oocysts through feces. It was classified in the coccidian subclass, phylum Apicomplexa, and the infectivity of the three parasitic stages (tachyzoite, cyst, and oocyst) was well characterized. Within intermediate hosts, the parasite undergoes only asexual development. Sexual reproduction strictly occurs only in domestic and wild cats. After the ingestion of cysts present in tissues of an intermediate host, the cyst wall is destroyed by gastric enzymes, where they differentiate into tachyzoites (Fig. 6.14A). Tachyzoites rapidly replicate by endodyogeny inside any kind of cell and disseminate throughout the organism (Fig. 6.14C). As a result of the conversion from tachyzoite to bradyzoite (Fig.6.14B), tissue cysts develop as early as 7 to 10 days postinfection and may remain throughout life in most hosts. These cysts predominantly harbor in the brain or musculature. Upon the ingestion of these tissue cysts by an intermediate host through raw or undercooked meat, cysts are ruptured as they pass through the digestive tract, causing the release of bradyzoites. The bradyzoites then infect the intestinal epithelium of the new host and differentiate back into the rapidly dividing tachyzoite stage for dissemination throughout the body[56].

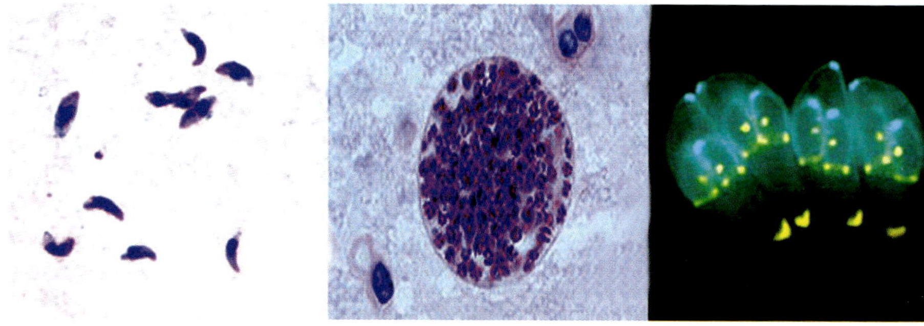

Fig. 6.14: (A) *Toxoplasma gondii* tachyzoites, (B) *T. gondii* tissue cyst in a mouse brain, individual bradyzoites can be seen within, stained with Giemsa; (C) Dividing *T. gondii* parasites (adapted from http://en.wikipedia.org/wiki/Toxoplasma_gondii

Diagnosis

Diagnosis can be made by direct observation of the parasite in stained tissue sections (Fig. 6.13 A), cerebrospinal fluid (CSF), or other biopsy material. Serologic testing is the routine method of diagnosis. Molecular techniques that can detect the parasite's DNA in the amniotic fluid can be useful in cases of possible mother-to-child (congenital) transmission. Diagnosis and treatment during pregnancy appears to reduce the effects on the baby.

A highly sensitive and specific method has been developed to detect and quantitate *Toxoplasma gondii* burden in animal tissue samples using *T. gondii* ITS1-derived primers from 18S rRNA gene. A fluorogenic probe has also been developed to diagnose more accurately by real-time PCR. To confirm the specificity of the assay, a panel of DNA samples from *T. gondii* and other common protozoa and host animal tissue were used as positive and negative controls respectively. This Toxo TaqMan assay was able to detect as low as 0.1 pg of *T. gondii* genomic DNA, which is equivalent to one *T. gondii* bradyzoite, and has a dynamic range of detection from 100 ng to 100 fg of *T. gondii* DNA[57].

Pathogenesis

Toxoplasmosis infection is asymptomatic in 80 % of people. The most common symptom of the disease is the sign of enlarged lymph nodes, especially around the neck. The illness may mimic glandular fever with other symptoms of muscle pain, intermittent fever and malaise. Dormant infection may persists for life and often reactivate in the immunosuppressed person. More serious disease can develop or reactivate in immunosuppressed patients involving different organs like brain, heart or eye. They may develop pneumonia and occasionally the disease may be fatal. Cerebral toxoplasmosis or chorioretinitis are frequent complications of AIDS when the lymphocyte CD4 cell count drops below 100 / cu mm. Acute toxoplasmosis in pregnant women can affect the unborn child. In early pregnancy brain damage as well as liver, spleen and eye disorders may occur. However, infection in late pregnancy may result in persistent eye infection through life. Toxoplasmosis if acquired after birth usually results in no symptoms or only cause a mild illness.

Epidemiology

Infection is often highest in areas of the world that have hot, humid climates and lower altitudes. The source of most infection are contaminated meat, especially pork and lamb. One survey of pigs in the US found a seroprevalence of 42 percent in breeder pigs and 23 percent in market pigs. Another study in the 1960s found *T. gondii* in 32 percent of pork chops and 4 percent of lamb chops in grocery stores. In humans the seroprevalence is 22.5 percent at the national level, according to a study of 17,658 people from 1988 to 1994. Since there is the extra concern of congenital infection in mothers, additional attention is given to detect the prevalence of the disease in women of child-bearing age. The seroprevalence in the group of 15 to 44 year old women was reported to be 15 percent[58]. The oocysts of this parasite are not killed by chemical and physical treatments currently applied in water treatment plants, including chlorination and ozone. The oocysts can remain viable for long periods of time in water and can resist freezing and moderately high water temperatures treatment. Toxoplasma oocysts can remain viable for extended periods of time in seawater[59]. Shellfish are filter feeders that concentrate *T. gondii.* Oocysts remained viable and were detected in various species of shellfish under natural conditions[60]. The consumption of oysters, clams, and mussels has shown to be a risk factor for acquiring Toxoplasma infection.

Treatment and Prevention

Healthy people can recover from toxoplasmosis without treatment. When an individual comes down with the disease, can be treated with a combination of drugs such as pyrimethamine and sulfadiazine and folinic acid. Management of maternal and fetal infection varies depending on the treatment center. In general during gestation period in women, spiramycin is recommended for the first and early second trimesters whereas pyrimethamine/sulfadiazine and leucovorin are recommended for late second and third trimesters with acute *T. gondii* infection diagnosed at a reference laboratory. Congenitally infected newborns are generally treated with pyrimethamine, a sulfonamide, and leucovorin for 1 year[61]. The parasites can remain within tissue cells in a less active phase. Persons with compromised immune systems need to be treated until they have improvement in their condition. For AIDS patients, continuation of medication for the rest of their lives may be necessary, or for as long as they are immunosuppressed. The serologic screening for the Toxoplasma antibody both for the donors and recipients should be made mandatory during organ transplantation in the countries endemic for the disease.

Preventive measures:

1. Food must be cooked to safe temperatures. Use of a food thermometer is advised to measure the internal temperature specially while cooking meat items.
2. Whole cuts of meat must be cooked to at least 145° F (63° C) and requires a rest time for three minutes minimum before consumption.
3. Ground meat should be cooked to at least 160° F (71° C); ground meats do not require a rest time.
4. For all Poultry (whole cuts and ground) must be cooked to at least 165° F (74° C), and rest for three minutes must be allowed before carving or consuming.

To prevent risk of toxoplasmosis from the environment:

- Consumption of safe drinking water.
- Using appropriate gloves while gardening in order to avoid any contact with soil contaminated with fecal material from cat containing *Toxoplasma*.
- Washing hands with soap and warm water after gardening or contact with soil or sand.
- Outdoor sandboxes should be covered.
- Use of dried canned or dried commercial food or well-cooked table food for feeding the cat.
- Avoid feeding cats with raw or undercooked meats.
- Domesticated cats should be kept indoors.

6.3.3 Malaria

Malaria is a mosquito-borne infectious disease of humans and other animals caused by parasitic protozoa belonging to the genus *Plasmodium*. The disease is transmitted via a bite from an infected female *Anopheles* mosquito. During the blood meal, the parasite from the saliva of the mosquito gets introduced into the host's circulatory system. Through the blood, the parasite travel to the liver to mature and replicate. Malaria causes symptoms that typically include fever and headache, however in severe cases the disease can progress to coma or death. The disease is widespread in tropical and subtropical regions around the equator. It covers much of Sub-Saharan Africa, Asia and the Americas. Five species of *Plasmodium* are known to infect and get transmitted by humans. The majority of deaths are caused by *P. falciparum* and *P. vivax,* while *P. ovale,* and *P. malariae* cause a generally milder form of malaria that is rarely fatal. *A zoonotic* species known as *P. knowlesi,* is prevalent in Southeast Asia and causes malaria in macaques but can also cause severe infections in humans.

Signs and Symptoms

Early symptoms include: fever, chills, headache, sweats, fatigue, nausea and vomiting. These symptoms resemble many other infections caused by bacteria, viruses, or parasites. In case of malaria, the initial chill is normally followed by high fever for few hours and then the patient sweats profusely. Symptoms may appear in cycles but often at different intensities and for different lengths of time. This typical pattern is very characteristic feature of malaria.

This cyclic pattern of malaria symptoms represents the life cycle of malaria parasites as they develop, reproduce, and eventually released from the red blood cells and liver cells in the human body. This cycle of symptoms is also one of the major indicators of infection with malarial parasite. Other common symptoms of malaria include: dry cough, muscle and back pain and often exhibit enlarged spleen. In severe cases, malaria can lead to impaired function of the brain or spinal cord and may lead to seizures, or loss of consciousness. Infection with the *P. falciparum* parasite is usually more serious and may become life-threatening. Other serious complications of falciparum malaria include:

- "Blackwater fever" – caused by the rupture of large numbers of red blood cells, releasing hemoglobin into the urine which gives the urine a darker color
- Kidney failure – due to the release of hemoglobin going through the kidneys
- Anaemia – a condition where all the red blood cells get destroyed
- Pulmonary oedema – accumulation of fluid in the lungs due to which breathing problems may arise.

Incubation period

It generally ranges from:

- 9 to 14 days for *Plasmodium* (*P.*) *falciparum.*
- 12 to 18 days for *P. vivax* and *P. ovale.*
- 18 to 40 days for *P. malariae.*
- 11 to 12 days for *P. knowlesi.*

Symptoms normally appear in 7 days. Sometimes, the time between exposure and signs of illness may be as long as 8 to 10 months with *P. vivax* and *P. ovale.* The incubation period may be longer if a person is under medication to prevent infection, a stage defined as chemoprophylaxis or has developed some immunity due to previous infection.

Diagnosis

The disease is commonly diagnosed by the microscopic examination of blood, utilizing blood films. Although blood is the sample most frequently used to make a diagnosis however both saliva and urine have been investigated as alternative and less invasive specimens. Malaria is usually confirmed by the microscopic examination of blood films or by antigen-based rapid diagnostic tests (RDT)[62,63.] Areas that cannot afford laboratory diagnostic tests often use only a history of subjective fever as the indication to treat for malaria. The sensitivity of the test using blood films ranges from 75–90% in optimum conditions, to as low as 50%. Commercially available RDTs are often more accurate than blood films at predicting the presence of malaria parasites. Over diagnosis of malaria may occur when it is misdiagnosed with non-malarial fever due to limited resources and expertise. Though polymerase chain reaction-based tests have been developed but not widely implemented in malaria-endemic regions due to their complexity.

Malaria rapid diagnostic tests (RDTs) are a useful tool in endemic malaria countries, where light microscopy is not feasible. In non-endemic countries they can be used as complementary tests to provide timely results in case of microscopy inexperience. A newly devised commercially available kit "The VIKIA Malaria Ag Pf/Pan™" performs similar to PCR based methods for the detection of *P. falciparum*. This is therefore considered as a valuable tool for detection of malaria in endemic and non-endemic regions[64].

Life Cycle and Transmission

During the life cycle of *Plasmodium*, (Fig. 6.15) a female Anopheles mosquito, defined as the definitive host transmits a motile infective form called the sporozoite to a vertebrate host such as a human, the secondary host. The mosquito thus acts as a transmission vector. A sporozoite travels through the blood vessels to liver cells known as hepatocytes where it reproduces asexually known as tissue schizogony producing thousands of merozoites. These infect new red blood cells and initiate a series of asexual multiplication cycles named as blood schizogony that produce 8 to 24 new infective merozoites. At this point the cells burst and the infective cycle begins anew. Other merozoites develop into immature gametocytes, which are the precursors of male and female gametes. When a mosquito bites an infected person, gametocytes get transmitted with the blood and they finally mature in the mosquito gut. During this stage in the mosquito, the male and female gametocytes of the parasite fuse and form a ookinete which is a fertilized, motile zygote. Ookinetes further develop into new sporozoites that migrate to the insect's salivary glands, ready to infect a new vertebrate host. The sporozoites carried through the saliva of the mosquito are injected into the skin of the host when the mosquito takes a subsequent blood meal.

Only female mosquitoes feed on blood; male mosquitoes feed on plant nectar, and thus do not contribute in the transmission of the disease. The females of the Anopheles genus of mosquito prefer to feed at night. Malaria parasites can also be transmitted by blood transfusions, although this is rare[65]. Some of the *P. vivax* sporozoites do not immediately develop into exoerythrocytic-phase merozoites, but instead produce hypnozoites that remain dormant for periods ranging from several months (7–10 months is typical) to several years. At the end of the dormancy, they may reactivate and produce merozoites. Hypnozoites are responsible for long incubation and late relapses in *P. vivax* infections[66].

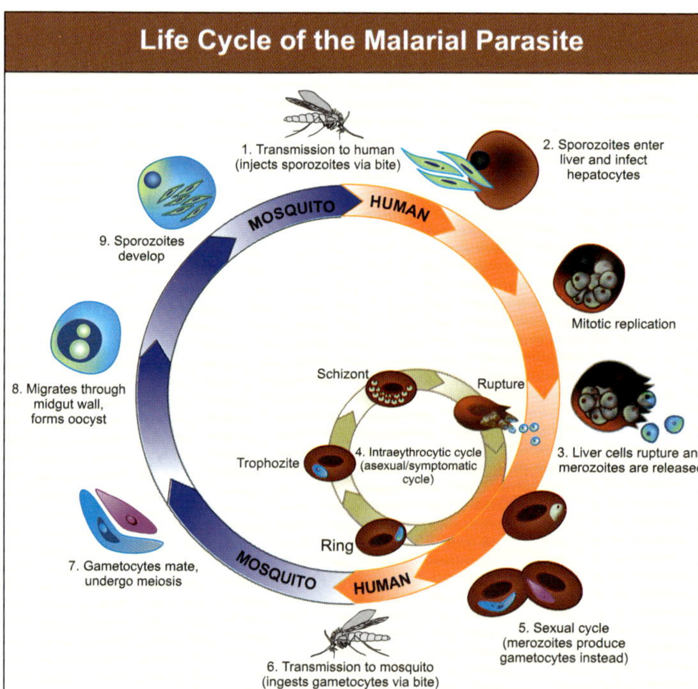

Fig. 6.15: The different stages of the Life Cycle of Malaria.

A mosquito causes infection by taking a blood meal. First, sporozoites enter the bloodstream, and migrate to the liver. They infect liver cells, where they multiply into merozoites, rupture the liver cells, and return to the bloodstream. Then, the merozoites infect red blood cells, where they develop into ring forms, trophozoites and schizonts that in turn produce further merozoites. Sexual forms are also produced, which, if taken up by a mosquito, will infect the insect and continue the life cycle.

Human Factors and Malaria

Evidences show that the first line of defense against malaria is provided by genetically controlled innate resistance. Host resistance to malaria therefore involves blood cell defect such as abnormal hemoglobins, Glucose-6-phosphate dehydrogenase deficiency, and Duffy antigens, which provide innate resistance. In malaria, as in other infections, innate immune responses lead into and stimulate adaptive immune responses through major histocompatibility complex genes, which regulate adaptive immune responses.

However the potent effect of genetically controlled innate resistance is reflected in the probability of survival of young children in endemic environment. The demonstration that sickle-cell heterozygotes have some degree of protection against *falciparum* malaria was the first example of genetically controlled innate resistance to human malaria, as recognized by experts on inherited factors affecting human infectious diseases[67]. Individuals who are negative for the Duffy blood group, possess red blood cells that show resistance to infection by *P. vivax.* Since the majority of Africans are Duffy negative, *P. vivax* is rare in Africa south of the Sahara, especially West Africa. In those areas, the *P. vivax* has been taken over by *P. ovale,* a very similar parasite that dominates and can infect Duffy-negative persons.

Epidemiology

Malaria is prevalent in tropical and subtropical regions because of the weather conditions favorable for continuous breeding of mosquitoes. Conditions like rainfall, consistent high temperatures and high humidity, along with stagnant waters in which mosquito larvae readily mature, providing them the right breeding ground for multiplication. Malaria is not just a disease commonly associated with poverty: some evidence suggests that it is also a cause of poverty and a major hindrance to economic development.

Malaria is prevalent in most of the tropical regions of the world, with *P. falciparum* causing the largest burden of disease, followed by *P. vivax. P. falciparum* predominates in Africa, New Guinea, and Hispaniola. *P. vivax* is more common in the Americas and the western Pacific[68]. The prevalence of these two species is approximately equal in the Indian subcontinent, eastern Asia, and Oceania. *P. malariae* is uncommon and is found only in sub-Saharan Africa. *P. ovale,* even less common, is restricted Africa and relatively unusual outside of Africa. This species comprises less than 1 percent of other isolates. Another species *P. knowlesi,* morphologically similar to *P. malariae,* has been identified by molecular methods in patients from Malaysia, the Philippines, Thailand, and Myanmar[69]. This species has not yet been proven to be transmitted from humans to mosquitoes. It is assumed that a monkey reservoir may be required to infect mosquitoes. Thc distribution of malaria in the world is represented in Fig. 6.16.

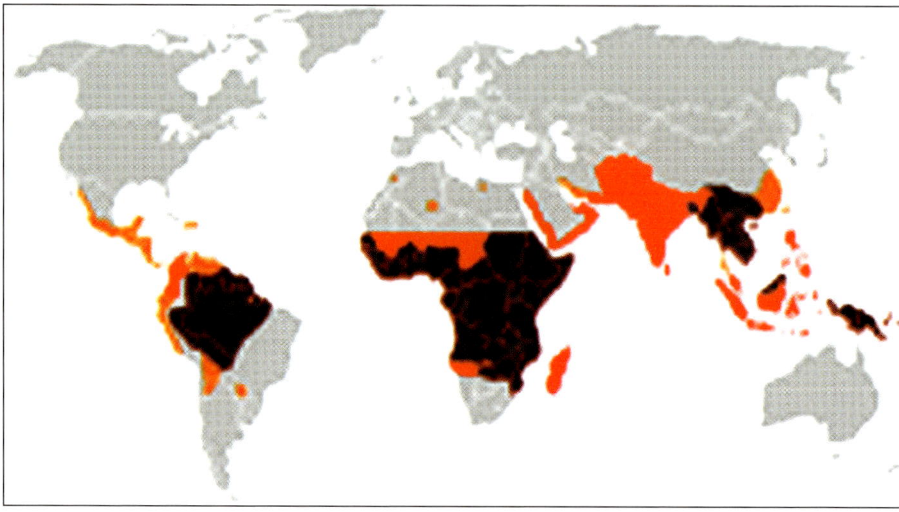

Fig. 6.16: Distribution of malaria in the world on the basis of chloroquine resistance: ■ Elevated occurrence of chloroquine- or multi-resistant malaria, ■ Occurrence of chloroquine-resistant malaria, □ *No Plasmodium falciparum* or chloroquine-resistance, □ No malaria (Adapted from "Malaria". US Centers for Disease Control and Prevention. April 15, 2010) with permission.

Study of the epidemiology of severe malaria in Africa has shown different epidemiological patterns for the two most frequent forms of this condition: cerebral malaria and severe malarial anaemia. Severe malarial anaemia is encountered frequently in young children where the transmission rate is high. In contrast, cerebral malaria predominates in areas of moderate transmission where it mostly found to be seasonal, and it is seen more frequently in older children[70]. Global efforts to control and eliminate malaria have saved an estimated 3.3 million lives since 2000, reducing malaria mortality rates by 45 percent globally and by 49 percent in Africa, according to the "World malaria report 2013" published by WHO. The large majority of the 3.3 million lives saved between 2000 and 2012 were from the 10 countries with the highest malaria burden. The affected population mostly includes the group of children aged less than 5 years. Over the same period, malaria mortality rates in children in Africa were reduced by an estimated 54 percent. http://www.who.int/mediacentre/news/releases/2013/world-malaria-report-20131211/en/)

A hospital-based studies on severe malaria in Thailand, Vietnam, Burma, Bangladesh, Indonesia, India, and Mozambique revealed consolidated information on mortality rate based on age structure (Fig. 6.17). There was no significant heterogeneity observed between countries.

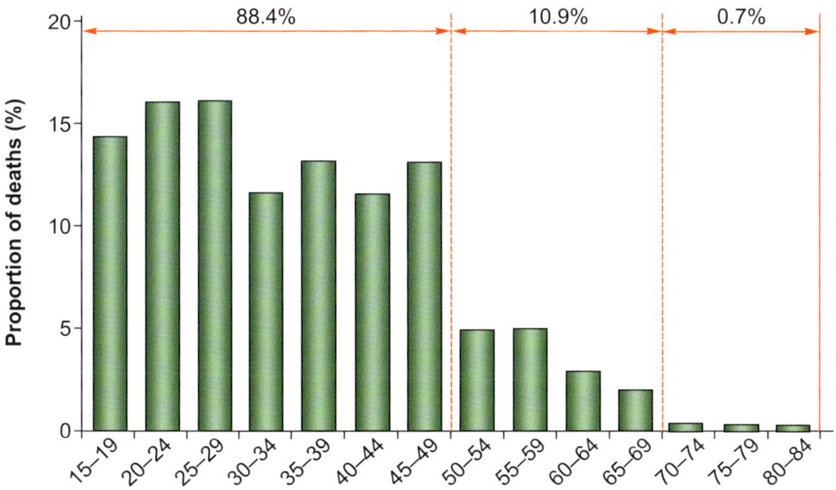

Fig. 6.17: Age structure, in 5-year intervals, of mortality from severe malaria in 2991 adults included in prospective studies in Thailand, Vietnam, Burma, Bangladesh, Indonesia, India, and Mozambique over the past 30 years(reproduced with permission, White et al, 2012[64])

Indian Scenario

Historically, the highest incidence of malaria in India occurred in the 1950s, with an estimated 75 million cases and 0.8 million deaths per year (World Health Organization, Country Office for India). The launch of the National Malaria Control Program (NMCP) in 1953 resulted in a significant decline in the number of reported cases to <50,000 and no reported mortality, by 1961. Despite its near elimination in the mid-1960s, malaria resurged to ~6.45 million cases in 1976. Since then, confirmed cases have gradually decreased to 1.6 million cases and ~1100 deaths in 2009. Recently, it has been suggested that the malaria incidence is between 9 and 50 times greater than reported with a ~13-fold under-estimation of malaria-related mortality[72,73]. The projected results show 205,000 malaria deaths per year in India before age 70 years (55,000 in early childhood, 30,000 at ages 5-14 years, 120,000 at ages 15-69 years); 1•8% cumulative probability of death from malaria before age 70 years.

The difficulty in making these estimations is further exacerbated by (*i*) epidemiological profiles that are highly variable, (*ii*) the transmission and overlap of multiple Plasmodium species and Anopheles vectors, (*iii*) increasing antimalarial drug resistance and insecticide resistance, and (*iv*) the impact of climate change on each of these variables. In the neighboring country, Bangladesh, most malaria cases were *P falciparum* (89•3%, 255 266), 10 • 3% (29, 367) were *P vivax,* and 0 • 4% (1089) were mixed infections (*P falciparum* plus *P vivax*)[74].

Treatment and Prevention

Malaria control program aim at reducing the transmission level to an extent so that it is no longer a public health problem. “Control” of malaria differs from “elimination” or “eradication of malaria.” “Elimination” program focuses at a local or regional level. Eradication program works for “global elimination” of the disease.

All cases of severe malaria are considered to have *P. falciparum* malaria. Following protocol needs to be followed for severe cases of malaria

- Admission to the hospital for proper evaluation, treatment and monitoring.
- Should be treated with injectable antimalarials (chloroquine, quinine, artemisinin) so as to ensure adequate absorption and plasma drug levels.
- Often to use two blood schizonticidal drugs, one fast acting and another slow acting, to ensure complete treatment.
- Choloroquine is considered to be the drug of choice when the patient hails from a zone of chloroquine sensitive area.
- For patients coming from chloroquine-resistant zones, combination of drugs are administered like malarone or oral quinine plus antibiotic such as doxycycline.
- All conditions like pregnancy, epilepsy, heart diseases should be carefully assessed before treating for malaria.
- In case of mixed infections by *P. falciparum* and *P. vivax,* Chloroquine 25 mg of salt/kg over 36-48 hours + Primaquine for 14 days are prescribed.

Treatment of malaria is also dependent on following host factors.

- Patient's age and weight should be recorded so as to administer adequate doses of anti malarial drugs.
- Functional capacity of the patient- independent, dependent, bed ridden etc. Patients with severe prostration are advised to be admitted to a hospital and monitored.
- Adequate hydration should be ensured.

Preventive measures are as follows:

- *Avoid outbreaks:* Travelers should avoid traveling in areas of known malaria outbreaks. The CDC Travelers' Health web page provides alerts and information on regional disease transmission patterns and outbreak alerts (http://www.cdc.gov/travel).
- *Be aware of peak exposure times and places:* Vectors for malaria are most active in twilight periods or in the evening after sunset. Risk of mosquito bite can be reduced by avoiding visits to outdoors during peak hours and adopting preventive measures.
- *Wear appropriate clothing:* Exposure of the body can be minimized by wearing long-sleeved shirts, long pants, proper shoes as to avoid mosquito bite. Tucking in shirts and wearing socks and closed shoes instead of sandals may reduce risk.
- *Use of mosquito repellants* Repellents or insecticides such as permethrin can be applied to clothing for added protection.
- *Bed nets:* When accommodations are not adequately screened or air conditioned, bed nets are essential to provide protection. Bed nets are most effective when they are treated with an insecticide or repellent such as permethrin.
- *Insecticides:* Aerosol insecticides, vaporizing mats, and mosquito coils can help to clear rooms or areas of mosquitoes. Proper care should be taken in order to avoid direct inhalation of spray or smoke.
- *Optimum protection can be provided by applying repellents.* The CDC recommended insect repellent should contain up to 50% DEET (N,N-diethyl-m-toluamide), which is the most effective mosquito repellent for adults and children over 2 months of age.

6.3.4 Leishmaniasis

Leishmaniasis is defined as a vector-borne neglected tropical disease associated with a spectrum of clinical manifestations, vary from self-healing cutaneous lesions to fatal visceral infections. The parasite is represented by 20 different species that are transmitted to humans by the bites of infected female sandflies. Parasites are categorized into three different groups: 1. *Leshmania donavani* causes a visceral leishmaniasis where the parasite invades the internal organs. 2. and 3. include *L. tropica* and *L. braziliensis* groups that grow preferentially at cooler temperatures and cause lesions of the skin or of the mucous membrane.

Pathogenesis

Signs and Symptoms

Leishmaniasis may be divided into the following types[75]:

1. **Cutaneous leishmaniasis (CL)** is the most common form, which causes an open sore at the bite sites, which normally takes few months to a year to heal leaving an unpleasant-looking scar on the site that sometimes resemble leprosy. The cutaneous form of the disease is caused by more than 15 types of *Leishmania*.
2. **Mucocutaneous leishmaniasis** causes both skin and mucosal ulcers with damage primarily of the nose and mouth
3. **Visceral leishmaniasis (VL)** also known as **kala-azar,** black fever, and **Dumdum fever.** This is considered to be the most serious form, and is potentially fatal if remains untreated.

Other consequences, which can occur anywhere from a few months to years after infection, include fever, damage to the spleen and liver, and anemia[76]. Leishmaniasis is considered one of the classic causes of a markedly enlarged spleen; may even become larger than the liver in severe cases. This disease is the second-largest parasitic killer in the world after malaria. The disease may be responsible for an estimated 500,000 infections each year worldwide[77]. The parasite migrates to the internal organs such as liver, spleen and bone marrow hence referred as '*visceral*' leishmaniasis. This stage leads to producing hepatosplenomegaly, oedema and anaemia. It is a slow but progressive illness, with bouts of irregularly recurring fever, and is invariably fatal, unless treated. Common signs and symptoms of the disease include fever, weight loss, mucosal ulcers, fatigue, anemia accompanied by swelling of the liver and spleen. According to the World Health Organization (WHO), of particular concern is the emerging problem of HIV/VL co-infection.

Diagnosis

The gold standard for diagnosis is visualization of the amastigotes in splenic aspirate or bone marrow aspirate. However, sampling is a technically challenging procedure and often not available in areas where visceral leishmaniasis is endemic.Serological testing is considered to be simple and accurate and therefore more frequently used in areas where leishmaniasis is endemic. The **K39** dipstick test is easy to perform and therefore even village health workers can be easily trained to adopt it. The other advantage is that it can be stored at ambient temperature and no additional equipment needs to be carried to remote areas. In highly endemic areas, not all infected individuals actually develop clinical disease or require treatment. Indeed, up to 32% of the healthy population may test positive, but not require treatment.

Life Cycle

Leishmania parasites have a complex life cycle that involves both vertebrate and invertebrate hosts and two developmental stages: promastigotes, capable of multiplication while residing in the lumen of the female sandfly, and amastigotes, also a proliferative form found inside several types of mammalian host cells. The life cycle of *Leishmania* involves two different hosts: a female sandfly and mammals (including humans and dogs), as summarized in Figure 6.18.

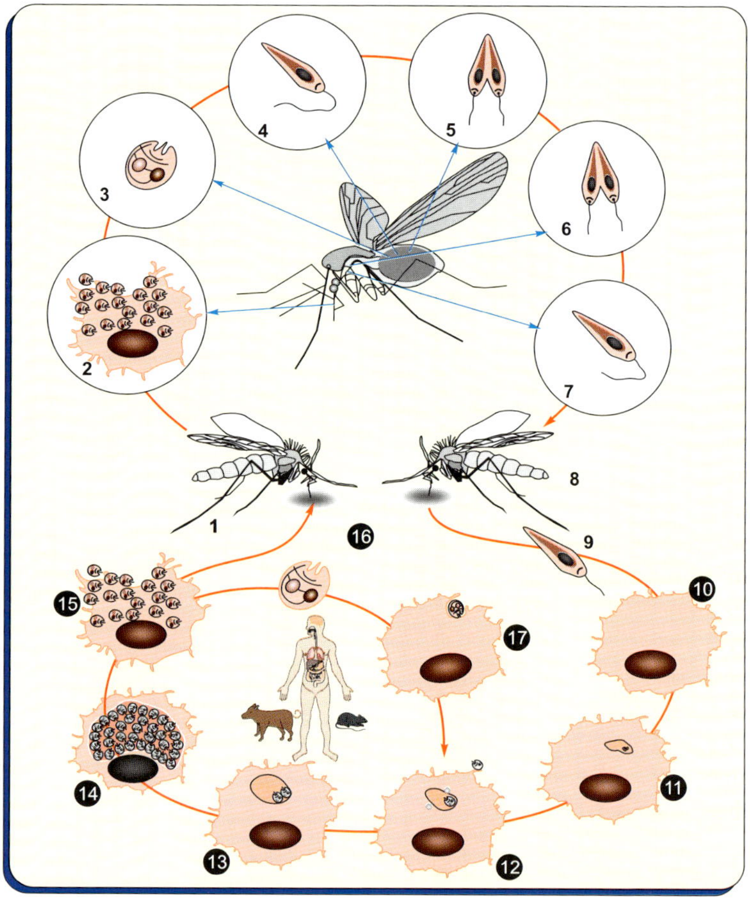

Fig. 6.18: The life cycle of Leishmania amazonensis—*The female sandfly (1) insect bites an infected mammal during the blood meal. Infected macrophages (2) with amastigote forms. (3) Amastigote form. (4) Amastigotes transform into procyclic promastigotes. (5) Procyclic promastigotes multiply in the midgut. (6) Promastigotes migrate toward the stomodeal valve in the anterior midgut and initiate cell division. (7) Promastigotes transform into infective metacyclic promastigotes. (8) The female sandfly releases the metacyclic promastigotes into a new mammalian host via regurgitation during the blood meal. (9) Metacyclic promastigotes. (10) Metacyclic promastigotes infect macrophages. (11) Metacyclic promastigotes transform into amastigotes. (12) Amastigotes attach to the membrane of the parasitophorous vacuole. (13) Amastigotes multiply in the vacuole. (14) Intense amastigote multiplication. (15) Amastigotes burst out of the cell. (16) Amastigote form. (17) An amastigote infects a macrophage. In the central portion of the figure is shown the important reservoirs involved in the maintenance of the parasite. (adapted from Teixeira et al (2013)*[78]

Transmission

All species of Leishmania are transmitted by small blood-sucking sandflies, known as *Phlebotomus* spp. belonging to the Old World and Lutzomyia spp. popular in the New World. In many geographic areas, even infected animals such as rodents or dogs, along with sand flies, maintain the cycle. However, in some parts of the world, infected people are involved to maintain the cycle; this type of transmission (human—sand fly—human) is called anthroponotic. In areas with anthroponotic transmission, effective treatment of individual patients can help control the spread of the parasite. Overall, infection in people is caused by more than 20 species of *Leishmania* parasites, which may involve about 30 species of phlebotomine sand flies. Selectivity is often encountered between the species of leishmania and its vector. The sand fly vectors are the most active during twilight, evening, and night-time hours. The identification of visceral disease determinants is required to understand the evolution of the disease and its survival mechanisms in order to develop better interventions for this largely neglected disease.

Epidemiology

Leishmaniasis is found in people in focal areas of more than 90 countries belonging to the tropics, subtropics, and southern Europe. The disease is prevalent from rain forests to deserts. Leishmaniasis usually is more common in rural than in urban areas, but it is also found in the peripheral regions of some cities. Climate and other environmental changes have the potential to expand the geographic range of the sand fly vectors and the areas in the world where leishmaniasis is found.

A comprehensive literature review on global incidence of leishmaniasis has been carried out by Alvar et al, 2012 as a part of WHO led effort. Based on these estimates, approximately 0.2 to 0.4 cases and 0.7 to 1.2 million VL and CL cases, respectively, occur each year. More than 90 percent of global VL cases occur in six countries: India, Bangladesh, Sudan, South Sudan, Ethiopia and Brazil. The Americas, the Mediterranean basin, and western Asia from the Middle East to Central Asia have been identified as three important areas involved in cutaneous leishmaniasis (CL) Estimated annual incidence of visceral leishmaniasis globally and in the Indian subcontinent are represented in Table 6.3 and Table 6.4[79].

Table 6.3: Global reported and estimated incidence of visceral leishmaniasis[79]

Region	Reported VL cases/year	Countries with 5 years of data	Estimated Annual VL incidence
Americas	3662	8/11 (73%)	4500 to 6800
Sub-Saharan Africa	1	3/11 (27%)	
East Africa	8569	5/8 (63%)	29,400 to 56,700
Mediterranean	875	21/26 (81%)	1200 to 2000
Middle East to Central Asia	2496	14/17 (82%)	5000 to 10,000
South Asia	42,623	3/6 (50%)*	162,100 to 313,600
Global total	58,227	54/79 (68%)	202,200 to 389,100

*3/3 (100%) of high burden countries (India, Bangladesh, Nepal) reported 5 years of data.

Table 6.4: Reported and estimated incidence of visceral leishmaniasis in the Indian subcontinent (adapted from Alvar et at 2012)[79]

*3/3 (100%) of high burden countries (India, Bangladesh, Nepal) reported 5 years of data. Reports are incomplete for Sri Lanka, Bhutan and Thailand.

Region	Reported VL cases/ year	Years of Report	Estimated Annual VL incidence
Bangladesh	6224	2004–2008	12,400 to 24,900
Bhutan	2	2005–2009	10 to 20[2]
India	34,918	2004–2008	146,700 to 282,800
Nepal	1477	2004–2008	3000 to 5900
Sri Lanka		No data	6 to 10[4]
Thailand	2	2006-2010	5 to 10[5]
Region	42,623		162,100 to 313,600

Prevention and Treatment

Prevention and control of leishmaniasis require a combination of intervention strategies since the transmission of the parasite takes place through complex biological system involving the human host, parasite, sandfly vector and in some cases an animal reservoir as well.

Key strategies include:

1. **Early diagnosis and effective case management :** effectively reduces the prevalence of the disease and prevents disabilities and death. Timely administration of highly effective and safe anti-leishmanial medicines particularly for VL.
2. **Vector control :** reduces the transmission of the disease. Control methods include insecticide spray, use of insecticide–treated nets and personal protection.
3. **Effective disease surveillance:** Early detection and treatment helps to monitor the spread and burden of disease.
4. **Control of reservoir hosts:** It is complex and should be tailored to the local situation.
5. **Social mobilization and strengthening partnerships**–mobilization and education of the community with effective communication strategies for reporting the cases in time to the health workers. Partnership and collaboration with various stakeholders and other vector-borne disease control programs is critical at various levels.

Drugs like Pentavalent antimony (Meglumine antimoniate, Sodium stibogluconate), Amphotericin B, Pentamidine isethionate are administered as injections. The association of pentavalent antimony with toxicity is well documented therefore these are to be used with great caution. Drugs like fluconazole, itraconazole, and ketoconazole are been used as oral drugs.

Liposomal amphotericin B is considered to be a highly effective and well tolerated drug. But due to high cost of this drug, it finds very limited use in developing countries. In these countries drugs containing pentavalent antimony are widely used. The parasite is also developing resistance against this drug in some of the developing countries like India. However in such areas, the drug miltefosine has been found to be effective. So far, there are no available vaccines that can prevent leishmaniasis.

6.3.5 Babesiosis

The causal organism of this disease is tick-transmitted hemoparasites of the protozoan genus Babesia belonging to the phylum Apicomplexa. After the trypanosomes, this parasite is considered as the second most common blood-borne parasites of mammals[80]. This parasite was discovered by Victor Babes at the end of the 19th century. He first discovered the presence of these microorganisms in erythrocytes of cattle in Romania and associated them with bovine hemoglobinuria or red water fever[81]. Seven distinct babesia parasites have been reported so far responsible to cause human babesiosis: *B. microti* and related organisms, *B. divergens, B. bovis, B. canis, B. duncani, B. venatorum,* and a novel type of Babesia sp. similar to ovine babesia which was named as KO1.

Babesia are transmitted by ixodid ticks (*Ixodes scapularis*) and infection of the host causes a host-mediated pathology and erythrocyte lysis, resulting in anemia, hyperbilirubinuria, hemoglobinuria, and possibly organ failure. Babesiosis can be a severe, life-threatening disease, particularly in people who do not have spleen, have a weak immune system such as cancer or AIDS patients, suffering from serious liver or kidney diseases and those who are elderly people.

The disease resembles malaria in some respects and at times even wrongly diagnosed. The parasite replicates in the RBCs and cause a prolonged illness. The common symptoms are of fever, chills and night sweats. is treated with a combination of two types of anti-parasite drugs, atovaquone (Mepron, Malarone) plus an erythromycin-type drug (azithromycin, clarithromycin, or telithromycin) are used to treat Babesiosis. Long-standing infections may need to be treated for several months, and relapses sometimes occur and must be retreated.

6.3.6 Schistosomiasis

The disease is caused by several species of trematode belonging to the genus *Schistosoma*, commonly known as **blood-flukes.** They are the members of the class Trematoda. These flukes are also known as parasitic flatworms which infect blood and internal organs of humans. Snails serve as the intermediate host between mammalian hosts. Although it has a low mortality rate, schistosomiasis often leads to chronic illness that can damage internal organs and, in children, impair growth and brain development. Schistosoma is known to exist in different forms. The urinary form of schistosomiasis is associated with increased risks for bladder cancer in adults. Schistosomiasis is the second most socioeconomically devastating parasitic disease after malaria

Species of *Schistosoma* that can infect humans:

1. *Schistosoma mansoni* and *Schistosoma intercalatum* causing intestinal schistosomiasis
2. *Schistosoma haematobium* causes urinary schistosomiasis
3. *Schistosoma japonicum* and *Schistosoma mekongi* cause Asian intestinal schistosomiasis

Signs and Symptoms

Some of the infections appear in a mild form in endemic areas leading to anemia. Acute schistosomiasis (Katayama's fever) is caused by *S. mansoni* and *S. japonicum* and manifests normally weeks after the initial infection. The main symptoms include: abdominal pain, cough, diarrhea, eosinophilia, fever, fatigue. The infection further may cause enlargement of liver and spleen, genital sores. These lesions make an individual more vulnerable to HIV infection. Skin symptoms are also common and that starts with mild itching and popular dermatitis of the feet and other parts of the body. Individuals contract with infection after swimming in polluted streams containing cercariae, the larval form of the parasite.

Occasionally a person may suffer from cerebral granulomatous disease where lesions are formed in the central nervous system caused by ectopic *S. japonicum* eggs in the brain. It may also lead to granulomatous lesions in the spinal cord due to deposition of ectopic eggs in the spinal cord from *S. mansoni* and *S. haematobium* infections. The local tissue damage leads to formation of granulomas as a defensive reactions mount by the human host to these foreign particles.

Transmission and Life Cycle

The disease is mainly transmitted through contaminated water bodies carrying eggs of schistosome through excreta or urine samples. When this contaminated water bodies enter water supplies with which humans come in contact, develop this infection. The Life cycle is depicted in the figure 6.19. Eggs are released with feces or urine (1). They enter the water body and under favorable conditions, the eggs hatch and release miracidia (2), they swim in water and may penetrate specific snail intermediate hosts (3), Sporocysts are formed inside the snail and undergo two generation of replication (4) and finally produces the cercariae (5). Upon release from the snail, the infective cercariae keep swimming using their forked tail and finally penetrate the skin of the human host (6), and finally get converted into schistosomulae by shedding their forked tail (7). The schistosomulae finally migrate to different parts of the body through portal veins (8, 9). As the worms mature turn into adults in humans, they harbor in the mesenteric veins draining different locations depending on the species of the schistosome (10). For instance, *S. japonicum* is more frequently found in the superior mesenteric veins draining the small intestine (A), and *S. mansoni* occurs more often in the superior mesenteric veins draining the large intestine (B). However, both species can occupy either location, and they are capable of moving between sites, so it is not possible to state unequivocally that one species only occurs in one location. *S. haematobium* most often occurs in the venous plexus of bladder (C), and sometimes observed in the rectal venules. The size of the females varies from 7 to 20 mm whereas males are slightly smaller. They deposit eggs in the small venules of the portal and around the urinary bladder. The eggs progressively migrate toward the lumen of the intestine when an individual is infected with *S. mansoni* or *S. japonicum*. The eggs migrate towards the bladder and ureters if the person is infected with *S. haematobium*. The eggs are finally released out of the body through feces or urine. The structure of paired adult worms is represented in Fig. 6.20. The females live in the groove on the ventral surface of the male schistosome. The males appear as split-body. In the mesenteric venules, schistosomes continuously fertilize and lay eggs. The males use the sucker to attach to the host.

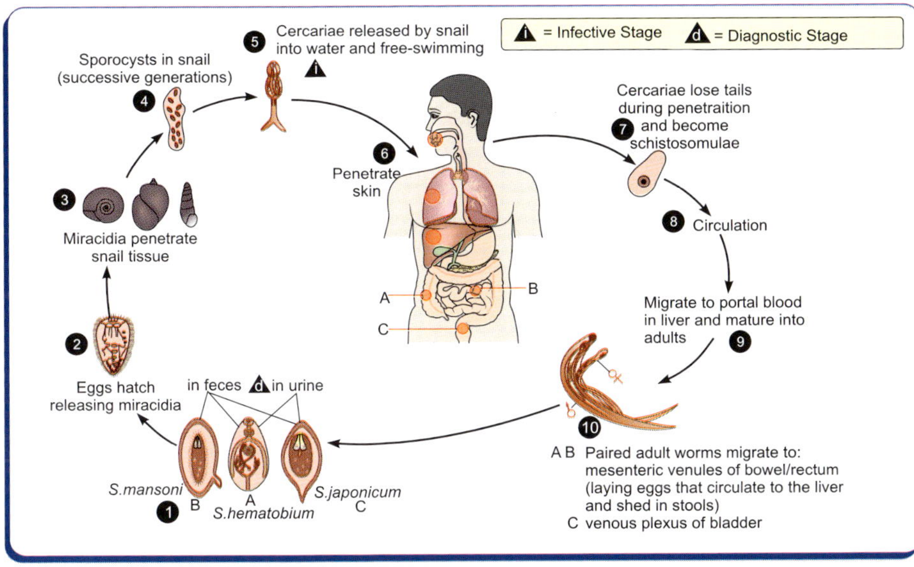

Fig. 6.19: Life cycle of Schistosoma

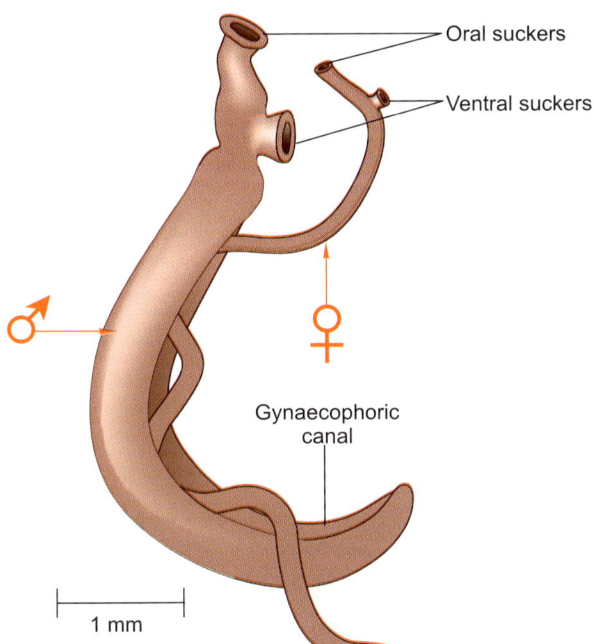

Fig. 6.20: Diagrammatic representation of male and female schistosomes [http://www.cdc.gov/parasites/schistosomiasis/biology.html]

Epidemiology

Among human parasitic diseases, schistosomiasis also referred as bilharziasis ranks second behind malaria in terms of socio-economic and public health importance in tropical and subtropical areas. The disease is endemic in most of the developing countries. People living in rural agricultural and peri-urban areas are more prone to the disease[82].

The disease is quite prevalent in tropical countries like Africa, the Caribbean, South eastern America, Southeast Asia and in the Middle East. A study conducted in 2010 revealed that around 238 million people were affected residing in the above countries, out of which, 85 percent were from Africa. An estimated 600 million people worldwide are at risk from the disease[83]. Worldwide study estimated 12,000 to 200,000 people die yearly due to schistosomiasis related disease. About 20 million people suffer from severe consequences because of the disease. In many areas, schistosomiasis infects a large proportion of children under 14 years of age.

Two groups can be clearly identified throughout Asia; the *S. japonicum* group found throughout eastern and south eastern Asia, including China, the Philippines and Malaysia, and the *S. indicum* group inhabiting the western and southern regions including India, Sri Lanka and Thailand[84]. Both *S. mansoni* and *S. haematobium* groups are found throughout Africa, often sharing overlapping geographical ranges. However in South America, *S. mansoni* is the predominant species, mainly restricted to areas of Brazil, Venezuela, Surinam and the Caribbean.

Treatment and Prevention

In countries where schistosomiasis causes significant disease, control efforts usually focus on improved sanitation that could reduce or eliminate transmission of this disease. Groups targeted for treatment are:

- the number of infections in people
- eliminating the snail population, required to maintain the parasite's life cycle.

Targeted groups are selected based on:

- children in endemic areas;
- adults considered to be at risk in endemic areas,
- people with occupations involving contact with infested water, such as fishermen, farmers, irrigation workers, and women carrying out domestic tasks using infested water.
- communities living in highly endemic areas.

The frequency of treatment is determined by the prevalence of infection in school-age children. In high transmission areas, treatment may have to be repeated every year and continue for couple of years. Monitoring the infection is essential to determine the impact of control interventions.

Praziquantel is the recommended drug for treating all forms of schistosomiasis. It is quite effective, safe and low-cost. Even though re-infection may occur after treatment, however the risk of developing severe disease reduces in case the treatment is initiated in childhood. It is a broad spectrum drug that attacks the adult worm in all species. Using only 2 doses of 40mg/kg given in a day can kill 96 percent of eggs. Side effects could include mild abdominal pain and fever.

6.3.7 Lymphatic Filariasis

Lymphatic filariasis is commonly known as elephantiasis. It is a painful and profoundly disfiguring disease as it buildup excess tissue in the affected area. The infection is usually acquired in childhood, but its visible manifestations occur later in life, causing temporary or permanent disability. The disease is caused by thread-like nematodes referred as filariae.

Filariae are long hair-like tissue-dwelling nematodes. These nematodes use insects as intermediate hosts. The life cycle of the parasite includes two stages- an obligatory maturation stage in a blood-sucking insect followed by a reproductive stage in the tissues or blood of a definitive host (Fig. 6.21)[85].

Lymphatic filariasis is a mosquito-borne parasitic disease that is caused by three species of filariae: *Wuchereria bancrofti* is the predominant species responsible for 90 percent of cases and is prevalent in the tropics and in some sub-tropical areas of the world. *Brugia malayi* is confined to restricted geographical areas like Southeast and Eastern Asia. The third species *Brugia timori* is restricted only to Timor and its adjacent islands. The symptoms of the disease caused by the three species of filariae are identical. *W. bancrofti* is the common cause of elephantiasis. Brugian filariasis does not characteristically include symptoms associated with genitalia or chyluria whereas Bancroftian filariasis is known to express these symptoms.

Mode of Transmission

Lymphatic filariasis gets transmitted by various species of mosquitoes belonging to the genera *Anopheles, Culex, Aedes, Ochlerotatus,* and *Mansonia.* The biting-time of the mosquito correlates with the periodicity of the microfilariae. In most parts of the world, the vectors are found to be nocturnal feeders and the microfilariae exhibit nocturnal periodicity. The numbers are also found highest in the blood samples around midnight. In the region of Central Pacific where the mosquito feeds mostly during the day, there is a diurnal periodicity where the highest number of microfilariae occurs at midday. In some areas of Southeast Asia, a sub periodic pattern is observed where microfilariae are found in the blood at all times but there is a nocturnal peak.

Diagnosis

The simplest method of detection of the parasite is to draw capillary blood from a finger-prick and prepare a thick film as carried out for malaria diagnosis (Fig. 6.22). However, recent faster diagnostic methods are now available.

1. **Membrane filtration method:** Venous blood drawn from the patient during night and filtered through millipore membrane filters. This technique enables an easy detection of microfilaria and also the load of infection can be quantified.
2. **Ultrasonography:** By employing this technique, one can record the movements of living adult filarial worms of *W. bancrofti* in the scrotal lymphatics of asymptomatic males with microfilaraemia.

3. **Lymphoscintigraphy:** In this technique, radiolabelled albumin or dextran is injected in the web space of the toes and the structural changes are imaged using a gamma camera[86].
4. **Immunochromatographic test (ICT):** This test is considered to be highly sensitive and specific filarial antigen detection assays for the diagnosis of *W. bancrofti* infection. This test is now available both as card test and in ELISA based format. This test shows positive result only at early stages of the disease when the adult worms remain alive and becomes negative once they are dead[87].
5. **DNA probes using Polymerase Chain Reaction:** These tests are of high specificity and sensitivity and are able to detect parasite DNA in humans as well as vectors in both bancroftian and brugian filariasis[88].

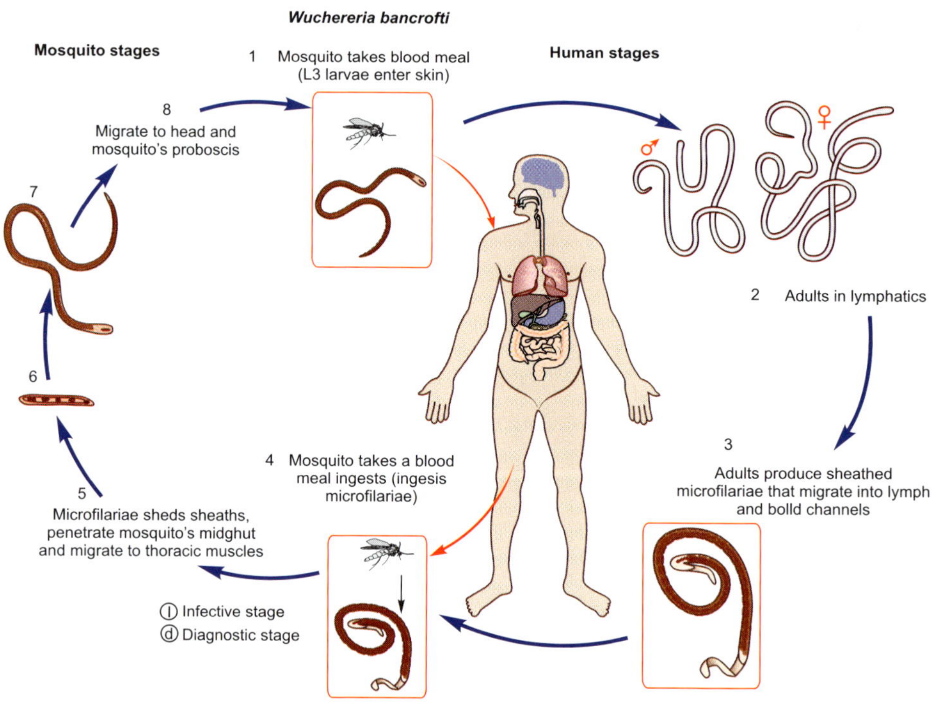

Fig. 6.21: Typical filaroid life cycle (reproduced with permission) (http://www.dpd.cdc.gov/dpdx)

During a blood meal, an infected mosquito introduces third-stage filarial larvae (L3) onto the skin of the human host, where they penetrate into the bit wound ❶. They develop in adults that commonly reside in the lymphatics ❷. The female worms measure 80 to 100 mm in length and 0.24 to 0.30 mmin diameter, while the males measure about 40 mm by .1 mm. Adults produce microfilariae measuring 244 to 296 μm by 7.5 to 10 μm, which are sheathed. The microfilariae migrate into lymph and blood channels moving actively through lymph and blood ❸. A mosquito ingests the microfilariae during a blood meal ❹. After ingestion, the microfilariae lose their sheaths and some of them work their way through the wall of the proventriculus and cardiac portion of the mosquito's midgut and reach the thoracic muscles ❺. There the microfilariae develop into first-stage larvae (L1) ❻ and subsequently into third-stage infective larvae (L3) ❼. The infective L3 larvae migrate to the mosquito's proboscis ❽ and can infect another human when the mosquito takes a blood meal ❶.

Pathogenesis

The filarial life cycle consists of 5 developmental stages in a vertebral host and an arthropod intermediate host and vector (Fig. 6.21). Adult female worms produce thousands of first-stage larvae or microfilariae that are ingested by a feeding insect vector. Some microfilariae have a unique daily circadian periodicity in the peripheral circulation. The arthropod vectors, mosquitoes and flies, also have a circadian rhythm in which they obtain blood meals. The highest concentration of microfilariae usually occurs when the local vector is feeding most actively.

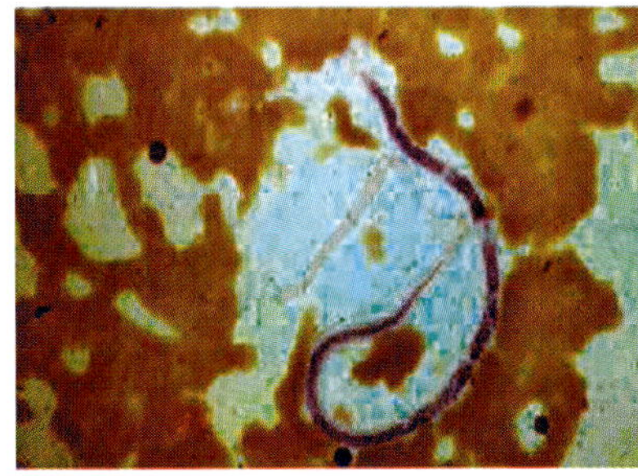

Fig. 6.22: Microfilaria of *W. bancrofti* in thin blood film attained with Giemsa stain and showing the sheath (X200) (adapted from www.tropmed.org/primer/chapter24.pdf with permission)

Microfilariae then undergo two developmental changes in the insect. Third-stage larvae then are inoculated back into the vertebral host during a blood meal. The final two stages of development completes in the human host. These larvae travel through the dermis and enter regional lymphatic vessels. Larvae further mature into full grown worms (20-100 mm in length) over a period of nine months. An average parasite can survive for about 5 years.

The following factors affect the pathogenesis of filariasis:[89,90]

1. The quantity of accumulating adult worm antigen in the lymphatics
2. The duration and level of exposure to infective insect bites
3. If the patient suffers from secondary bacterial and fungal infections
4. The degree of host immune response

Filarial infection generates significant inflammatory immune responses that participate in the development of symptomatic lymphatic obstruction. Increased levels of immunoglobulin E (IgE) and immunoglobulin G4 (IgG4) secondary to antigenic stimulation of Th2-type immune response often arising from dead worms have been demonstrated.

Prenatal exposure seems to be an important determinant in conferring greater immune tolerance to parasite antigen. Thus, individuals from endemic areas are often asymptomatic until late in disease when they develop high worm burden. Those who migrate from nonimmune areas, normally mounts a brisk immune responses and show severe early clinical symptoms when they get infected[91].

Until quite recently it was generally accepted that three groups of people are found in filarial- endemic areas: Those who are exposed, but with no evidence of disease – so called "endemic normals", those with "asymptomatic microfilaraemia", and those with "chronic pathology" - lymphodema, elephantiasis, or hydrocele. In some areas, most notably in Asia, another manifestation of filariasis arises which is referred as "Tropical pulmonary eosinophilia(TPE)"[92]. These individuals demonstrate marked eosinophilia, extreme levels of serum IgE, and high titres of antifilarial IgG and IgE[93]. Circulating microfilariae can not be detected in these individuals however, adult worms may be detected by ultrasound. Clinically TPE is characterized by nocturnal coughing and asthma. Pulmonary infiltration gives rise to a characteristic X-ray picture called "snow flake" lung. Signs and symptoms resolve quite quickly after treatment with antifilarial drugs.

Geographical Distribution and Epidemiology

Filariae have a specific geographic distribution. For example, *W bancrofti* is found in sub-Saharan Africa, Southeast Asia, India, and the Pacific Islands. *B malayi* is found in similar locations, but it is absent from sub-Saharan Africa. *B timori* is concentrated to the Timor island of Indonesia.

It has been observed that in endemic areas, the prevalence of microfilaremia increases with age, as adult worms are gradually acquired over period of many years. Lymphatic filariasis is first contracted in childhood. Individuals in endemic areas are mostly exposed to the parasite by the third or fourth decade of life. The proportion of infected individuals remains constant[94]. As in case of most helminths, adult filarial parasites replicate in a secondary host. The adult worm burden in an individual cannot increase unless the host is exposed to additional microfilaria. It is also observed that infected individuals cannot sustain higher levels of parasitemia once they leave the endemic area.

Because the mosquito vector is inefficient in non endemic areas, a relatively prolonged stay in an endemic area is usually required to acquire the infection. Disorganized urbanization is adding to the vector population and hence to the increased incidence and prevalence of such diseases in developing countries. Lymphatic filariasis affects 119 million people living in 73 countries and is found throughout the tropics and subtropics. Estimates suggest that more than 40 million infected individuals are seriously incapacitated and disfigured due to lymphatic filariasis[95]. The prevalence of microfilaremia in endemic communities increases with age.

Indian Scenario

A study from India reveals that India alone accounts for 40 percent of the global prevalence of filariasis. It is also estimated that more than \$840 million is lost each year due to treatment costs and missed working days[96]. Another study in India noted that patients with chronic filariasis lose around 29 days of work per year due to complications of infection, highlighting the considerable burden the disease on individuals and on the community[97]. Many studies showed that the prevalence of Microfilaria in females was lower than that in males in South East Asia, and particularly India[98]. Gender-specific estimates indicated that prevalence of *W. bancrofti* infection in males was 10 percent more than that in females[99]. The disease rate was found to be age-dependent and steadily increased from age 10 onwards[100,101]. Prevalence of chronic manifestations was significantly higher than acute manifestations in both the sexes. Prevalence of lymphangitis was

62. Abba K, Deeks JJ, Olliaro P, Naing CM, Jackson SM, Takwoingi Y, Donegan S, Garner P (2011) Rapid diagnostic tests for diagnosing uncomplicated P. falciparum malaria in endemic countries. In Abba, Katharine. Cochrane Database of Systematic Reviews (7): CD008122.

63. Kattenberg JH, Ochodo EA, Boer KR, Schallig HD, Mens PF, Leeflang MM (2011) Systematic review and meta-analysis: Rapid diagnostic tests versus placental histology, microscopy and PCR for malaria in pregnant women. Malaria J. 10: 321.

64. Eibach D, Traore B, Bouchrik M, Coulibaly B, Coulibaly N, Siby F, Bonnot G, Bienvenu AL, Picot S. (2013) Evaluation of the malaria rapid diagnostic test VIKIA malaria Ag Pf/Pan™ in endemic and non-endemic settings. Malaria J. 12:188.

65. Owusu-Ofori AK, Parry C, Bates I (2010) Transfusion-transmitted malaria in countries where malaria is endemic: A review of the literature from sub-Saharan Africa. Clinical Infectious Diseases 51: 1192–8.

66. White NJ (2011) Determinants of relapse periodicity in Plasmodium vivax malaria. Malaria Journal 10: 297.

67. Alcaïs A, Abel L, Casanova JL (2009) Human genetics of infectious diseases: between proof of principle and paradigm. J Clin Invest 119: 2506–2514.

68. Guerra CA, Gikandi PW, Tatem AJ, et al (2008) The limits and intensity of Plasmodium falciparum transmission: implications for malaria control and elimination worldwide. PLoS Med 5:e38

69. White NJ (2008) Plasmodium knowlesi: the fifth human malaria parasite. Clin Infect Dis 46:172.

70. Greenwood BM (1997) The Epidemiology of malaria. Ann Trop Med Parasitol. 91:763-9.

71. White N.J., Dondorp A.M., Faiz A., Mishra S., Hien T.T (2012) New global estimates of malaria deaths. The Lancet, 380 :559-560.

72. Hay SI, Gething, PW, Snow RW (2010) India's invisible malaria burden. Lancet 376, 1716–1717.

73. Dhingra N, Jha P, Sharma VP, Cohen AA, Jotkar RM, Rodriguez PS, Bassani DG, Suraweera W, Laxminarayan R, Peto R (2010) Million Death Study Collaborators. Adult and child malaria mortality in India: a nationally representative mortality survey. Lancet. 376:1768-74.

74. Haque U., Overgaard H.J., Clements A.C.A., Norris D.E., Islam N., Karim J., Roy S., Glass G.E (2014).Malaria burden and control in Bangladesh and prospects for elimination: An epidemiological and economic assessment. The Lancet Global Health 2 (2) e98-e105.

75. James WD, Berger TG et al. (2006). Andrews' Diseases of the Skin: clinical Dermatology. Saunders Elsevier. pp. 422–428.

76. Barrett, MP; Croft, SL (2012) Management of trypanosomiasis and leishmaniasis Br Med Bull 104: 175–96.

77. Desjeux P. (2001) The increase of risk factors for leishmaniasis worldwide. Trans. Royal Soc. Trop. Med. Hyg. 95: 239–43.

78. Teixeira DE, Benchimol M, Rodrigues JCF, Crepaldi PH, Pimenta PFP, et al. (2013) The Cell Biology of Leishmania: How to Teach Using Animations. PLoS Pathogen 9(10): e1003594.

79. Alvar J, Vélez ID, Bern C, Herrero M, Desjeux P, et al. (2012) Leishmaniasis Worldwide and Global Estimates of Its Incidence. PLoS ONE 7(5): e356

80. Telford 3rd S.R, Gorenflot A, Brasseur P, Spielman, A.(1993) Babesial infection in human and wildlife. In: Kreier, J.P. (Ed.),Parasitic Protozoa, vol. 5. Academic Press, San Diego, CA, pp. 1–47.

81. Babes V (1888) Sur l'hemoglobinurie bacterienne du boeuf. Compt. Rend. Acad. Sci. Ser. III Sci. Vie. 107, 692–694.

82. Oliveira G, Rodrigues NB, Romanha AJ, Bahia D (2004) Genome and Genomics of Schistosomes. Canad. Jr. Zool. 82: 375–90.

83. Vos T, Flaxman AD, Naghavi M, Lozano R, Michaud C, Ezzati M, Shibuya K, Salomon JA et al. (2012). Years lived with disability (YLDs) for 1160 sequelae of 289 diseases and injuries 1990-2010: a systematic analysis for the Global Burden of Disease Study 2010. Lancet 380 (9859): 2163–96.

84. Agatsuma T (2003) Origin and evolution of Schistosoma japonicum. Parasitol Int 52:335-340.

85. Roberts LS, Janovy J. Nematodes: Filaroidea, the filarial worms (1996) In: Schmidt GD, Roberts LS. Foundations of Parasitology 5th. Edition, Boston. WCB Publishers.

86. McCarthy J. (2000) Diagnosis of lymphatic filarial infection. In:Lymphatic Filariasis, ed. Nutman TB, Imperial College Press, London, 127-41.

87. Weil G, Lammie PJ ,Weiss N. (1997) The ICT filariasis test: A rapid format antigen test for diagnosis of bancrofitian filariasis. Parasitol Today 13: 401-4.

88. McCarthy J. Diagnosis of lymphatic filarial infection. In:Lymphatic Filariasis, ed. Nutman TB, Imperial College Press, London, 2000; 127-41.

89. Baird JB, Charles JL, Streit TG, Roberts JM, Addiss DG, Lammie PJ (2002) Reactivity to bacterial, fungal, and parasite antigens in patients with lymphedema and elephantiasis. Am J Trop Med Hyg.66:163-9

90. Lamb TJ, Le Goff L, Kurniawan A, Guiliano DB, Fenn K, Blaxter ML, et al. (2004) Most of the response elicited against Wolbachia surface protein in filarial nematode infection is due to the infective larval stage. J Infect Dis.189:120-7.

91. Ottesen EA. (1992) The Wellcome Trust Lecture. Infection and disease in lymphatic filariasis: an immunological perspective. Parasitology. 104 Suppl:S71-9.

92. Dasgupta A. (1984) Cursory overview of lymphatic filariasis. Ind J Pathol Microbiol 27: 273-280.

93. Melrose WD (2002) Lymphatic filariasis – new insights into an old disease.32: 947-960.

94. Witt C, Ottesen EA. Lymphatic filariasis: an infection of childhood (2001) Trop Med Int Health. 6:582-606.

95. Global Programme to eliminate lymphatic filariasis: progress report on mass drug administration, 2010.(2011) Wkly Epidemiol Rec.86:377.

96. Ramaiah KD, Das PK, Michael E, Guyatt H (2000) The economic burden of lymphatic filariasis in India. Parasitol Today.16:251.

97. Babu BV, Swain BK, Rath K.(2006) Impact of chronic lymphatic filariasis on quantity and quality of productive work among weavers in an endemic village from India. Trop Med Int Health. 11:712.

98. Babu BV, Swain BK, Rath K (2006) Impact of chronic lymphatic filariasis on quantity and quality of productive work among weavers in an endemic village from India. Trop Med Int Health.11:712.

99. WHO: Lymphatic filariasis infection and disease: Control strategies. Report of a consultative meeting held at the Universiti Sains Malaysia, Penang, Malaysia. Geneva: World Health Organization, mimeographed document WHO/TDR/CTD/FIL/PENANG 1994;94:1-30.

100. Pani SP, Yuvaraj J, Vanamail P, Dhanda V, Michael E, Grenfell BT, et al. (1995) Episodic adenolymphangitis and lymphoedema in patients with bancroftian filariasis. Trans R Soc Trop Med Hyg 89:72-4

101. Surendran K, Pani SP, Soudarssanane MB, Srinivasa DK, Bordolai PC, Subramanian S (1996) Natural history, trend of prevalence and spectrum of manifestations of bancroftian filarial disease in Pondicherry (South India). Acta Trop 61:9-18.

102. Pani SP, Das LK, Balakrishnan N, Sadanandane C, Rajavel AR, Subramanian S, et al. (1989) A study on the clinical manifestations of Bancroftian filariasis in Pondicherry. Indian Med Gazette CXXII:111-5.

106. Sabesan S, Vanamail P, Raju K, Jambulingam P.(2010) Lymphatic filariasis in India: Epidemiology and control measures. J Postgrad Med 56:232-8.

104. Andrade LA, Medeiros Z, Pires ML, et al.(1995) Comparative efficacy of three different diethylcarbamazine regimens in lymphatic filariasis. Trans Royal Soc Trop Med Hyg 89:319-21.

105. Emilio Palumbo (2008) Filariasis: diagnosis, treatment and prevention. Acta Biomed 279: 106-109.

106. Shenoy RK, John A, Babu BS, et al. (2000) Two-year follow-up of the microfilaraemia of asymptomatic Brugian filariasis, after treatment with two, annual, single doses of invermectin, diethylcarbamazine or albendazole in various combinations. Ann Trop Med Parasitol 94: 607-14.

7

Nervous System Infections

Chapter Outline

7.1 Bacterial infections
7.2 Viral infections
7.3 Fungal infections
7.4 Protozoal infections

7.1 Bacterial Infections of the Nervous System

Some of the most devastating infectious diseases are those that affect the vital organs like brain and spinal cord of our nervous system. Any damage to these organs due to infections can lead to serious consequences like deafness, blindness, learning disabilities, paralysis and even death. Both these organs are protected from pathogens that may be circulating in the blood stream because of blood-brain barrier. Only pathogens of special virulence characteristics can invade this defence barrier with serious consequences. Drugs used for treatment can not cross the blood-brain barrier unless they are lipid soluble.

7.1.1 Meningitis–Mode of Action, Laboratory Diagnosis and Prevention

An inflammation of the meninges is called meningitis. Meningococcal disease manifests most commonly as meningitis and/or as bacteremia. In some occasions, pneumonia and septic arthritis are encountered. Invasive meningococcal infections are caused by the bacterium *Neisseria meningitidis*, a gram-negative diplococcus. Humans are known to be the only reservoir of *N. meningitidis*.

Pathogenesis

Meningitis represents an infection of the membrane covering the brain and spinal cord. The typical symptoms of meningitis include the sudden onset of high fever and headache. Infected fluid from the meninges passes into the spinal cord. This leads to symptoms including stiff neck, fever and rashes. Inflammation of the meninges affects the function of central nervous system. Meningococcal meningitis is a consequence of bacteria entering the cerebrospinal fluid (CSF) and affect the meninges - the membranes that line the brain and spinal cord.

Meningococcal septicaemia or meningococcaemia, is defined as a form of sepsis. The causative organism is *Neisseria meningitidis*[1] (Fig. 7.1). It is caused by the release of toxins into the blood that break down the walls of blood vessels. Often a rash develops under the skin due to blood leakage that may leave red or brownish pin prick spots.

There are at least 13 serogroups of *N. meningitidis,* classified according to the immunologic reactivity of their polysaccharide capsules (A, B, C, D, 29E, H, I, K, L, W-135, X, Y, and Z). Of these, five are clinically relevant, being responsible for nearly all disease worldwide (A, B, C, Y, and W-135). The incubation period can vary from one to ten days; the average incubation period is two to four days. The patient remains infectious as long as meningococci are present in respiratory/ oral secretions:

Modes of Transmission

The principal route of meningococcal transmission is person-to-person contact via respiratory droplets or oral secretions from an asymptomatic carrier or individual with invasive disease. The bacteria may also get transmitted by a vehicle contaminated with saliva (*e.g.,* a cigarette, food utensils, or water bottle) however the risk of such transmission is very low. Up to 15 percent of the population may be asymptomatic nasopharyngeal carriers of strains of *N. meningitidis.* Less than 1 percent of these individuals will progress to invasive disease.

Diagnosis

The gold standard of diagnosis is isolation of *N. meningitidis* from sterile body fluid. Initial diagnosis of meningococcal meningitis can be made by clinical examination followed by a lumbar puncture showing a purulent spinal fluid. The bacteria can sometimes be seen in microscopic examination of the spinal fluid. Diagnosis relies on culturing the organism on a chocolate agar plate or a blood agar plate enriched with a growth supplement. Further biochemical analysis can differentiate at species level. All Neisseria show a positive reaction when tested for oxidase. When test for carbohydrate utilization is carried out, *N. meningitidis* will oxidize the glucose and maltose but not sucrose. The diagnosis is supported or confirmed by growing the bacteria from specimens of spinal fluid or blood, by agglutination tests or by polymerase chain reaction (PCR). The disease has a high fatality risk approaching 15 percent within 12 hours of infection therefore, it is crucial to initiate testing as quickly as possible. It is advised to initiate antibiotic therapy instead of waiting for the results of the identification tests. The identification of the serogroups and susceptibility testing to antibiotics are important to define control measures.

Epidemiology

Many countries in the developing world suffer from repeated epidemics of meningococcal disease. Annual attack rate is estimated to 1 to 3 per 100,000 of the population. In temperate regions, an increase in the number of cases observed during winter and spring. Serogroups B and C together account for a large majority of cases in Europe and the Americas. Major African epidemics are associated with *Neisseria meningitidis* serogroup A. Even in Asian countries the same serogroup is the cause of meningococcal disease.

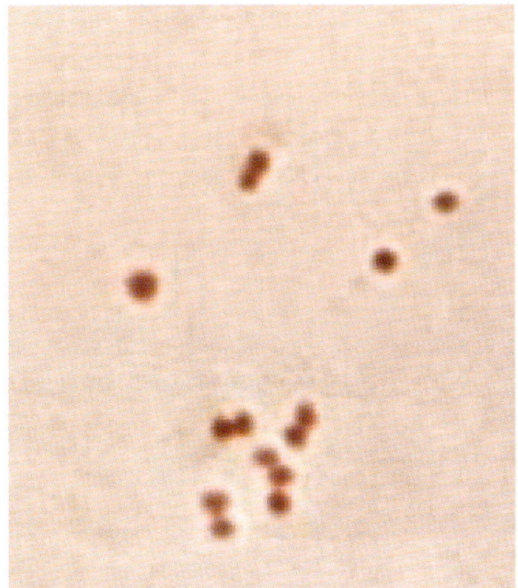

Fig. 7.1: *Neisseria meningitidis* Gram–ve diplococcal Bacteria

In major African epidemics, the attack rate ranges from 100 to 800 per 100,000 population, but individual communities have reported an average rate of 1 per 100. In comparison, the average attack rate in industrialized nations is estimated as 1 to 3 per 100,000 population.

Meningococcal meningitis occurs in small clusters throughout the world with seasonal variation and accounts for a variable proportion of epidemic bacterial meningitis. The largest burden of meningococcal disease occurs in an area of sub-Saharan Africa known as the meningitis belt, which extends from Senegal in the west to Ethiopia in the east. The risk of meningococcal disease increases during dry season when dusty winds blow due to which upper respiratory tract gets infected, making an individual more prone to the disease.

Several risk factors have been associated with the development of epidemics in the meningococcal belt. They include: 1. Immunological susceptibility of the population, 2. Demographic conditions like travel or displacement of population from one place to the other, 3. Socioeconomic conditions like poor sanitation, overcrowding 4. Climatic conditions like sudden drought and dust storms etc.

Indian Scenario

Incidence of endemic meningococcal disease in India is low, but that occasional epidemics of meningococcal disease have been recorded for a long time. The larger epidemics have affected mainly the cities of northern India and have almost universally been caused by meningococci belonging to serogroup A[2] (Fig.7.2). These epidemics have showed some characteristic feature, including a marked seasonality, which are similar to those of epidemic meningococcal A disease observed in Africa.

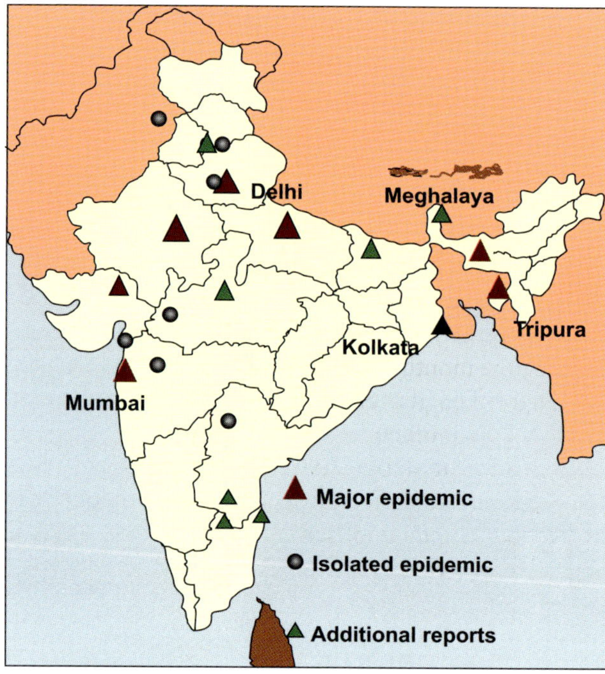

Fig. 7.2: The distribution of Indian epidemics over time and place (1910-2000) Adapted from Sinclair et al, 2010[2]

Since 1930s, India has experienced three documented periods of increased meningococcal disease with cases always reported first in New Delhi with an apparent periodicity of approximately 20 years, and with small isolated outbreaks in between[3].

Treatment and Prevention

Meningococcal disease is potentially fatal and should always be considered as a medical emergency. Children 2–10 years of age are at high risk for meningococcal disease due to certain chronic medical conditions and those who travel to or reside in countries with hyperendemic or epidemic meningococcal disease should receive primary immunization. A patient could receive treatment with a number of effective antibiotics. A range of antibiotics can treat the infection, including penicillin, ampicillin, chloramphenicol and ceftriaxon [http://www.who.int/mediacentre/factsheets/fs141/en/]. Appropriate treatment can reduce most common types of bacterial meningitis. With treatment, the risk of death from meningitis reduces below 15 percent, although the risk is higher among the elderly.

There are three vaccines available for the prevention of meningitis. The most common vaccine is Meningococcal conjugate vaccine (MCV4) is normally administered. This vaccine is licensed as Menactra (sanofi pasteur) and Menveo (Novartis) and is administered in the age group between 2-55 years. Another vaccine known as Meningococcal polysaccharide vaccine (MPSV4; Menomune [sanofi pasteur]), is administered in the age group above 55 years. These vaccines are 85 to 100 percent effective in preventing all the four kinds of meningococcus bacteria belonging to types A, C, Y, W-135. Nearly, 70 percent of the diseases are caused by these four types. However these vaccines fail to prevent the infection caused by type B that accounts for about one-third of cases in adolescents. Meningitis A vaccine (MenAfriVac) manufactured by Serum Institute of India, Pune will be introduced as a part of the routine immunization program. The vaccine has been successfully implemented in those aged 1-29 years. This vaccine has been now been approved by WHO in January, 2015 for use in infants. The vaccine had a dramatic impact in breaking the cycle of meningitis A epidemic in sub-Saharan African populations when introduced four years back. The WHO has approved the use of a 5 microgram dose of the vaccine for children, which will be administered in their nine months along with a vaccine against measles. A booster dose is recommended when the child is 12-18 months old to confer lifelong protection. MenAfriVac manufactured in Serum Institute, India has an additional advantage since it is an heat stable vaccine therefore it can remain outside the cold chain at temperatures 40°C for up to four days without any change in the potency [http://www.thehindu.com/sci-tech/science/meningitis-indian-vaccine-will-protect-infants-also/article6789358.ece]

7.1.2 Mycobacterium Leprae

Mycobacterium leprae is the causal organism for leprosy or Hansen's disease. Cases of leprosy have been recorded as early as 600 B.C. The organism was first discovered by the Norwegian physician Gerhard Armauer Hansen in 1873, while he was culturing the bacteria present in the skin nodules of patients with leprosy. In 2004, The World Health Organization reported that there were 407,791 new cases of leprosy. Leprosy is a serious world issue in countries like Brazil, India, Democratic of Congo, Tanzania, Nepal, Mozambique, Madagascar, Angola and the Central African Republic. *Mycobacterium leprae* is considered as an intracellular bacterium that may infect different types of cells like nerve, skin and mucosal cells.

Pathophysiology

Leprosy is a chronic disease and clinically characterized by one or more of the three important symptoms: 1. hypopigmented or erythematous skin patches with definite loss of sensation, 2. thickened peripheral nerves, and 3. detection of acid-fast bacilli on skin smears or biopsy material. *M. leprae* primarily infects the principal glia of the peripheral nervous system (Schwann cells) leading to nerve damage and finally leads to the development of disabilities. Optical nerves may get damaged leading to blindness. In case the patient is not treated in time, leprosy causes blindness in 3.2 percent of patients[4]. The disabilities of the eyes, hands, and feet due to neuropathy are often not reversible and may require lifelong care and rehabilitation. In addition to humans, leprosy has also been observed in nine-banded armadillos and three species of primates[5].

This bacterium cannot be cultured in vitro conditions, therefore in the laboratory, *Mycobacterium leprae* is cultured on the feet of mice. Mycobacteria are known for their notoriously slow growth with the doubling time of 14 days. *M. leprae* has not yet been successfully cultured in vitro. Mycobacteria have a unique lipid that makes up their membranes that gives them their unique characteristic. The mycolic acids are very large lipids chains ranging from 60 to 80 carbons long. Covalent bonds link these lipids to one another forming a very thick waxy coat. The optimum temperature of growth is recorded at 30°C. The bacilli tend to grow mainly in the extremities of the body. The bacteria can survive ingestion by the macrophages and eventually invades cells of the myelin sheath of the peripheral nervous system. The destruction of Schwann cells by immune-mediated reactions cause nerve damage.

With the recent progress in the molecular biology techniques, molecular markers have been developed for faster diagnosis of leprosy. The genome of *M. leprae* has been sequenced in totality. Interestingly, the genome presents less than 50 percent coding capacity with a large number of pseudogenes. *M. leprae* genes help to define the minimal gene set necessary for in vivo survival of this mycobacterial pathogen as well as genes potentially required for infection and pathogenesis seen in leprosy[6] .

Classification of Disease and Host Response

Patients are classified as having either paucibacillary or multibacillary Hansen's disease. Paucibacillary Hansen's disease is milder and characterized by one or more hypopigmented skin macules. Multibacillary Hansen's disease is associated with symmetric skin lesions, nodules, plaques, thickened dermis and frequent involvement of the nasal mucosa resulting in nasal congestion and epistaxis. Signs of leprosy are painless ulcers, skin lesions of hypopigmented macules (flat, pale areas of skin), and eye damage due to dryness with reduced blinking. At a later stage, large ulcerations, loss of digits, skin nodules, and facial isfigurement may develop (Fig. 7.3)

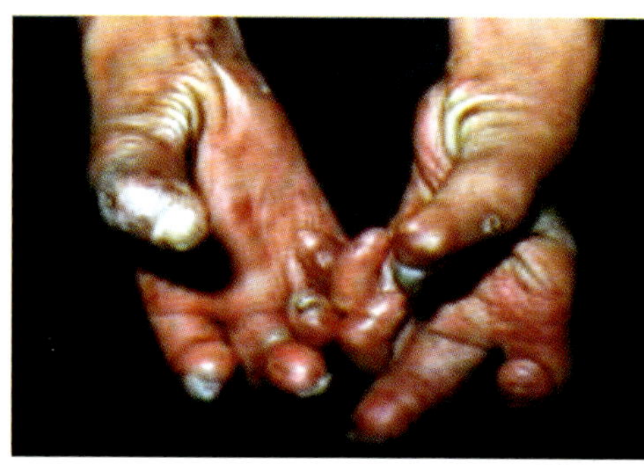

Fig. 7.3: Lepromatous (progressive) leprosy: Deformation of hand into clawed form and considerable necrosis of tissue occurs

With the progression of the disease, transition between different clinical forms can be identified. Clinical, histopathological, and immunological criteria identify five forms of leprosy: tuberculoid polar leprosy (TT), borderline tuberculoid (BT), midborderline (BB), borderline lepromatous (BL), and lepromatous polar leprosy (LL). On the basis of the progression of the disease, patients are divided into two groups for therapeutic purposes: paucibacillary (TT, BT) and multibacillary [midborderline (BB), BL, LL][7].

Polyneuritic Leprosy (PL) is a separate type, manifesting with only neural signs without any evidence of skin lesions. This form is mostly well recognized in the Indian subcontinent. This form is prevalent both in males and females of all age groups. The affected nerves are thickened or may become tender, or both. Localized involvement of the nerves lead to nerve abscesses[8] .The innate immune response to the infection is inflammation. The inflammation in the neural cell causes neuropathy and serious nerve damage. Nerve damage doesn't occur until an immune response happens. The nerve damage quickly progresses to numbness, paralysis and deformity.

Human genetic factors influence the acquisition of leprosy and the clinical course of disease. Single-nucleotide polymorphism (SNP) association studies showed a low lymphotoxin-α (LTA)-producing allele as a major genetic risk factor for early onset leprosy[9].

Transmission

The exact transmission pattern of the leprosy bacillus is uncertain but patients with leprotamous leprosy shed large numbers of bacteria in the environment through their nasal secretions and the exudates (oozing matter) of their skin lesions.

Epidemiology

Of all leprosy patients, lepromatous ones (LL/BL), who have a large load of *M. leprae* (AFB), are considered more infectious. There are several notable challenges that limit the study of *M. leprae* by molecular epidemiology. There is a substantial time period, as long as 2–3 years, between infection and the manifestation of leprosy symptoms making it difficult to identify the precise time and source of the primary infection. Though *M. leprae* is considered to be primarily a parasite of humans, however some evidences have indicated possible non-human sources of the agent. Thus it has been suggested that *M. leprae* may be harbored in soil, in water, on plants, or in various animal species including amoeba, insects, fish, primates and armadillos[10]. If non-human sources exist, their recognition may help to explain patterns of infection and disease in human populations.

It is now known, out of the abbreviated genome of *M. leprae*, it is most unlikely that leprosy bacilli can replicate in any extra-cellular environment[11]. Molecular epidemiological studies are called for to resolve the issue of multiple reservoirs of *M. leprae*. Recently, a cost-effective panel of variable-number tandem-repeat (VNTR) markers and SNP analysis has been developed. Use of this panel has allowed the genotyping of 475 *M. leprae* strains from six different countries[12]. It would likewise benefit establishing a culture of data sharing and maintenance of a publicly accessible strain typing database.

Indian Scenario

Currently, seven states in India, namely Bihar, Chhattisgarh, Jharkhand, Maharashtra, Orissa, Uttar Pradesh and West Bengal contribute to 73.03 percent of the total cases in India and 74.9 percent of the new cases[13]. In India, during the 1980s prior to the introduction of multidrug therapy, the southern states of Tamil Nadu and Andhra Pradesh had the highest prevalence. Today, the endemic and hyperendemic areas are located in the northern and eastern part of India. Several reasons could be cited for south Indian states to have reached elimination faster: earlier implementation of multi drug therapy (MDT) and better coverage, and timely release from treatment (RFT) are the major contributing factors[14].

Table 7.1: Detection of new cases in leprosy in 2007 and the number of new cases detected previously (adapted from Prasad et al, 2010[15])

	1993	2002	2003	2004	2005	2006	2007
India	456,000	473,658	367,143	260,063	169,709	139,252	137,685
Global	590,933,	620,638	514,718	407,791	299,036	250,017	254,525

There is a dramatic fall in the prevalence rate (PR) of leprosy, but the new case-detection rate (NCDR) has not been reduced concomitantly. It is the operational efficiency of National Leprosy Eradication Programme (NLEP) that has led to a significant reduction in the NCDR in Andhra Pradesh and Tamil Nadu[16].

Table 7.2: Global leprosy situation at the beginning of 2008. South-east Asia statistics is given below [Available from: http://www.who.int/wer][17]

	Registered prealence	No. of new cases detected –2007	No. of new MB cases	No. of new female cases	No. of new cases among childred	No. of new cases with grade-2 deformity	No. of relapses 2007
India	87,228	137,685	64,980	47,537	12,976	3,477	—
Total	120,967	171,552	87,532	53,445	15,763	6,332	119

Percent of cure rates for Paucibacillary (PB) and Multibacillary (MB) are not known.

Treatment and Prevention

Multidrug therapy (MDT) combining all three drugs-clofazimine, rifampicin and dapsone was first recommended by a WHO Expert Committee in 1981. These three anti-leprosy drugs are still used in the standard MDT regimens. None of them is used alone because of the risk of developing resistance[17]. The development of simulation models by Meima et al (1999)[18] have been useful in understanding the multiple factors involved in both the transmission as well as incidence of the disease. They utilized seven leprosy intervention scenarios with a micro simulation model (SIMCOLEP) to predict future leprosy trends. The baseline scenario consisted of passive case detection, MDT, contact tracing and BCG vaccination of infants.

Close contacts with patients with untreated, active, predominantly multibacillary disease, and persons living in countries with highly endemic disease are the major risk factors for the disease. On the basis of earlier studies on the prevention of deformities following recommendations were made[19]. Steroids were recommended to treat reactions and nerve function impairments (NFIs) of recent onset; the expected recovery rate for NFI is approximately 60 percent.

1. MB patients and those with existing NFIs would be carefully monitored for new nerve function loss, as they are the groups at the greatest risk.
2. Assessment of nerve function using standard methods every month during MDT is recommended.

7.1.3 Tetanus

Tetanus also known as 'lock jaw'is a medical condition characterized by a prolonged contraction of skeletal muscle fibers. Tetanus is often associated with rust, especially rusty nails. Objects that accumulate rust are often found outdoors, or in places that harbor anaerobic bacteria. The disease causing organism is a Gram-positive, rod-shaped, endospore-forming, obligate anaerobic bacterium *Clostridium tetani*. The patient suffers from tetanospasmin, due to the production of a neurotoxin by the bacterium. The rough surface of rusty metal merely provides a prime habitat for *C. tetani* endospores to reside in, and the nail is responsible in the formation of a wound thus help to deliver endospores deep within the body at the site of the wound. The endospores can survive for a long time in oxygen free environment but begins to metabolise and cause infection once they encounter an adequate environment.

Symptoms

Tetanus often begins with mild spasms in the jaw muscles due to which the jaws get locked and thus the mouth fails to open, a stage also defined as 'trismus'. The spasms can also affect other organs like the chest, neck, back, abdominal muscles, and buttocks. Tetanus affects skeletal muscle, a type of striated muscle used in voluntary movement. Sometimes the spasms affect inspiratory and expiratory muscles that help with breathing thus causing breathing problems during attack. Because of the prolonged muscular action, the patient suffers from sudden, powerful, and painful contractions of muscle groups, which is called "tetany". Repeated episodes results in fractures and muscle tears. Other symptoms of the disease are drooling, excessive sweating, fever, elevated blood pressure, hand or foot spasms. The patient also suffers from swallowing difficulty, frequent episode of rapid heart rate and uncontrolled urination or defecation.

The frequency of the spasms may last for several minutes. Due to frequent spasms, the body gets deshaped into a characteristic form called opisthotonos. Spasms may continue for several weeks and recovery is slow and often may take months. Sympathetic over activity (SOA) is common in severe tetanus and manifests as labile hypertension, tachycardia, dysrhythmia, peripheral vasculature constriction, profuse sweating, fever, increased carbon dioxide output, increased catecholamine excretion and late development of hypotension. In severe cases, death may occur within four days (Figure 7.4)

Fig. 7.4: An advanced case of tetanus. Spasms known as ophisthotonos results in a fractured spine

Transmission

Tetanus is infectious but not contagious. The disease does not transmit from person to person. The disease can be prevented by administration of vaccine. Since *C. tetani* spores are ubiquitous in nature, tetanus is considered as an international health problem. The disease occurs almost exclusively in persons without vaccination or inadequately immunized[20]. The *C. tetani* spores inhabit more often hot, damp climates and soil rich in organic matter. This is particularly true with manure-treated soils, as the spores are widely distributed in the intestines and feces of many animals such as horses, sheep, cattle, dogs, cats, rats, guinea pigs, and chickens. Spores can be introduced into the body through any wound.

Pathogenesis

Tetanus affects skeletal muscle, a type of striated muscle used in voluntary movement. The tetanus toxin initially binds to peripheral nerve terminals. It is transported within the axon and across synaptic junctions until it reaches the central nervous system. There it becomes rapidly fixed to gangliosides at the presynaptic inhibitory motor nerve endings, and is taken up into the axon by endocytosis. The effect of the toxin is to block the release of inhibitory neurotransmitters glycine and gamma-Aminobutyric acid (GABA) across the synaptic cleft, which is required to check the nervous impulse. If nervous impulses cannot be checked by normal inhibitory mechanisms, the generalized muscular spasms characteristic of tetanus are produced. The toxin appears to act by selective cleavage of a protein component of synaptic vesicles, synaptobrevin II, and this prevents the release of neurotransmitters by the cells[21].

Epidemiology

Tetanus occurs worldwide but is most frequently encountered in densely populated regions of tropical countries. According to the report of 2010, it caused about 61,000 deaths whereas the figure was 272,000 in 1990[22]. Tetanus particularly in the neonatal form, remains a significant public health problem in non-industrialized countries. As per report, 59,000 newborns worldwide died in 2008 as a result of neonatal tetanus[23].

Treatment and Prevention

Tetanus toxoid was first produced in 1924, and tetanus toxoid immunizations were used extensively in the armed services during World War II. Tetanus cases among this population declined from 70 in World War I (13.4/100,000 wounds and injuries) to 12 in World War II (0.44/100,000). Of the 12 case-patients, half had received no prior toxoid. Almost all reported cases of tetanus were the individuals who were either not vaccinated or were partially immunized without the required booster dose in the preceding 10 years. The tetanus vaccine is a toxoid, an inactivated toxin that stimulates the formation of antibodies that neutralize the toxin produced by the bacteria.

Effective vaccines for tetanus have been available since the1940s.The routine vaccination of children includes combination of vaccines used against diphtheria, tetanus and pertussis: DTaP, Tdap, DT, and Td. Two of these (DTaP and DT) are given to children younger than 7 years of age, and two (Tdap and Td) are given to older children and adults. Children should get 5 doses of **DTaP,** one dose at each of the following ages: 2, 4, 6, and 15-18 months and 4-6 years. **DT** does not contain pertussis, and is used as a substitute for DTaP for children who cannot tolerate pertussis vaccine. Often, one encounters minor reactions due to the vaccine, including pain and redness at the injection site, headache, fatigue or a feeling of discomfort for some time.

Being fully immunized is the best tool to prevent tetanus. Tetanus vaccines are recommended for people of all ages, with booster shots throughout life. Worldwide, there are estimated 1 million cases annually with at least half occur in newborns. In many parts of the world, the severed umbilical cords of infants are dressed with materials such as soil, clay and even cow dung. In such cases the mortality rate from tetanus is about 50 percent in developing areas.

Tetanus is a medical emergency requiring:

- Hospitalization and breathing aid if required
- Immediate treatment with human tetanus immune globulin (TIG) (or equine antitoxin)
- Tetanus vaccine
- Drugs to control muscle spasms
- Aggressive wound care
- Antibiotics

7.1.4 Botulism

Botulism was first described as a clinical disease in the early 1800s, when it was known as a sausage disease. It is a form of food poisoning and is caused by a neurotoxin produced by the bacterium *Clostridium botulinum*, an obligately anaerobic, endospore-forming gram-positive rod, found mostly in soil, and many aquatic sediments. It is a rare and potentially fatal paralytic illness.

Signs and Symptoms

Symptoms typically appear within a day or two. The muscle weakness of botulism characteristically starts in the muscles supplied by the cranial nerves. A group of twelve nerves controls various functions like eye movements, the facial muscles and the muscles controlling chewing and swallowing. Double vision, drooping of both eyelids, loss of facial expression and swallowing problems may therefore occur, as well as difficulty with talking (Fig. 7.5). The weakness then spreads to the arms starting in the shoulders and proceeding to the forearms and legs again from the thighs down to the feet.

Individuals suffering from botulism undergo a progressive flaccid paralysis for 1 to 10 days. Severe botulism leads to difficulty in breathing (dyspnea) in some cases can lead to respiratory failure. Often it also results in a depressant effect on the brain. This may lead to coma and eventually death if untreated. In addition to affecting the voluntary muscles, it can also cause disruptions in the autonomic nervous system. The patient experiences dryness in mouth due to decreased production of saliva, postural hypotension. He patient also suffers from decreased blood pressure on standing with resultant light headedness and often faces risk of blackouts and eventually constipation due to decreased peristalsis[22].

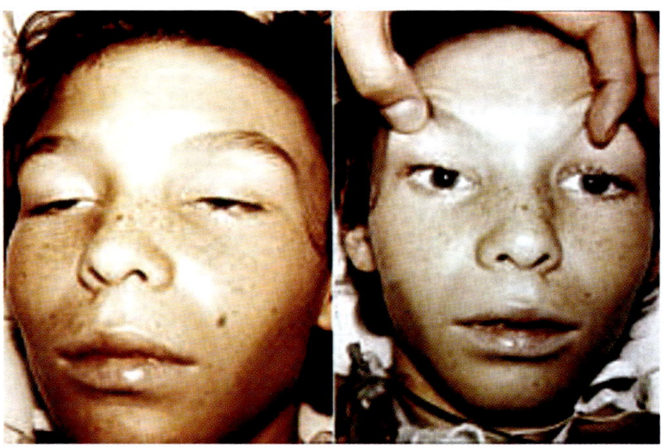

Fig. 7.5: A 14-year-old with botulism showing the symptoms of weakness of eye muscles and the drooping eyelid

Source: en.wikipedia.org/wiki/Botulism

Diagnosis

The disease is diagnosed by inoculating mice with samples from patient serum, stool or vomitus specimens. Different sets of mice are immunized with type

A, B, or C or E antitoxin. All the mice are then inoculated with the test toxin and their survival rate is counted. The toxin in food can similarly be identified by mouse inoculations. Botulinum toxin can also be detected by a variety of techniques, including enzyme-linked immunosorbent assays (ELISAs) and electrochemiluminescent (ECL) tests.

Mechanism of Action

The botulinum toxin is a protein produced under anaerobic conditions by the bacterium *Clostridium botulinum*. Botulinum toxin is broken into 8 neurotoxins (labeled as types A, B, C [C1, C2], D, E, F, G & H), which are antigenically and serologically distinct but structurally similar. Human botulism is caused mainly by types A, B, E, and rarely F. Types C and D are known to cause toxicity only in other animals.

Some types produce a characteristic putrefactive smell and digest meat (types A and some of B and F); these are said to be proteolytic; type E and some types of B, C, D and F are non proteolytic and can go undetected because there is no strong odor associated with them. Type A toxin is probably the most virulent. In untreated cases, the mortality rate is 60-70%. The type A endospore is the most heat-resistant of all *C. botulinum* strains.

When the bacteria are under stress, they develop spores, which are inert. Their natural habitats are in the soil, in the silt that comprises the bottom sediment of streams, lakes and coastal waters and ocean, while some types are natural inhabitants of the intestinal tracts of mammals (*e.g.,* horses, cattle, humans), and are present in their excreta. The spores can survive in their inert form for many years

Toxin is produced by the bacteria when environmental conditions are favorable for the spores to replicate and grow, but the gene that encodes for the toxin protein is actually carried by a virus or phage that infects the bacteria. Unfortunately, little is known about the natural factors that control phage infection and replication within the bacteria.

The spores require warm temperatures, a protein source, an anaerobic environment, and moisture in order to become active and produce toxin. In the wild, decomposing vegetation and invertebrates combined with warm temperatures can provide ideal conditions for the botulism bacteria to activate and produce toxin that may affect feeding birds and other animals. Spores are not killed by boiling. However, fortunately botulism is uncommon because special, rarely obtained conditions are necessary for botulinum toxin production from *C. botulinum* spores. The physiological conditions include anaerobic, low-salt, low-acid, low-sugar environment at ambient temperatures[24]. Botulinum inhibits the release of acetylcholine within the nervous system. This chemical is required to form a bridge across synapses where nerve cell axons and dendrites connect with each other.

Kinds of Botulism

Infant botulism: The most common form encountered in Western countries is infant botulism. This occurs in small children who are colonized with the bacterium during the early stages of their lives. The bacterium then releases the toxin into the intestine, which is absorbed into the bloodstream. The consumption of honey during the first year of life has been identified as a risk factor for infant botulism. Lethal dose has been described as few as 2000 bacteria.

Wound botulism: Wound botulism results if the wound is contaminated with the bacteria. These bacteria then secrete the toxin into the bloodstream. This has been found to be common method of contracting with the disease during infected intravenous injections by drug users since 1990. This is especially observed in people using contaminated black tar heroin and those injecting heroin into the skin rather than the veins.

Food botulism: Foodborne botulism results from contaminated food in which *C. botulinum* spores have been allowed to germinate in low oxygen conditions. This typically occurs in home-canned food substances and fermented uncooked dishes. Symptoms usually appear 12–36 hours after eating, but can also appear within 2 hours to 10 days. Improperly preserved food is the most common cause of food-borne botulism. Consumption of pickled fish preserved in low saline and smoked fish stored at high temperature are also potential source of infection. Intoxication occurs as a result of ingesting **pre-formed** bacterial toxins in food. Immunocompromised individuals, young children, and elderly individuals may suffer from more serious symptoms.

Treatment and Prevention

The treatment of botulism relies heavily on supportive care. Antibiotics cannot be administered for treatment because the toxin is preformed. Antitoxins aimed at neutralizing A,B and E toxins are available and are usually administered together. The deadly toxin of botulism (Botox) finds some therapeutic uses for a number of medical conditions, such as chronic headaches. It is also useful for relieving painful muscle contractions in cases like cerebral palsy, Parkinson disease and multiple sclerosis. Injections in the area of facial wounds prevent muscle movement during healing. It has been approved to control involuntary eye lid twitching (blepharospasm), crossed eyes (strabismus) and also prevents armpit sweating.

While the botulinum spores can survive in boiling water, the toxin is heat-labile, meaning that it can be destroyed at high temperatures. Heating food to a typical cooking temperature of 80°C (176°F) for 10 minutes before consumption can greatly reduce the risk of foodborne illness. Since temperatures higher than boiling (212°F) are necessary to kill the spores, a pressure cooker is suggested for home-canning purposes (CDC 2014).

7.2 Viral Infections of the Nervous System

1. Encephalitis
2. Poliomyletis
3. Human rabies

7.2.1 Encephalitis

It refers to brain inflammation caused by a virus. It is a rare disease that occurs in approximately 0.5 per 100,000 individuals. The disease is prevalent in children, the elderly, and immunocompromised people like those with HIV/AIDS or cancer. During the disease condition, the patient shows symptoms of loss of function of the brain and spinal cord. The inflammation of the brain results in changes in neurological condition including mental confusion and seizures.

Pathogenesis

Most often encephalitis is caused due to a viral infection. Many types of viruses may cause this disease. Exposure to viruses can occur through:

- Respiratory droplets from an infected person
- Contaminated food or drink
- Mosquito, tick, and other insect bites
- Skin contact

Different viruses located in different locations can be responsible for transmission. Many cases will tend to cluster in a certain season. Encephalitis caused by the herpes simplex virus is the leading cause of more severe cases in all ages including newborns. Interestingly, a number of viruses for which vaccines are now available are also responsible for encephalitis. These include:

- Measles
- Mumps
- Polio
- Rabies
- Rubella
- Varicella (chickenpox)

Other viruses that are known to cause encephalitis include:

- Adenovirus
- Coxsackievirus
- Cytomegalovirus
- Eastern Equine Encephalitis Virus
- Echovirus
- West Nile virus
- Japanese encephalitis virus (JEV)

Symptoms

The virus causes inflammation of brain tissue. The swelling in the brain tissue develops leading to cerebral edema that destroys nerve cells causing bleeding in the brain (intracerebral hemorrhage) followed by brain damage. Besides low fever, headache and loss of appetite, few emergency symptoms are also noticed. They are:

- Loss of consciousness, poor responsiveness, stupor, coma
- Muscle weakness or paralysis
- Seizures
- Severe headache
- Sudden change in mental functions: Memory loss (amnesia), impaired short-term or long-term memory

Diagnosis

Clinical examination of infected individual shows: Abnormal reflexes, Mental confusion, Mouth ulcers, Muscle weakness, Neck stiffness, Skin rash and Speech problems.

Following tests are required to be done for complete diagnosis:

- Brain MRI
- CT scan of the head
- Culture of cerebrospinal fluid (CSF), blood, or urine (however, this test is rarely useful)
- Electroencephalogram (EEG)
- Lumbar puncture and CSF examination
- Serological tests that detect antibodies to a virus
- Final confirmation by subjecting virus DNA to polymerase chain reaction (PCR)

The immunoglobulin M (IgM) antibody capture ELISA (MAC ELISA) for serum and cerebrospinal fluid (CSF) has become the accepted standard for rapid diagnosis of Japanese encephalitis[25]. However, because these ELISAs require sophisticated equipment, their use has been confined largely to a few academic or referral centers. Since most patients are seen in rural hospitals with limited facilities, there is a need for a simple and reliable diagnostic test which is appropriate for such settings.

Epidemiology

Some viruses are transmitted by animals while others by human-to-human transmission only. Some are vector-borne and therefore geographically limited, while air-borne transmitted viruses are distributed worldwide. Few of them are prevented by individual or mass vaccination, and even fewer can be treated with specific drugs.

Scenario in Asian Countries

Japanese encephalitis virus (JEV): JEV is a mosquito-borne flavivirus. It is a single stranded RNA virus and shows close relationship with West Nile, St. Louis encephalitis, yellow fever, and dengue viruses. JEV is the most frequent cause of encephalitis clusters in Asian countries with an estimated 50,000 cases/year[26]. Currently, poliomyelitis is almost completely eradicated however JEV now stands as a leading cause of viral neurological infection and disability in children belonging to Asian countries. Pigs and birds are the virus reservoir, and the vectors are *Culex* mosquitoes. The virus is present from Korea to Australia and its western limit is Pakistan. The most affected countries of Asia are China, Japan, South Korea, Nepal, northern Vietnam, and northern India where temperate weather prevails with spell of warm weather followed by rains. In Asia, the disease is frequent around rice fields which are favorable sites for breeding of the vector. The disease tends to increase in water logging areas due to flooding of rivers. JEV infections are encountered almost in all seasons in South Asia, while they occur as summer outbreaks in northern areas. The virus first appeared in Australia in 1995[27]. JEV encephalitis can be identified with a unique pattern showing high frequency of seizures up to 80 percent and Parkinson like syndromes[28]. Other than the above symptoms, patients may also experience behavioral disorders. Encephalitis is not the only neurological syndrome caused by JEV but some patients develop polio-like acute flaccid paralysis as a major clinical presentation of the infection.

Indian Scenario

Currently disease is reported from the states of Andhra Pradesh, Assam, Bihar, Goa, Haryana, Karnataka, Kerala, Mahrashtra, Manipur, Tamil Nadu, Uttar Pradesh, West Bengal and Nagaland. Till 2007, 103389 JEV cases and 33729 deaths and Case Fatality Rate of 32.62 percent have been reported since 1978. Government of India launched vaccination campaign in highly endemic states of Assam, Karnataka, West Bengal and Uttar Pradesh in 2006 and in Andhra Pradesh, Bihar, Haryana, Maharashtra, Tamil Nadu in 2007 and 2008 respectively which has resulted in reduced incidence of JE in these states[29]. In India, the virus has been isolated from more than 15 species of mosquitoes belonging to genera Culex, Aedes and Anopheles. *Cx. tritaeniorhynchus* and *Cx. vishnui* are considered as the main vectors (refer to http://www.icmr.nic.in/pinstitute/niv/JAPANESE%20 ENCEPHALITIS.pdf).

Treatment and Prevention

CDC's Advisory Committee on Immunization Practices (ACIP) regarding the prevention of Japanese encephalitis (JE) among travelers, CDC recommended use of Inactivated Japanese encephalitis virus vaccine: recommendations of the Advisory Committee on Immunization Practices [ACIP]. MMWR 1993; 42 [No. RR-1]. It is recommended only for travelers from non endemic countries visiting endemic areas on a short trip.

These vaccines are not been used in endemic areas and currently the patients are provided with supportive care like rest, nutrition, sufficient intake of fluids to help the body fight the infection, and to relieve symptoms. Medications given to reduce the symptoms include:

1. Antiviral medications, such as acyclovir (Zovirax) and foscarnet (Foscavir) to treat herpes encephalitis or other severe viral infections.
2. Anti-seizure medications (such as phenytoin) to prevent seizures
3. Steroids such as dexamethasone to reduce brain inflammation
4. Sedatives to treat irritability or restlessness
5. Acetaminophen for fever and headache

If required, interventions like physical therapy and speech therapy are given when the disease is under control. Prevention from mosquito bite can reduce the transmission from an infected person to a healthy individual. Application of an insect repellant containing the chemical, DEET (N,N-diethyl-meta-toluamide) can prevent mosquito bite in case of adults but the same is not recommended for infants younger than 2 months. The Environmental Protection Agency (EPA) recommends two products-DEET and picaridin to repel mosquitoes. Removal of standing water sources such as old tires, cans, gutters can prevent the infection.

7.2.2 Poliomyletis–Pathogenesis, Epidemiology and Prevention

Poliovirus (PV) is the causal agent of paralytic poliomyelitis. This is considered as an acute disease of the central nervous system (CNS) resulting in flaccid paralysis, a condition termed as polio or infantile paralysis. It is an acute viral infectious disease and attacks only humans. It is often referred as inflammation of the grey matter of the spinal cord, although a severe infection can extend into the brainstem and even higher structures, resulting in polio *encephalitis*. Poliomyelitis was first recognized as a distinct metabolic condition by Michael Underwood in 1789. Its causative agent, poliovirus was identified in 1908 by Karl Landsteiner[30]. Children under five years of age are more prone to poliovirus attack. First major outbreaks were reported from Europe and United States in early 19th century.

Pathogenesis

The different steps involved in the pathogenesis of Poliomyelitis are represented in Fig. 7.6. The virus preferentially replicates and destroys motor neurons within the spinal cord, brain stem, or motor cortex. This leads to the development of paralytic poliomyelitis. This may manifest in different forms like the spinal, bulbar, and bulbospinal. The variations in these forms are seen based on the region of CNS affected. They also vary with the amount of neuronal damage and inflammation that occurs. Spinal polio considered to be the most common form of paralytic poliomyelitis that results from viral invasion of the motor neurons of the anterior horn cells, or the ventral (front) grey matter section in the spinal column. These neuronal cells are responsible for movement of the muscles, including those of the trunk, limbs, and the intercostal muscles. Due to the destruction of neuronal cells, lesions are formed within the spinal ganglia[31]. However spinal polio is rarely fatal. Individuals suffering from Bulbar polio face the danger of death because of respiratory involvement. The patients with this kind of paralytic polio suffer from suffocation since the muscles involved in breathing do not function. Mortality rate can be reduced by providing artificial respiratory support to the patients from time to time.

Poliovirus, the causative agent of poliomyelitis is a human enterovirus, a member of the family of Picornaviridae. Poliovirus is composed of a RNA genome enveloped in a protein coat called capsid (Fig. 7.7). The genome is a single-stranded positive-sense RNA genome that is about 7500 nucleotides long[32].

Symptoms

When a person becomes infected with poliovirus, the virus comes in contact with the cells of pharynx or intestinal mucosa. The virus enters the cell with the help of immunoglobulin like receptor known as polio virus receptor present on the cell membrane and then begins to replicate within the cells that line the back of the throat, nose, and intestines. Infected individuals develop polio symptoms like fever, sore throat, and nausea between 7 to 10 days. This period is often referred as "polio incubation period." The incubation period varies from 4 days to as long as 35 days. About 0.5 percent of infected children suffer from muscle weakness due to which they are unable to carryout normal movement of leg. The most affected organ is leg but at times it may also affect muscles of neck, head or diaphragm. In some individuals, even after recovery post polio symptoms may develop leading to further weakening of their limbs. Transmission normally takes place through fecal-oral route. Many infected persons remain asymptomatic but may shed the virus through feces and may contaminate food and drinking water if proper hygienic practices are not followed.

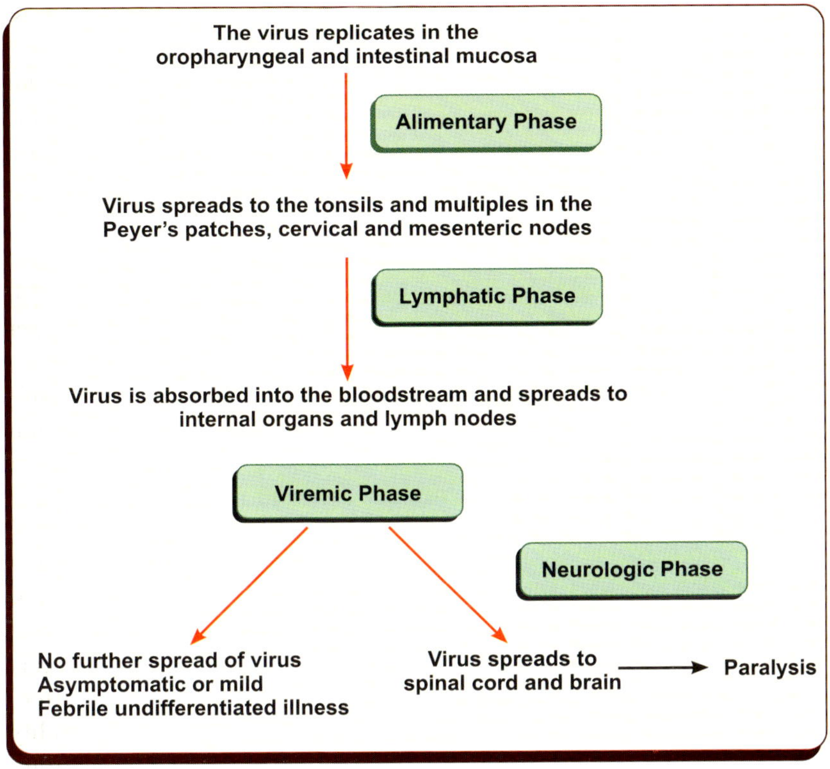

Fig. 7.6: Steps in the Pathogenesis of Poliomyletis

Diagnosis

The diagnosis of polio is a clinical one. History of exposure with no history of previous vaccination is the initial hint. Often for diagnosis, a spinal tap or a lumbar puncture known as LP is done by the insertion of a needle into the fluid within the spinal canal to obtain CSF fluid. This test is very crucial to distinguish polio from meningitis since initial symptoms are quite similar in both the cases. After that, viral cultures are carried out in the laboratory from the samples taken from throat washings, stools, or CSF fluid. Measurement of polio

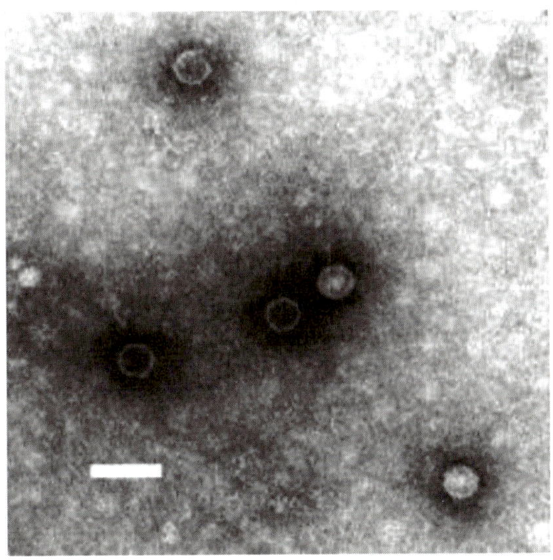

Fig. 7.7: Transmission electron micrograph of Poliovirus virions. Scale bar: 50 nm.
Source: en.wikipedia.org

antibodies is carried out for diagnostic purpose. A positive culture of poliovirus from a patient experiencing acute flaccid paralysis is further confirmed by genetic fingerprinting or more recently by PCR amplification to determine "wild type", the virus encountered in nature or "vaccine type" derived from a strain of poliovirus used for generation of polio vaccine[33]. It is important to determine the source of the virus because for each reported case of paralytic polio caused by wild poliovirus, an estimated 200 to 3,000 other contagious asymptomatic carriers exist[34].

Epidemiology

Polio was one of the most dreaded childhood diseases of the 20th century. Polio epidemics have crippled thousands of people in the past, mostly young children. Polio had existed for thousands of years quietly as an endemic pathogen. However from 1880s, major epidemics have been reported from Europe followed by the United States. By 1910, much of the world experienced a dramatic increase in polio cases and epidemics became regular events. Most of the cases were seen primarily in cities during the summer months. However, with the help of widespread polio vaccination programs, global incidence has decreased by 99 percent since 1988. The disease is currently endemic only in parts of Pakistan, Nigeria and Afganistan. In any country, a single case of poliomyelitis is now considered a public health emergency. This would initiate an extensive supplementary immunization response over a large geographic area. In 2013 the Center for Disease Control received reports of 183 cases of polio in Somalia, 14 in Kenya and eight cases in the Somali Region of Ethiopia. Due to ongoing civil war, 90 polio cases have been reported in Syria perhaps due to poor sanitary conditions and lack of safe drinking water.

Polioviruses have three antigenically distinct types, giving no cross immunity:

- Type I : "Leon"; the commonest in epidemics
- Type II : "Berlinhide"; the prevailing type in endemic areas.
- Type III : "Lansing"; occasionally causes epidemics.

Polioviruses are present in the environment in three forms:

Wild polioviruses: field isolates and reference strains of polioviruses known to have circulated persistently in the community.

Oral poliovirus vaccine strains: attenuated polioviruses approved for use in oral vaccines by national control authorities.

Vaccine derived polioviruses: mutated progeny of approved poliovirus vaccine strain

Indian Scenario

India recorded 741 cases of polio in 2009 - nearly half the number of global cases. There were few endemic pockets in the country like in the state of Uttar Pradesh and Bihar. But in a remarkable turn of events, India reported just one case of the crippling disease in 2011 and went on to record zero cases in over 12 months since then. India has reported its first case of vaccine derived polio virus (VDPV) infection of 2012 from the Murshidabad district of West Bengal. According to the current reports available, India has not reported any case of polio past three years and therefore India has been declared as a polio free country in the year 2014. [http://www.unicef.org/india/health_3729.htm]

Treatment and Prevention

There is no cure for polio, only treatment to alleviate the symptoms. Heat and physiotherapy is used to stimulate the muscles. Often, antispasmodic drugs are given to relax the muscles. These treatments can improve mobility but unable to reverse permanent polio paralysis.

Polio can be prevented through effective immunization program requiring couple of booster doses. Polio vaccine when administered following a prescribed schedule almost always protects a child for life. Two polio vaccines are commonly used throughout the world for poliomyelitis. The first was developed by Jonas Salk in 1952; the second was an oral vaccine developed by Albert Sabin. These 2 vaccines have eradicated polio from most countries and have reduced the worldwide incidence of polio.

The Salk vaccine, or inactivated poliovirus vaccine (IPV), is based on 3 virulent reference strains—Mahoney, MEF-1, and Saukett. The Salk vaccine provides immunoglobulin G-mediated immunity in the bloodstream, which prevents infection from progressing to viremia and protects the neurons. The Salk vaccine is 60%–70% effective against poliovirus 1 and is 90% effective against both poliovirus 2 and 3[35]. A major concern about OPV is its ability to revert to a form that can cause paralysis. Outbreaks of vaccine-associated paralytic poliomyelitis (VAPP) have been reported in many countries of the world[36]. In 2006, around 1600 cases of vaccine-induced polio were recorded in India, according to the Indian Medical Association Sub-Committee on Immunization report on the Polio Eradication Initiative[37]. It was noted that these cases were reported during repeated mass-immunization campaigns in which repeated doses of OPV were administered, suggesting that OPV has lost its effectiveness in providing herd immunity[38].

Dose schedule: Children should receive doses of IPV or IPV combined with hepatitis B vaccines at two months, four months and six to 18 months. A booster dose is of IPV (not a combination vaccine) is given at four to six years.

7.2.3 Human rabies–Pathogenesis, Symptoms, Diagnosis, Transmission and Epidemiology, Treatment and Prevention

Rabies is considered to be a fatal viral disease that causes acute encephalitis in warm-blooded animals including humans. Dogs are considered to be the most important host of rabies virus. High rate of fatality is seen in children infected with rabies The disease is mainly transmitted to humans by infected animals like dogs, cats and ferrets. The disease gets contracted through infected saliva that enters the body through a bite or broken skin. The virus travels through the nerves from the wound to the brain, causing inflammation.

Pathogenesis

Rabies is an acute infection of the central nervous system. The disease is caused by rabies virus (RV) which is a member of the rhabdovirus family. This is a negative-stranded RNA virus. Neuroinvasiveness and neurotropism are the main features that define the pathogenesis of rabies[39a]. Though several elements of the RV genome are involved in the pathogenecity (multigenic trait), however the RV glycoprotein plays a major role. This glycoprotein controls the rate of virus uptake and also helps in trans-synaptic spreading of the virus. Neuroinvasiveness of the RVstrains isolated from infected road dogs are found to be higher than the culture adapted RV strains perhaps due to their ability to evade immune system[39a].

Following inoculation, the virus replicates in the striated or connective tissue at the site of inoculation and enters the peripheral nerves through the neuromuscular junction (Fig. 7.8). It then spreads to the CNS in the endoneurium which is a delicate layer of connective tissue around the myelin sheath of nerve fiber. Some structural abnormalities are seen in infected neurons. The nature of the profound disorder is still not understood. The most significant factor underlying the lethal outcome of an RV infection appears to be the neuronal dysfunction due to drastically inhibited synthesis of proteins required in maintaining neuronal functions[39b].

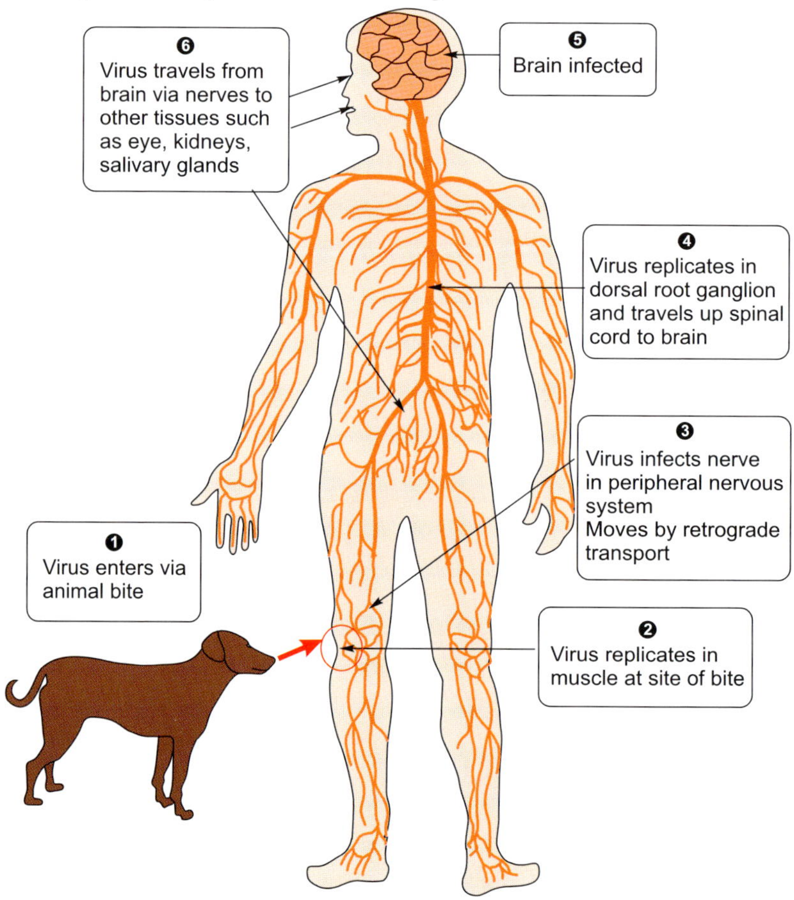

Fig. 7.8: Different steps in the Pathogenesis of Rabies virus (***Source:*** pathmicro.med.sc.edu)

Symptoms

Once the virus starts multiplying in the brain, the infected animals begin to show the first signs of rabies. Within a short period of time, usually within 3 to 5 days, the virus is able to cause sufficient damage to the brain. The incubation time may be from few weeks to months. At the later stage of the disease, the virus now travels through salivary glands to the saliva. At this stage, the typical signs of rabies become apparent in the animal and may lead to death of the animal. In case, the same animal infects a human through a bite, then it is very important to monitor the movement and fate of the animal. Animals with furious rabies usually have an excitation phase lasting several days when the animal turns vicious, restless and start biting other animals or humans more frequently. The infected human should be treated immediately.

Another name for the disease rabies is *hydrophobia*, which literally means "fear of water," a symptom shared by half of all people infected with rabies. Other symptoms include fever, depression, confusion, painful muscle spasms, sensitive to touch, loud noise, and light, extreme thirst, painful swallowing, excessive salivation, and loss of muscle tone.

Diagnosis

After the onset of symptoms, blood tests and cerebrospinal fluid (CSF) analysis tests are conducted. CSF is collected from the patient by a procedure called a lumbar puncture in which a needle is used to withdraw a sample of CSF from the area around the spinal cord. However, the CSF test is not the ultimate diagnosis however other potential causes can be ruled out. A positive diagnosis can be made by looking into the inclusion bodies known as Negri bodies that can be detected in the cytoplasm of neurons.

The two most common diagnostic tests are the fluorescent antibody test and isolation of the rabies virus from an individual's saliva or throat culture. The direct fluorescent antibody (dFA) test involves taking a small sample of skin biopsy from the back of the neck of the patient that produces the antigen and testing it against fluorescently- labeled anti-rabies antibody. Another diagnostic procedure involves taking a corneal impression in which a swab or slide is pressed lightly against the cornea of the eye to determine whether viral material is present. Immunofluorescent antibody (IFA) staining of the epithelial cells on the corneal impression test is found to be quite reliable and can be carried out by Opthalmologists on a routine basis for suspected rabies patients.

Molecular diagnostic studies carried out in India on Rabies virus (RV) isolates revealed three amino acid changes (amino acid 462G [aa462G], aa465H, and aa468K) that distinguished the Indian RVs from RV isolates in other parts of the world. Analysis of the data indicated that the dog rabies virus variants are the major circulating viruses in India that transmit the disease to other domestic animals and humans as well[40].

Transmission and Epidemiology

Rabies is widely distributed disease across the globe, with only few exceptions in countries located mainly in islands and peninsulas that are free of the disease.

Some of the bat species have been shown to harbor rabies or rabies-related viruses in Africa, Australia, central and south-east Asia, Europe and most of the Americas. However, canine rabies predominates in most of the developing countries of central and south Americas, Africa and Asia. More than 90 percent of cases of human rabies are transmitted by dogs; most deaths occur in Asia and Africa. [http://www.who.int/rabies/epidemiology]

Indian Scenario

Rabies is one of the oldest recognized diseases affecting humans and one of the most important zoonotic diseases in India. Rabies causes 50,000 – 60,000 deaths worldwide and it is estimated that every year about 20,000 are from India that amounts to 36 percent of these deaths. The Rabies situation in India has shown some improvement, but the scenario is still not very encouraging. The number of human rabies cases has declined from 30,000 estimated in the 1990s to 17,000 in 2003. WHO-APCRI (Association for Prevention and Control of Rabies in India) survey found 28-million pet dog population in India and annual animal bite incidence to be 17 per 1000 population. Even today, 96 percent of rabies transmission in India is by dogs, mainly due to high number of stray dogs and lack of appropriate practices in maintaining pet dogs free of rabies[41].

Treatment and Prevention

Since rabies is practically 100 percent fatal therefore, rabies vaccines and immunoglobins are considered to be life saving drugs. Rabies post exposure prophylaxis is highly effective if given promptly. Various steps of the treatment includes wound cleansing, immunization with a modern cell-culture vaccine followed by an administration of human rabies immunoglobulin (HRIG)[42].

The following preventive measures must be observed in order to avoid contacts of humans with infected animals:

- Domesticated animals including household pets must be vaccinated against rabies before adopting as a pet.
- In case a pet is bitten by an animal suspected to have rabies, one must contact a veterinarian immediately and notify the local animal control authorities.
- Domestic pets with current vaccinations should be revaccinated immediately
- In case of unvaccinated dogs, cats, or ferrets, animals are usually euthanized (put to sleep).

Further information about domestic pets and rabies is available on the American Veterinary Medical Association (AVMA) web site.

- While handling pets, one must use rubber gloves since rabies is transmitted through saliva.
- Persons whose occupation involves frequent contact with wild animals or non-immunized domestic animals must take preventative vaccination against rabies.
- Travelers visiting countries where the disease is prevalent must take appropriate precautions for protecting themselves.
- Facilities for the surveillance and diagnosis of animal rabies must be available easily.

7.3 Fungal infections of the Central Nervous system

7.3.1 Cryptococcal Meningoencephalitis—Pathogenesis, Symptoms, Diagnosis, Epidemiology, Treatment and Prevention

Cryptococcosis is considered to be one of the major opportunistic pathogen worldwide. The management of the disease in specific regions depends on the medical resources available to clinicians. The infection is caused by the encapsulated yeast *Cryptococcus neoformans*. The incidence of the disease has increased over past 20 years due to overuse of immunosuppressive therapies since HIV has come in an epidemic form[43].

Pathogenesis

Cryptococcus neoformans, the causal organism of cryptococcal encephalitis can infect any organ in the body, but frequently spread in the lung and the CNS (Fig.7. 9 A and B). The usual entry of the fungal infection is through the lungs and symptoms range from asymptomatic colonization to severe pneumonia. Meningitis is the most frequent manifestation of cryptococcosis. Infection of the subarachnoid space (referred as the anatomic space between the arachnoid membrane and pia mater) is accompanied by involvement of the brain parenchyma, and therefore it is known as meningoencephalitis.

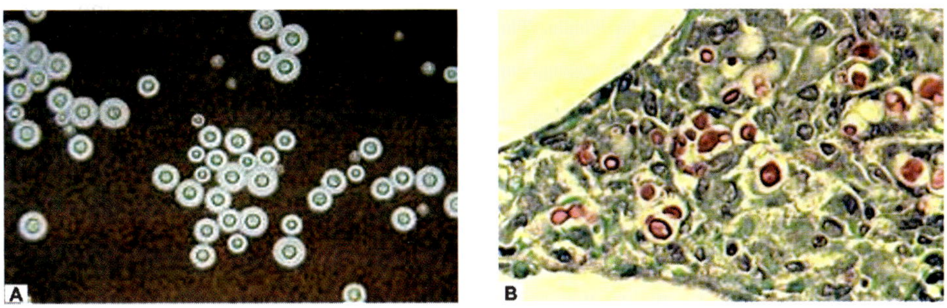

Figs. 7.9A: *Cryptococcus neoformans* in a culture plate; **B.** *C. neoformans* seen in the lung of a patient with AIDS. The inner capsule of the organism stains red
Source: www.mycology. adelaide.edu.au; en.wikipedia.org

Symptoms

Patients with meningoencephalitis usually suffer from headache, fever, malaise and altered mental status over several weeks. Often patients do not show up with typical symptoms of the disease however may show few symptoms like meningism, papilloedema, cranial nerve palsies and other focal neurological deficit like double vision etc. They may also show depressed conscious level and sometimes oversensitive to light (photophobia).

In HIV-infected patients, the disease is associated with severe immunosuppression, usually CD4 counts fall below <100 cells/μl. When compared with HIV-negative patients, higher cryptococcal antigen titer is observed in the serum and a poor CSF inflammatory response (white blood cell count <20/μl) in HIV-positive patients.

Diagnosis

In HIV-associated cryptococcal infection, high organism load is seen and therefore, Indian ink preparations of CSF are usually positive. Cryptococccal antigen testing of either CSF or serum has a high sensitivity and specificity. On the other hand, in non-HIV-associated disease, especially in apparently immunocompetent patients, cultures and antigen tests of CSF may sometimes be negative and the diagnosis is hard and requires higher volume of CSF fluid obtained through multiple lumbar punctures. CSF protein is usually elevated and CSF glucose may be low. Indian ink examination is positive in 70–90 percent of AIDS patients but in only ~50 percent of non-AIDS patients. MRI scans are more sensitive for detection of multiple enhancing nodules within the brain parenchyma, meninges, basal ganglia and midbrain. In AIDS patients there is often cortical atrophy and there may be coexistent pathology[44].

In a recent study, it was found that *C. neoformans*-derived microvesicles (CnMVs) can enhance the traversal of the blood-brain barrier (BBB) by *C. neoformans in vitro*. The immunofluorescence imaging demonstrated that CnMVs can fuse with human brain microvascular endothelial cells (HBMECs), the constituents of the BBB[45].

Taxonomy

Cryptococcus neoformans and *C. gattii* belong to basidiomycetes group and they are encapsulated yeasts. *C. neoformans* and *C. gattii* can be sub classified into four serotypes. The serotypes are based upon capsular agglutination reactions and are designated A, B, C, or D. The serotype A cryptococci is considered as a separate variety based upon genotypic differences[46]. Serotype A cryptococci are therefore now considered as variety *grubii*; serotype D are classified under the variety *neoformans* and they are divided into four molecular types (VNI, VNII, VNIII, VNIV). Serotypes B and C are now considered as a separate species called *Cryptococcus gattii* and further divided into four cryptic species identified by molecular types (VGI, VGII, VGIII, VGIV), and several subtypes. Various molecular techniques have been applied like PCR fingerprinting, AFLP, and MLST to subtype *C. neoformans* and *C. gattii* strains.

Epidemiology

Worldwide, *C. neoformans* and *C. gattii* infections cause an estimated one million cases of cryptococcal meningitis per year among people with HIV/AIDS, resulting in nearly 625,000 deaths (Centers for Disease Control and Prevention, CDC, Atlanta, USA,http://www.cdc.gov/). The greatest burden of disease occurs in sub-Saharan Africa, where mortality is estimated to be 50 to 70 percent[47].

Reservoir: *C. neoformans* may be found in humans and various domestic and wild animals. Soil and decaying vegetation is considered to be a reservoir for serotypes A and D, while serotypes B and C are found in trees. *C. neoformans* is associated with various environmental niches, especially avian guano[48]. *C. neoformans* can be isolated from various sources like soil, bird guano, blood, urine, and specimens from bone marrow, brain, CSF, eye, respiratory sites, skin, and mucous membrane.

A detailed analysis of the geographical distribution of the major molecular types in each continent has been described and represented on thematic maps that may serve as an useful tool to start new epidemiological surveys on the basis of the present knowledge[49].

Indian Scenario

The tropical climate of the Indian subcontinent offers a suitable environment for *Cryptococcus neoformans*, and the onslaught of the AIDS pandemic since the early 1990s has led to a sharp increase in the number of reported cases of cryptococcosis in the past decade[50]. The wide variety of presentations of the disease seen in India suggests the possibility of strain variation. Clinical isolates studied were predominately *C. neoformans* var. *grubii* isolates (serotype A) and a small number of *C. neoformans* var. *gattii* isolates (serotype B) and one *C. neoformans* var. neoformans strain (serotype D). *C. neoformans* var. *neoformans* is found primarily in temperate climates and appears to have lower thermotolerance, which may reduce its fitness in the tropics[51]. The study also indicated that human cryptococcosis can result from simultaneous infection with two strains, which may complicate treatment and contribute to failure.

Treatment

Treatment aims at bringing down the fungal burden in the CSF or blood to the point of sterile cultures. High-dose amphotericin B, with or without 5-flucytosine, followed by fluconazole is the standard treatment for acute cryptococcal meningitis. Though the duration of intravenous amphotericin B induction treatment is arbitrary, it is generally given at dose of 0.5-0.7 mg/kg/day for 2 weeks, followed by consolidation therapy of fluconazole 200 mg twice for 8 weeks. The dose of fluconazole can be reduced to a maintenance dose of 200 mg daily after 10 weeks of standard therapy with amphotericin B and fluconazole. It is customary to give amphotericin B for a longer period if response to treatment is unsatisfactory. For very mild disease with normal mental state, some authorities suggested that amphotericin B might be omitted and fluconazole given alone. Response to acute antifungal treatment needs to be monitored against clinical, biochemical and microbiological parameters. During amphotericin B therapy, complete blood picture, renal and liver function tests need to be carried out twice per week. The treatment response is best gauged by clinical assessment and CSF findings. Therefore examination of CSF has to be repeated after 2-4 weeks of therapy.

Maintenance Treatment

Chronic suppressive treatment for life is required after induction treatment of cryptococcal meningitis as relapse rates are quite high (50-60 percent). Moreover, life span reduces for patients not receiving such maintenance[52]. Life-long suppressive therapy with fluconazole is thus indicated after acute treatment, which may be substituted later by itraconazole if fluconazole is not tolerated. In case of some difficult cases, a higher than usual (1.5-2 times) dose of fluconazole are required to be administered for might be required for maintenance. Despite maintenance, relapse may occur especially in patients with persistent severe immunosuppression. It is also reported that, *C. neoformans* can develop resistance to flucocytosine when used alone [53].

Prevention

The pathogen has been classified under Risk group 2[54]. Containment Level 2 facilities, equipment, and operational practices for work should be strictly followed involving infectious or potentially infectious materials, animals. All procedures that may produce aerosols, or involve high concentrations or large volumes should be conducted in a biological safety cabinet (BSC). The use of needles, syringes, and other sharp objects should be strictly limited to avoid any injury.

C. neoformans is susceptible to 1 percent sodium hypochlorite, iodine, phenolic disinfectants, and formaldehyde however its susceptibility to 70 percent ethanol is questionable. This organism can also be killed by moist heat of 121°C for a minimum of 15 minutes. Person-to-person transmission is very rare.

Animals that may be sources of *C. neoformans* should be treated with caution, particularly by people who are immunosuppressed. Although no cases of mammal-to-human transmission have been reported, people handling animals with cryptococcosis should also use appropriate precautions and wear protective clothing such as gloves and a mask.

7.4 Protozoal Infection

Chances of invading central nervous system and causing infections is rare by a protozoan, however, in some cases if it enters may cause a devastating disease.

7.4.1 African Sleeping Sickness- Causal Organism, Mode of Action, Symptoms, Diagnosis, Pathogenesis, Epidemiology, Treatment and Prevention

Sleeping sickness is also known as trypanosomiasis, caused by single-celled protozoal parasite that leads to a serious infection in the brain and the meninges (the covering of the brain and spinal cord). The disease develops at a slow pace and in case the treatment is delayed it usually proves fatal. These dangerous parasites are carried by the tsetse fly (*Glossina species*), prevalent in rural Africa. The fly and the parasite are found in a broad belt around the Equator. According to the World Health Organisation (WHO) report, major outbreaks have been reported in 2005 from Angola, the Democratic Republic of Congo and Sudan.

Causal Organism and Mode of Action

Sleeping sickness is caused by *Trypanosoma brucei*, a flagellate (organism with a propulsive tail) that lives in the blood of the host. The parasite exists in two sub-species. The flagellate reproduces in the bloodstream, and the symptoms develop gradually as the burden of parasites increases. The organism is known to migrate into the central nervous system producing typical symptoms.

T. brucei gambiense is the cause of a persistent infection that lasts several years and ultimately the patient enters a state of coma from which the patient cannot return back to normal state. Hence the name 'sleeping sickness' is popularized. There are three different sub-species of *T. brucei*, which significantly differ in their epidemiology causing different intensity of trypanosomiasis[55].

1. *T. brucei gambiense*—Causes slow onset of chronic trypanosomiasis in humans. This sub species is most common in central and western Africa. Humans are the only reservoir for this sub-species.
2. *T. brucei rhodesiense*—Causes fast onset of acute trypanosomiasis in humans. Most common in southern and eastern Africa. Domestic livestock and wild animals are thought to be the primary reservoir.
3. *T. brucei brucei*—Causes animal African trypanosomiasis, along with several other species of trypanosoma. *T. b. brucei* does not infect human due to its susceptibility to lysis by Trypanosome Lytic Factor-1 (TLF-1)[56].

Symptoms, Diagnosis and Pathogenesis

T. brucei gambiense: The incubation period varies from a few days to several weeks or months. A local infection may occur with a large 2 to 10 cm red area or sore at the site where the fly originally bit. The patient develops swelling of lymph nodes, general illness with fever, muscle pain and joint pain. Symptoms of infection in the brain begin in a few months to two years later. The symptoms observed are confusion, reversed diurnal rhythm where the patient is found sleeping during the day and waking at night. They also show symptoms of dementia and convulsions. The patient increasingly shows signs of mental impairment and finally loses bladder control, and often develops other forms of paralysis. Terminal patients finally go into irreversible state of coma.

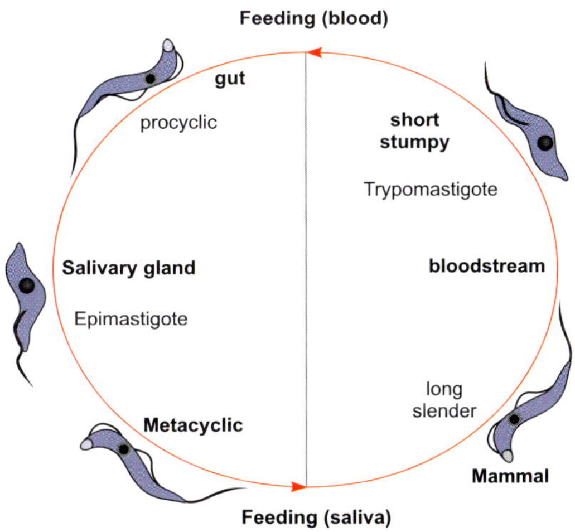

Fig. 7.10: Life cycle of *Trypanosoma brucei*

T. brucei rhodiense: This parasite is more aggressive and the stages all develop more rapidly. The disease usually proves fatal within 9 to 12 months and therefore requires immediate medical attention. Once the parasite enters the CNS, the patient starts showing up the symptoms like loss of appetite and mental disturbances. The disease rarely becomes chronic because the parasite starts damaging the vital organs like kidneys and destroys the heart muscles with a fatal outcome.

The parasite is capable of escaping the initial host defense mechanisms by presenting extensive antigenic variation of parasite surface glycoproteins. These glycoproteins are known as major variant surface glycoprotein (VSG). Diagnosis relies on recognition of the flagellate on a blood smear. Motile organisms may be visible in the buffy coat when a blood sample is spun down. The symptoms of East African trypanosomiasis develop more quickly (starting 1 month after bite) than the symptoms of West African trypanosomiasis, which can begin months to a year after the first bite. Notably, individuals infected with the East African form are more likely to experience cardiac complications and develops CNS disease more quickly, within weeks to a month.

The CNS manifestations of behavioral changes, day time, night time insomnia, stupor, and coma result in death if the patient remains untreated. In West African trypanosomiasis, the patient remains in an asymptomatic state before the onset of fevers, rash that results in inflammation of cervical lymph nodes. In case the patient is not examined and treated in time, symptoms further progress to weight loss, asthenia, itching sensations and finally develops CNS disease. Death at this point is usually due to aspiration or seizures caused by CNS damage.

Life cycle: The life cycle of Trypanosomes depends on two hosts: mammalian and arthropod. The parasite lives in the midgut of the fly as a procyclic form from where it migrates to the salivary glands and then inject the parasite to the mammalian host on biting. The parasite lives within the bloodstream where it can re infect the fly vector while a fly bites an infected mammalian host. Later, during a *T. brucei* infection the parasite may migrate to other areas of the host. A *T. brucei* infection may also get transmitted from human to human via bodily fluid exchange, primarily during blood transfer.

Two specific cellular forms are manifested by *Trypanosoma brucei* at different stages of the life cycle (Fig. 7.10):

a. **Epimastigote**—represented by a basal body anterior of nucleus, with a long flagellum attached along the cell body
b. **Trypomastigote**—represented by a basal body posterior of nucleus, with a long flagellum attached along the cell body. These names are derived from the Greek *mastig*-meaning whip. Thus it refers to whip-like flagellum present in trypanosomes. *T. brucei* is found as a trypomastigote in either as a slender procyclic or stumpy metacyclic forms. The procylic form finally differentiates to the proliferative epimastigote form in the salivary glands of the insect.

Fig. 7.11: Close up of a Tsetse fly taking a blood meal
(***Source:*** 250 × 184 – cdc.gov)

During a blood meal, an infected tsetse fly injects metacyclic trypomastigotes into the skin tissue of the mammalian host (Fig. 7.11). The parasites eventually enter the lymphatic system and then pass into the bloodstream. Inside the host, they transform into bloodstream trypomastigotes, and then are carried to other sites throughout the body. When they reach other blood fluids like lymph, spinal fluid, they start replicating by binary fission. The entire life cycle of African

Trypanosomes is represented by extracellular stages. The tsetse fly becomes infected with bloodstream trypomastigotes while taking a blood meal on an infected mammalian host (Fig. 7.12). Inside the midgut of the fly, the parasites transform into procyclic trypomastigotes and continue to multiply by binary fission and leave the midgut and get transformed into epimastigotes. The epimastigotes finally reach the salivary glands and continue multiplication by binary fission. The duration of the cycle in the fly lasts for 3 weeks. Humans are the main reservoir for *Trypanosoma brucei gambiense*, but at times also reported in animals.

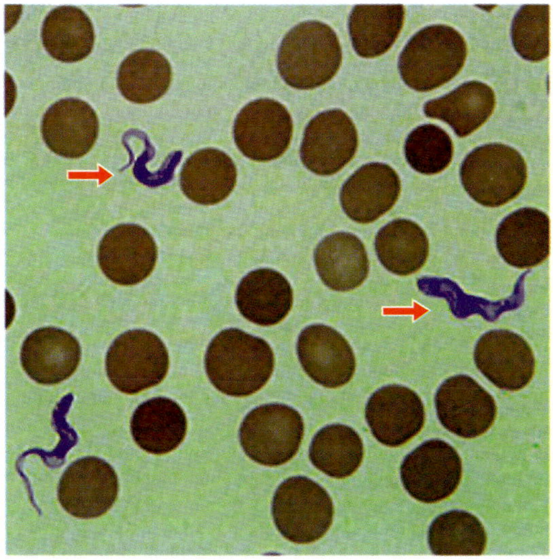

Fig. 7.12: Trypomastigotes of *T. brucei* sp. in a blood smear stained with Giemsa (***Source:*** www.cdc.gov/dpdx)

Epidemiology

In Central African Republic, Chad, Congo, Côte d'Ivoire, Guinea, Malawi, Uganda, and United Republic of Tanzania, sleeping sickness remains an important public health problem[57]. On an average, 50 new cases per year are reported in countries such as Burkina Faso, Cameroon, Equatorial Guinea, Gabon, Kenya, Mozambique, Nigeria, Rwanda, Zambia, and Zimbabwe[58]. Interestingly, the transmission of *T. brucei* has stopped due to which no new cases of African trypanosomiasis have been reported for several decades in countries such as Benin, Botswana, Burundi, Ethiopia, Gambia, Ghana, Guinea Bissau, Liberia, Mali, Namibia, Niger, Senegal, Sierra Leone, Swaziland, and Togo. However, even now sleeping sickness threatens millions of people in 36 countries of sub-Saharan Africa. Unfortunately in many endemic countries, due to lack surveillance and diagnostic expertise it is difficult to assess current situation. During recent epidemic periods, the prevalence of sleeping sickness has reached 50 percent in several villages in the Democratic Republic of Congo, Angola, and Southern Sudan. Sleeping sickness was considered as the first or second greatest cause of mortality in those communities. The mortality rate was higher than HIV infection and AIDS. With the reinforcement of surveillance mechanism and preventive measures, the number of new cases reported throughout the continent had substantially reduced between 1998 and 2004.

Treatment and Prevention

Antitrypanosomal treatment is indicated for all persons diagnosed with African trypanosomiasis. Choice of therapy depends on the infecting subspecies of the parasite as well as on the disease stage. The treatment regimen prescribed by CDC is represented in the Table 7.3 (www.cdc.gov/parasites/sleepingsickness/health_ professionals. It has been observed that the whole range of so far prescribed medicines leave some adverse effects on the body. The drug melarsoprol causes skin reactions, gastrointestinal upset and often leads to peripheral neuropathy. Intravenous injections of melarsoprol are painful and can cause bulging and tenderness of the vein. Adverse reactions to Suramin are also reported, but usually mild and reversible. These include drug rash, nephrotoxicity, and peripheral neuropathy. Suramin administration results in a hypersensitivity reaction in few cases therefore, a small test dose is usually given to check the reaction prior to the full first dose.

Pentamidine is generally well tolerated, but at times leave some adverse reactions like hypoglycemia, injection site pain, diarrhea, nausea and vomiting. Data on the use of pentamidine in pregnant women are limited, and risk to the embryo-fetus is unknown. Pentamidine should be used during lactation only if the potential benefit of therapy to the mother justifies the potential risk to the infant. Intravenous and intramuscular pentamidine have a similar safety profile in children age 4 months and in adults. Pentamidine is recommended as a medicine for the treatment of 1st stage African trypanosomiasis infection.

Adverse effects of eflornithine include bone marrow suppression, gastrointestinal symptoms, and often seizures. Eflornithine is not effective against *T. b. rhodesiense* and it is not recommended for treating the East African form of the disease. Eflornithine should be used with caution in breast-feeding women. Eflornithine should be used during pregnancy only if the potential benefit justifies the potential risk to the fetus. Eflornithine is not approved by the Food and Drug Administration (FDA) for use in pediatric patients. Eflornithine is listed for the treatment of 1st stage African trypanosomiasis in *Trypanosoma brucei gambiense* infection on the WHO Model List of Essential Medicines for Children that are intended for the use of children up to 12 years of age. Antigenic variation remains the major obstacle to developing a trypanosome vaccine.

Table 7.3: Drugs effective against various stages of *T. brucei* infection

Species	Drug of choice	Adult Dosage	Pediatric Dosage
T. b. rhodesiense, hemolymphatic stage	Suramin	1 gm IV on days 1,3,5,14, and 21	20 mg/kg IV on days 1, 3, 5, 14, and 21
T. b. rhodesiense, CNS involvement	Melarsoprol	2–3.6 mg/kg/day IV × 3 days. After 7 days, 3.6 mg/kg/day × 3 days. 3rd series of 3.6 mg/kg/d after 7 days.	2–3.6 mg/kg/day IV × 3 days. After 7 days, 3.6 mg/kg/day × 3 days. 3rd series of 3.6 mg/kg/d after 7 days.
T. b. gambiense, Hemolymphatic stage	Pentamidine	4 mg/kg/day IM or IV × 7–10 days	4 mg/kg/day IM or IV × 7–10 days
T. b. gambiense, CNS involvement	Eflornithine	400 mg/kg/day in 4 doses × 14 days	400 mg/kg/day in 4 doses x× 14 days

Trypanosomes sequentially express a series of surface antigens that are capable of inducing protective immunity. The immune response against each variant is rapid and highly effective in destroying any trypanosomes that possess that particular antigen. Thus, parasitaemia appears and falls in waves with each parasite population carrying different surface antigens[59].

7.4.2 Amebic Meningoencephalitis

There are two organisms involved in amebic encephalitis. *Naegleria fowleri*, is a protozoan that causes a neurological disease, known as primary amebic meningoencephalitis (PAM). A similar neurological disease granulomatous amebic encephalitis (GAE) is caused by a species of free living *Acanthamoeba*.

Pathogenesis of PAM

Naegleria fowleri propagates in warm, stagnant bodies of freshwater, during the summer months, and enters the central nervous system after blowing in of infected water. The parasite attaches itself to the olfactory nerve. The organism initially infects the nasal mucosa and then migrates into the olfactory bulbs of the forebrain[60]. There it tends to multiply by feeding on nerve tissue. During the post infection period (3-7 days), the patient develops typical symptoms of parosmia, a state when the patient fails to recognize the natural or intrinsic smell of a particular odor due to impairment of olfactory function and tends to experience unpleasant odors. This state finally progress to a state of anosmia when the patient loses the sense of smell due to consumption of olfactory bulbs that get replaced with necrotic lesions. The fatality rate is found to be nearly 100 percent. The high mortality rate of this disease is due to lack of typical symptoms at an early stage. In addition to this, culturing of the parasite from CSF is often not carried out in time for positive confirmation. The parasite also demonstrates rapid late-stage propagation through the nerves of the olfactory system.

The slow moving ameboid cells can be isolated only from the cerebrospinal fluid of a patient. In low nutrient conditions, the trophozoite forms rapidly motile cell with two flagella (Fig. 7.13). The trophozoite encysts during cold or dry conditions and reemerges when conditions improve. The protozoan has a spherical cyst stage and an ovoid flagellated stage which is considered to be infective stage of the parasite when it swims rapidly in aquatic habitat.

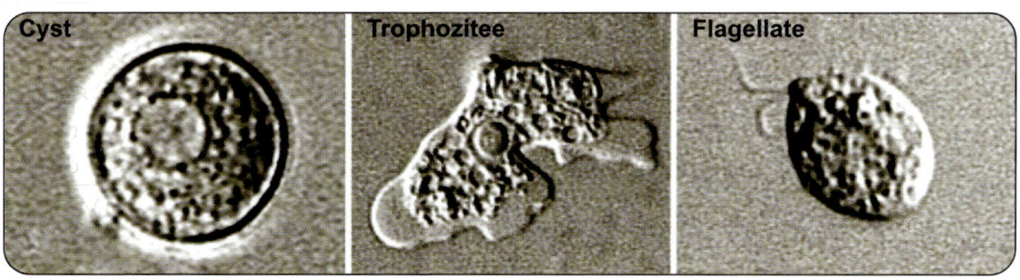

Fig. 7.13: Electron micrograph of *Naegleria fowleri* showing three stages of its life cycle. Adapted from (http://www.cdc.gov/parasites/naegleria)

Diagnosis of this disease is often made at autopsy. Very few survivors are treated with antifungal drug amphotericin B.

7.4.3 Granulomatous Amebic Encephalitis (GAE)

This is a neurological disorder caused by a species of Acanthamoeba. It is a chronic disease and becomes fatal within few weeks or months after infection. GAE can present with: focalparalysis, seizures, brainstem symptoms, and other neurological problems, some of which often resemble glioma (especially brainstem glioma), or other brain diseases. Therefore, diagnosis of the disease remains a problematic issue. The symptoms are usually caused by inflammatory necrosis of brain tissue due to the amoebic infiltrates.

Acanthamoebae are the causative agents of *Acanthamoeba* keratitis and granulomatous amebic encephalitis (GAE). This condition predominantly occurs in immunocompromised individuals. In GAE the skin and the olfactory neuroepithelium probably function as portals of entry and leads to inflammation at these primary foci. Multiple lesions are formed in the brain and other organs, especially lungs. Descriptions of approximately 150 cases of GAE caused by *Acanthamoeba* have been published worldwide and only less than 10 of these patients have survived[61]. The life cycle of *Acanthamoeba* consists of two stages: an actively feeding, dividing trophozoite and a dormant cyst (Fig. 7.14). The trophozoite varies in size from 25 to 40 μm and feeds on bacteria, algae, and yeast present in the environment. Uptake of food by trophozoites can occur by pseudopod formation and phagocytosis. In addition, the other distinguishing features of the trophozoite were the presence of spiny surface projections called acanthopodia. (Fig. 7.14), a prominent contractile vacuole in the cytoplasm that controls the water content of the cell. A double-walled wrinkled cyst composed of an ectocyst and an endocyst ranges in size from 13 to 20 μm.

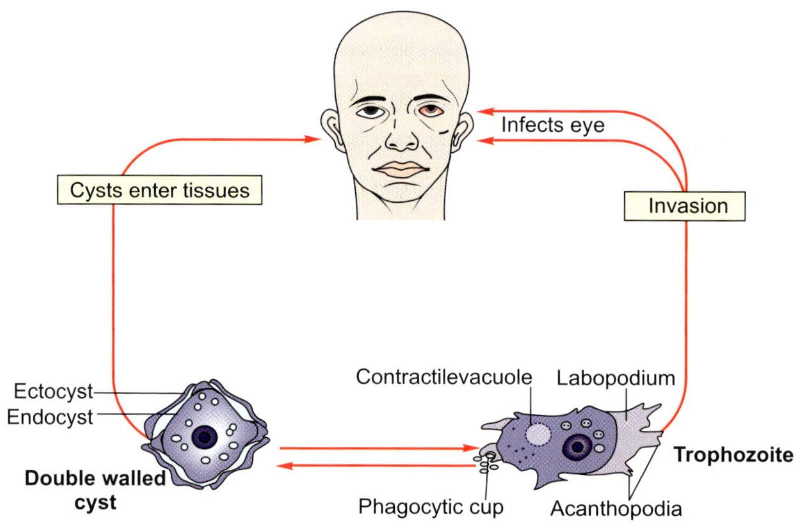

Fig. 7.14: *Acanthamoeba castellanii*: Free living trophozoites and cysts cause infection

REFERENCES

1. " Meningococcal" Meningitis Research Foundation 2008.
2. Sinclair D, Preziosi MP, Jacob John T, Greenwood B (2010) The epidemiology of meningococcal disease in India. *Trop Med Int Health.* 15:1421-35.
3. Verghese T (1990) *Meningococcal meningitis. National Institute of Communicable Diseases,* 13–20.
4. Walker, S., Lockwood, D (2007) "Leprosy". *Clinics in Dermatology.* 25: 165-172.

5a. Rojas-Espinosa O and Løvik M(2001) *Mycobacterium leprae* and *Mycobacterium lepraemurium* infections in domestic and wild animals. *OIE Revue Scientifique et Technique 20:* 219–251.

5b. Pedrini, Sílvia Cristina Barboza, Rosa, Patrícia Sammarco, Medri, Ísis Meri, Mourão, Guilherme, Bagagli, Eduardo, & Lopes, Carlos Alberto de Magalhães. (2010). Search for *Mycobacterium leprae in wild mammals. Brazilian Journal of Infectious Diseases.* 14: 47-53.

6. Bhat RM and Prakash C (2012) Leprosy: An Overview of Pathophysiology, Interdisciplinary Perspectives on Infectious Diseases; Vol 2012, Article ID 181089.
7. WHO, 'Chemotherapy of leprosy for control programmes' (1982) World Health Organisation Technical Report Series, vol. 675, pp. 1–33.
8. Sehgal VN, Tuli SM, and B. Dube B (1967) Leprotic nerve abscesses in northern India. Int. Jr. Leprosy and Mycobacterial Diseases, 35: 60–64.
9. Alcas A, Alter A, Antoni G et al. (2007) Stepwise replication identifies a low-producing lymphotoxin-α allele as a major risk factor for early-onset leprosy. *Nature Genetics* 39:517–522.
10. Truman R, Fine PE (2010) Environmental sources of *Mycobacterium leprae*: issues and evidence *Lepr Rev.* 81:89-95.
11. Cole ST, Eiglmeier K, Parkhill J et al (2001) Massive gene decay in the leprosy bacillus. *Nature*, 409:1007–1011.
12. B.G. Hall, S.J. Salipante (2010)Molecular epidemiology of *Mycobacterium leprae* as determined by structure-neighbor clustering. *J. Clin. Microbiol.,* 48:1997–2008.
13. Sundar Rao PS, Jayakumar S (2006) Trends in new case-detection rates atthe Leprosy Mission Trust India centers. *Indian J Lepr.* 78:187-94.
14. Noordeen SK (2005) The Chennai conference on leprosy discusses progress & prospects with leprosy in the region. *Bull Lepr Elim Alliance* 5: 10–12.
15. Prasad PVS and Kaviarasan PK (2010) Leprosy therapy, past and present: can we hope to eliminate it? *Indian J Dermatol* 55: 316–324.
16. Subramanian M, Showkath Ali MK, Thorat DM, Muthukumar M, Sathiskumar E, Ramadoss C, Ali Khan M (2003) Leprosy situation in endemic states of India and prospects of elimination of the disease. *Indian J Lepr.* 75:335-45.
17. WHO study group. Chemotherapy of leprosy for control programs. Report of WHO study (2008) World Health Organ *Tech Rep Ser.* 1982:675.
18. Meima A, Gupte MD, van Oortmarssen GJ, Habbema JD (1999) SIMLEP: a simulation model for leprosy transmission and control. *Int. J. Lepr. Other Mycobact. Dis.* 67: 215–236.
19. Report of the International Leprosy Association Technical Forum. Lepr Rev. (2002)73:513–6.
20. Wells, C. L., Wilkins, T. D. (1996) Clostridia: Sporeforming Anaerobic Bacilli. In Baron, S., *et al. Baron's Medical Microbiology.* Univ of Texas Medical Branch. ISBN 0-9631172-1-1.

21. http://textbookofbacteriology.net/themicrobialworld/Tetanus.htm
22. Lozano, R (2012). "Global and regional mortality from 235 causes of death for 20 age groups in 1990 and 2010: a systematic analysis for the Global Burden of Disease Study 2010.". *Lancet* 380 (9859): 2095–128.
23. Black, RE; Cousens, S; Johnson, HL; Lawn, JE; Rudan, I; Bassani, DG; Jha, P; Campbell, H; Walker, CF; Cibulskis, R; Eisele, T; Liu, L; Mathers, C; Child Health Epidemiology Reference Group of WHO and, UNICEF (2010). "Global, regional, and national causes of child mortality in 2008: a systematic analysis.". *Lancet* 375 (9730): 1969–87.
24. International Commission on Microbiological Specifications for Foods. Clostridium botulinum. In: Micro-organisms in foods 5: characteristics of microbial pathogens. London, UK: Blackie Academic & Professional; 1996:68–111.
25. 28 b. Innis B L. Japanese encephalitis. In: Porterfield J S, ed. Exotic viral infections. London, United Kingdom: Chapman & Hall; 1995. pp. 147–174
26. Stahl JP, Mailles A, Dacheux L, Morand P (2011) Epidemiology of viral encephalitis in 2011. *Med Mal Infect.* 41:453-64.
27. Mackenzie JS (2005) Emerging zoonotic encephalitis viruses: lessons from Southeast Asia and Oceania J Neurovirol, 11: 434–440.
28. Kumar R, Tripathi P, Singh S, Bannerji G (2006) Clinical features in children hospitalized during the 2005 epidemic of Japanese encephalitis in Uttar Pradesh, India Clin Infect Dis, 43: 123–131.
29. Dhillon GP, Raina VK (2008) Epidemiology of Japanese encephalitis in context with Indian scenario. J Indian Med Assoc 106:660-3.
30. Paul JR (1971). A History of Poliomyelitis. Yale studies in the history of science and medicine. New Haven, Conn: Yale University Press. pp. 16–18.
31. Mueller S, Wimmer E, Cello J (2005). Poliovirus and poliomyelitis: a tale of guts, brains, and an accidental event. Virus Res 111 : 175–93.
32. Ryan KJ, Ray CG (editors) (2004). Sherris Medical Microbiology (4th ed.). McGraw Hill.
33. Chezzi C (1996). Rapid diagnosis of poliovirus infection by PCR amplification. J Clin Microbiol 34 : 1722–5.
34. Gawande A (2004) The mop-up: eradicating polio from the planet, one child at a time". The New Yorker: 34–40.
35. John TJ. (2004) The golden jubilee of vaccination against poliomyelitis. Indian J Med Res 119:1-17.
36. Virendra B (2008) Problems with the oral polio vaccine. Nat Med 14:9.
37. Diamond B (2005) Global polio campaign doomed to fail, experts warn. Nat Med. 11:1260.
38. Shahzad A (2009) Time for a worldwide shift from oral polio vaccine to inactivated poliovaccine. Clin Infect Dis. 49:1287-8.
39. Dietzschold B, Schnell M, Koprowski H (2005) Pathogenesis of rabies. Curr Top Microbiol Immunol.292:45-56.
40. Nagarajan T, Mohanasubramanian B, Seshagiri EV, Nagendrakumar SB, Saseendranath MR, Satyanarayana ML, Thiagarajan D, Rangarajan PN and Srinivasan VA (2006) Molecular Epidemiology of Rabies Virus Isolates in Ind. J Clin Microbiol. 44: 3218–3224.
41. Sudarshan MK (2007) The Changing Scenario of Rabies in India: Are We Moving Towards its Prevention and Control? Ind. J. Public Health 51: 145-147.
42. Centers for Disease Control and Prevention. Human rabies prevention—United

States, 1999: recommendations of the Advisory Committee on Immunization Practices (ACIP) (1999) Morb Mortal Wkly Rep 48(RR-1):1-21.

43. Casadevall A, Perfect JR (1998) *Cryptococcus neoformans*. Washington, DC: ASM Press.

44. Bicanic T and Harrison TS (2004) Cryptococcal meningitis. Br Med Bull 72: 99-118.

45. Huang SH, Wu CH, Chang YC, Kwon-Chung KJ, Brown RJ, Jong A (2012) Cryptococcus neoformans-Derived Microvesicles Enhance the Pathogenesis of Fungal Brain Infection. PLoS One. 7: e48570.

46. Franzot SP, Salkin IF, Casadevall A (1999) Cryptococcus *neoformans* var. *grubii*: separate varietal status for *Cryptococcus neoformans* serotype A isolates. J Clin Microbiol.37:838.

47. Hakim JG, Gangaidzo IT, Heyderman RS et al.(2000) "Impact of HIV infection on meningitis in Harare, Zimbabwe: a prospective study of 406 predominantly adult patients," AIDS, 14:1401–1407.

48. Lin, X. (2009). *Cryptococcus neoformans*: Morphogenesis, infection, and evolution. Infect Genet Evol. 9:401-416.

49. Cogliati M, Global Molecular Epidemiology of *Cryptococcus neoformans* and *Cryptococcus gattii*: An Atlas of the Molecular Types (2013) Scientifica Article ID 675213, http://dx.doi.org/ 10.1155/2013/675213.

50. Banerjee U, Datta K, Majumdar T, Gupta K (2001) Cryptococcosis in India: the awakening of a giant? Med. Mycol. 39:51-67.

51. Jain N, Wickes BL, Keller SM, Fu J, Casadevall A, Jain P, Ragan MA, Banerjee U, Fries BC (2005). Molecular Epidemiology of Clinical *Cryptococcus neoformans* Strains from India. J Clin Microbiol.43: 5733–5742.

52. Zuger A, Louie E, Holzman RS, et al (1986) Cryptococcal disease in patients with the acquired immunodeficiency syndrome. Diagnostic features and outcome of treatment. Ann Intern Med. 04:234-40.

53. Cryptococcus, Histoplasma, Coccidioides, and Other Systemic Fungal Pathogens. In K. J. Ryan, & C. G. Ray (Eds.), Sherris Medical Microbiology: An introduction to infectious diseases (2004) (4th ed., pp. 669-684). New York, USA: McGraw-Hill.

54. Human Pathogens and Toxins Act. S.C. (2009) c. 24. Government of Canada, Second Session, Fortieth Parliament, 57-58 Elizabeth II, 2009.

55. Barrett MP, Burchmore RJ, Stich A et al. (2003). "The trypanosomiases" Lancet 362:1469–80.

56. Stephens NA, Kieft R, Macleod A, Hajduk SL (2012) Trypanosome resistance to human innate immunity: targeting Achilles' heel. Trends Parasitol.28:539-45.

57. Simarro PP, Cecchi G, Franco JR, Paone M, Fèvre EM, Diarra A, et al (2011) Risk for human african trypanosomiasis, Central Africa, 2000-2009. Emerg Infect Dis.17:2322-4.

58. Kohagne TL, M'eyi MP, Kamkuimo RG, Kaba D, Louis JF, Mimpfoundi R.(2011) Transmission of human African trypanosomiasis in the Komo-Mondah focus, Gabon. Pan Afr Med J. 8:36.

59. Vickerman, K (1978) Antigenic variation in trypanosomes. Nature, 273: 613-617.

60. Cervantes-Sandoval I, Serrano-Luna Jde J, García-Latorre E, Tsutsumi V, Shibayama M (2008). Characterization of brain inflammation during primary amoebic meningoencephalitis. Parasitol. Int. 57: 307–13.

61. Schuster FL,Visvesvara GS (2004) Free-living amoebae as opportunistic and non-opportunistic pathogens of humans and animals. Int. J. Parasitol. 34:1001-1027.

8

Nosocomial Infections

Chapter Outline

Hospital Acquired Infections, Mode of Transmission and Preventive Measures

8.1 Ventilator-associated pneumonia (VAP)

8.2 Bloodstream infection

8.3 Hospital acquired urinary tract infection (HAUTI)

8.4 Other antibiotic-resistant nosocomial infections

Hospital Acquired Infections, Mode of Transmission and Preventive Measures

The term nosocomial infection or hospital-acquired infection is applied to any clinical infection that was neither present nor was in its incubation period when the patient entered the hospital. Nosocomial infections may also make their appearance after discharge from the hospital, if the patient was in the incubation period at the time of discharge. Nosocomial infections, even in this modern era of antibiotics, continue to remain an important and formidable consequence of hospitalization. It has been estimated that about 3.5 percent of patients leave the hospital after having acquired infections. Such infections include fungal and bacterial infections and are aggravated by the reduced resistance of individual patients.

Site of infections are the followings:

- Urinary tract infections
- Surgical site infections
- Respiratory tract infections
- Blood stream infections
- Skin infections
- Gastrointestinal tract infections
- Central nervous system infections

8.1 Ventilator-Associated Pneumonia (VAP)

It occurs in people who are receiving mechanical ventilation. The infection manifests when the microorganisms invade the lower respiratory tract and lung parenchyma. A diagnosis of ventilator-associated pneumonia is made when the patient has a new diagnosis of pneumonia after having mechanical ventilation initiated. VAP should be suspected in any person on mechanical ventilation exhibiting increasing numbers of white blood cells on blood testing, and new shadows (infiltrates) on a chest x-ray as is indicative of pneumonia. Blood cultures may reveal the microorganisms causing VAP. It primarily occurs because the endotracheal tube allows free passage of bacteria into the lower segments of the lung in a person who often has underlying lung or immune problems. Often, bacteria colonize the endotracheal or tracheostomy tube. Bacteria may also be brought down into the lungs with procedures such as deep suctioning or bronchoscopy[1]. Between 8 and 28 percent of patients receiving mechanical ventilation are affected by VAP[2]. VAP normally develops within first few days after the procedure.

Involvement of Microbes

Bacterial flora responsible for VAP is different from that of the more common community-acquired pneumonia (CAP). There are several bacteria which are particularly important to cause VAP because of their resistance to commonly used antibiotics. These bacteria are referred to as multidrug resistant (MDR). VAP is especially common in people who have acute respiratory distress syndrome (ARDS).

- *Pseudomonas aeruginosa* is the most common MDR Gram-negative bacterium causing VAP. Evidences show that member of naturally occurring *P. aeruginosa* strains shows resistance to many antibiotics. They are also capable of transferring these resistance genes to disease causing strains. It has been known to acquire resistance to every antibiotic except for polymyxin B. Resistance is typically acquired through upregulation or mutation of a variety of efflux pumps which pump antbiotics out of the cell. Resistance may also develop through loss of an outer membrane porin channel (OprD).
- *Klebsiella pneumoniae* has natural resistance to some beta-lactam antibiotics such as ampicillin. Resistance to cephalosporins and aztreonam may arise through induction of a plasmid-based extended spectrum beta-lactamase (ESBL) or plasmid-based ampC-genes.
- *Serratia marcescens* has an ampC gene which can be induced by exposure to antibiotics such as cephalosporins. Thus, in a culture the sensitivity of the bacteria may change due to the administration of antibiotics.
- *Enterobacter* as a group also have an inducible ampC gene. Susceptible strains of Enterobacter may also develop resistance by acquiring plasmids.
- *Citrobacter* also possess similar inducible ampC gene that are encoded in plasmid.
- *Stenotrophomonas maltophilia* often colonizes tracheal tubes. The organism is also responsible for causing pneumonia. The organism exhibits resistance to wide variety of antibiotics but often sensitive to co-trimoxazole.

- *Acinetobacter* are now considered as important organism colonizing in VAP patients and may be resistant to carbapenems such as imipenem and meropenem
- *Burkholderia cepacia* responsible for nosocomial infections in cystic fibrosis patients often show resistance to multiple antibiotics
- *Methicillin-resistant Staphylococcus aureus* (MRSA) is of increasing concern for VAP in a hospital setting. It has been observed that nearly 50 percent of *Staphylococcus aureus* isolates from intensive care units exhibit resistance to methicillin. This resistance is conferred by the presence of mecA gene which encodes an alternative penicillin binding protein, PBP 2a. It has now been documented that horizontal transfer plays an important role in the dissemination of the *mecA* gene in the *S. aureus* population.

Mode of Transmission

The drug-resistant Gram-negative bacteria mostly threaten only hospitalized patients whose immune systems are weak. These bacteria can survive for a long time on various surfaces like instruments used in the hospital. The organism makes an entry in the body through wounds, catheters, and ventilators.

Different modes of transmission include:

1. *Contact transmission:* by direct physical contact
2. *Droplet transmission:* when droplets containing microbes from the infected person are propelled to a short distance through air and deposit on the body of the host. Droplets are generated from the infected person mainly by coughing, sneezing, and talking or during the performance of certain procedures such as bronchoscopy etc.
3. *Air-borne transmission:* Dust particles or evaporated droplets of small-particle residue of 5 μm or smaller in size containing infectious microorganisms may remain suspended in the air for long periods of time. These are normally dispersed widely by air currents and may be inhaled by a susceptible host. Common microorganisms that are transmitted by airborne transmission include *Legionella, Mycobacterium tuberculosis* and the rubella and varicella viruses.
4. *Common vehicle transmission:* This applies to microorganisms transmitted to the host by contaminated items, such as food, water, medications, surgical devices, and equipment.
5. *Vector borne transmission:* This occurs when vectors such as mosquitoes, flies, rats, and other organisms transmit microorganisms from an infected person to a healthy individual.

Risk Factors

Factors that predispose a patient to nosocomial infections have been identified:

1. Patients admitted to the hospital with an advanced age or babies with a history of premature birth or having a history of immunodeficiency present a general risk. Specific risks are presented by patients with diseases like chronic obstructive pulmonary disease that increases the chance of respiratory infections.

2. The invasive devices used through various surgical procedures like intubation tubes, catheters, surgical drains, and tracheostomy tubes planted in the patient's body all bypass the natural lines of defense against pathogens and thus provide an easy route for infection.
3. Patients' treatment with immunosuppressive drugs, antacid drugs, antimicrobial therapy and recurrent blood transfusions have been identified as risk factors.

Treatment

Treatment suitable for the diagnosed organism needs to be administered. When VAP is first suspected, the bacteria causing infection is typically not known and broad-spectrum antibiotics are given until the diagnosis is completed and the sensitivities of the bacteria towards the drug are determined. Risk factors for infection with an MDR strain include following factors like 1. Ventilation for more than five days; 2. duration of residence in a nursing home; 3. duration of treatment in a hemodialysis clinic ; 4. prior antibiotic use if any past 90 days before infection.

Possible empirical therapy combinations include following drugs [adapted from wikipedia.org/wiki/Ventilator-associated_pneumonia]

- vancomycin/linezolid and ciprofloxacin,
- cefepime and gentamicin/amikacin/tobramycin
- vancomycin/linezolid and ceftazidime
- ureidopenicillin plus β-lactamase inhibitor such as piperacillin/tazobactam or ticarcillin /clavulanate
- a carbapenem (e.g., imipenem or meropenem)

Before administering an antibiotic to a VAP patient, following information should be available to the clinician: 1. thorough knowledge of their mode of action, 2. previous history of antibiotic consumption by the patient, 3. local antibiotic resistance bacterial profile and 4. prevalence of local pathogens in the ICU where the patient is undergoing treatment.

Therapy is typically changed once the causative bacteria are known and continued until symptoms resolve within a fortnight. For patients with VAP not caused by aggressive Gram-negative bacilli like Acinetobacter, *Pseudomonas aeruginosa* use of short-course antimicrobial treatments are suggested for a week or ten days[3]. Indicators of response to treatment include normalization of temperature, leucocyte count and C-reactive protein (CRP) levels.

Prevention

In order to prevent VAP infection, following measures are mandatory: 1. limiting exposure to resistant bacteria, 2. discontinuing mechanical ventilation as early as possible, 3. maintaining hygienic practices like proper hand washing while attending the patient, 4. sterile technique for invasive procedures, and 5. isolation of individuals with known resistant organisms. Resistant bacteria also spread the same ways as any communicable disease. Therefore, instead of sanitization, sterilization of all the equipments, plastic wares and glass wares is recommended through exposing them to chemicals, ionizing radiation, dry heat, or autoclaving.

One more important aspect to prevent nosocomial infections is limiting the amount of sedation that a ventilated person receives. Other recommendations for preventing VAP include raising the head of the bed to at least 30 degrees and placement of feeding tubes beyond the pylorus of the stomach. Use of antiseptic mouth washes such as chlorhexidine may also reduce the incidence of VAP.

8.2 Bloodstream Infection

Central venous catheters (CVC) are life saving medical devices that are inserted in the patient's body during surgery for delivering fluids, medications etc. Catheter related blood stream infection (CRBSI) is the most common cause of health-care associated infection in the blood stream. On the basis of CDC report, around 12-25 percent of patients who acquire CRBSI often develops complications and are required to stay in the hospital for a prolonged period or else expire.

Nearly, 30 percent of hospital-acquired bloodstream infections in ICUs are due to gram-negative organisms[4] although this proportion is lower when hospital-wide data are examined[5]. Given an adequate portal of entry, almost any gram-negative organism can cause bloodstream infection; however, the most common organisms include Klebsiella species, *Escherichia coli,* Enterobacter sp., and *P. aeruginosa*. However, organisms showing resistance against extended-spectrum cephalosporins and carbapenems is an emerging problem. For example, of all the bloodstream isolates of *Klebsiella pneumoniae* from hospitals throughout the United States, 27.1 percent were resistant to third-generation cephalosporins and 10.8 percent were resistant to carbapenems[5]. Higher rates of resistance are also reported from different parts of Europe.

The most recent challenge has been the spread of carbapenemase-producing Enterobacteriaceae. The β-lactamase responsible for this phenotype, also known as *K. pneumoniae* carbapenemase, or KPC, confers reduced susceptibility to all cephalosporins (including cefepime), monobactams (aztreonam), and the carbapenems[6]. Carbapenemase producing Enterobacteriaceae have now been identified in hospitals in different parts of the world, including South America, Israel, China, but less commonly from Europe. The genetic relatedness of the strains responsible for outbreaks within and between countries highlights the importance of strict infection control to prevent ongoing dissemination. These β-lactamases are encoded on mobile genetic elements, mostly plasmids and transposons which function as vehicle to spread these genes among other gram-negative genera lacking β-lactamases. Furthermore, they often coexist with other resistance genes, including the most widespread of the ESBLs (the *bla* CTX-M-15 gene), aminoglycoside plasmid mediated quinolone-resistance genes (*qnrA* and *qnrB*), thus leaving the physician with few therapeutic options. As has been described for the nonfermenting gram-negative organisms, *K. pneumoniae* strains that are resistant to all currently available antibiotics, including the polymyxins, have also been reported[7].

Management Options

Patients who present at the hospital with suspected bloodstream infection and who have health care–associated risk factors should be treated initially with broad-spectrum empirical antibiotics, pending the results of blood cultures. Detailed guidelines for the management of central vascular catheter–related bloodstream infections have recently been published[8].

Preventive measures for Central Venous Catheter–Associated Bloodstream Infection (adapted from Yokoe et al, 2008)[8]

Multifactorial step-wise interventions are suggested to reduce Catheter associated bloodstream infection:

- Health care staff involved in central venous catheterization should be educated regarding the safety measures to be followed before the procedure
- Catheter-insertion checklist should be followed to ensure adherence to infection-prevention practices
- Hand-hygiene procedures must be adopted during catheter insertion or manipulation
- The attending staff must avoid use of the femoral vein in adults
- Use of an all-inclusive catheter kit and maximal sterile-barrier precautions during catheter insertion is recommended
- Use of chlorhexidine-based antiseptic is recommended for skin preparation in patients >2 months of age
- After insertion, catheter hubs that include needleless connectors and injection ports must be disinfected before accessing the catheter. Any nonessential catheters must be removed
- Appropriate maintenance of catheters require - change of transparent dressings and performing site care with a chlorhexidine antiseptic every 5 to 7 days; treat after every two days for gauze dressings or more frequently if dressing is soiled, loose, or damp
- Antimicrobial ointments are to be used for hemodialysis catheter insertion sites
- Surveillance for bloodstream infection must be performed on a regular basis
- In hospitals with suboptimal infection control practices, ICU patients >2 month of age should be given bath on a regular basis with a chlorhexidine preparation
- Antiseptic-impregnated or antimicrobial impregnated catheters should be used for adult patients,
- Chlorhexidine-containing sponge dressings in patients >2 mo of age, and
- Antimicrobial locks are recommended for central venous catheters

8.3 Hospital Acquired Urinary Tract Infection (HAUTI)

Nosocomial urinary tract infections (UTIs) account for up to 40 percent of all hospital-acquired infections. Gram-negative organisms predominate in hospital-acquired urinary tract infections and are mostly associated with urethral catheterization. Within 48 hr of catheterization, it is estimated that the risk of bacteriuria increases by 5 to 10 percent per day. Majority of cases of bacteriuria are asymptomatic, and the most effective management is the removal of the catheter rather than antibiotic treatment. However, in few cases, local and systemic complications ensue, and antibiotic treatment should be initiated even for asymptomatic bacteriuria in patients who need to undergo urologic surgery or implantation[9]. Such therapy should also be considered in immunocompromised patients. Recent data indicate the prevalence of microorganisms involved in urinary tract infection are in the following order- *E. coli* > *P. aeruginosa* > *Klebsiella species* > *Enterobacter species* > *A. baumannii*[5]. *Uropathogenic E. coli* strains infect the urinary tract through a range of mechanisms, including specialized structures like adhesins, fimbriae; by the formation of biofilm or by avoiding host responses.

Emergence of Resistance to Antibiotics

The emergence of resistance to quinolones and extended spectrum cephalosporins remains a considerable challenge, since these agents are often used as first-line therapy.

Extended spectrum beta-lactamases (ESBLs) are often plasmid mediated derived from mutations in the classic TEM and SHV genes by one or more amino acid substitution around the active site. Detection of TEM and SHV genes by molecular methods in ESBL producing bacteria and their pattern of antimicrobial resistance can provide useful information about its epidemiology and risk factors associated with these infections. Traditionally, the SHV-type and TEM-type ESBLs have predominated among hospital-acquired organisms. Frequency of antibiotic resistance was high in isolates having both TEM and SHV genes. The epidemiology of ESBLs is changing. A newer group of ESBLs have arisen and spread quickly on a global basis. The CTX-M family of enzymes emerged by initial transfer of the chromosomal beta-lactamase gene from *Kluyvera* spp. to conjugative plasmids that have readily spread among different members of the *Enterobacteriaceae* family and certain other gram-negative bacteria[10]. Among various members of CTX-M family of genes, CTX-M-15 has frequently been found to be associated with a uropathogenic *E. coli* clone known as sequence type 131[11]. Unfortunately, the plasmids carrying these ESBL genes often carry resistance determinants targeting fluoroquinolones as well. Therefore, in order to reduce the incidence of HAUTI urgent measures should be taken to prevent the dissemination of drug-resistant gram-negative organisms and at the same time the hospital care staff must adhere to the recommended guidelines and keep a close surveillance to reduce morbidity[12]. Compliance to recommended catheter associated urinary tract infections preventive practices can reduce the catheter associated urinary tract infections considerably in ICU setting[13]:

- Practicing decontamination of hand,
- Use of protective equipment while carrying out the procedure,
- use of disposable gloves,
- cleaning of urethral meatus prior to catheter insertion using sterile saline,
- assessment of catheter need,
- to follow aseptic urine sampling technique, and
- draining- bag should be positioned correctly
- proper transportation of health care wastes

Routine surveillance for nosocomial infections should be made mandatory in health care units.

8.4 Other Antibiotic-Resistant Nosocomial Infections

Infections with hospital- and community-acquired methicillin-resistant *Staphylococcus aureus* (MRSA) have rapidly spread worldwide[14]. Increased rates of MRSA infections are observed with the administration of glycopeptides, cephalosporins, and especially quinolones[15]. Among a limited selection of anti-MRSA drugs, vancomycin was introduced in 1991 for the treatment of MRSA infection. Because MRSA-infected patients are consistently threatened by possible mixed infections with gram-negative bacteria such as *Pseudomonas aeruginosa*, β-lactam antibiotics often have been used together with vancomycin, and this treatment has been successful for over 15 years[16].

Community-acquired MRSA (CA-MRSA) has now emerged as an epidemic that is responsible for rapidly progressive, fatal diseases, including necrotizing pneumonia, severe sepsis and necrotizing fasciitis[17].

A new type of vancomycin-resistant MRSA strain of *Staphylococcus aureus* appeared with the simultaneous use of vancomycin and β-lactam antibiotics. This organism exhibited elevated vancomycin resistance and has been designated as β-lactam-induced vancomycin-resistant MRSA (BIVR)[18]. Emergence of BIVR were reported from samples of a burn patient and a renal failure patient when both the patients were initially treated with a combination of vancomycin and imipenem[19]. Therefore, appropriate precautions need to be taken to prevent the nosocomial spread of BIVR.

Vancomycin Resistant Enterococci

Few bacterial strains of the genus Enterococcus are reported to show resistance to the antibiotic vancomycin. High-level vancomycin-resistant *E. faecalis* and *E. faecium* clinical isolates were first documented in Europe in the late 1980s[20,21] and subsequently VRE led to several outbreaks of hospital-acquired infections around the world. The risk factors for acquisition and mortality due to nosocomial infection with vancomycin resistant *Enterococcus faecium* (VREF) in orthotopic liver transplant (OLT) recipients have been found to be quite high. Infection with VREF is a predictor of morbidity and mortality in OLT patients[22]. Steps should be taken to regularly screen the enterococcal isolates for vancomycin and high level aminoglycoside resistance and to treat the enterococcal infections effectively to limit the spread of multidrug resistant enterococcal infections.

For further details, one may consult "Prevention of hospital-acquired infections- A practical guide 2nd edition" published by World Health Organization Department of Communicable Disease, Surveillance and Response

http://www.who.int/csr/resources/publications/whocdscsreph200212.pdf

Table 8.1 show results of the International Infection Control Consortium (INICC) surveillance study from January 2003 through December 2008 in 173 intensive care units (ICUs) in Latin America, Asia, Africa, and Europe.

Table 8.1: Antimicrobial resistance rates in the ICUs of the International Nosocomial Infection Control Consortium (Reproduced with permission[24])

	No. of pathogenic isolated tested, pooled	Resistance percentage, %	No. of pathogenic isolated tested, pooled	Resistance percentage, %	No, of pathogenic isolated tested, pooled	Resistance percentage, %
Pathogen, antimicrobial	(CLAB)	(CLAB)	(VAP)	(VAP)	(CAUTI)	(CAUTI)
Staphylococcus aureus						
OXA	761	84.1	715	77.5	43	74.4
Enterococcus faecalis						
VAN	115	8.7	277	0.72	277	2.9
Pseudomonas aeruginosa						
FQs	963	50.0	963	49.8	188	56.4
PIP or PTZ	703	78.0	1525	35.1	277	37.9
AMK	304	31.0	990	30.4	185	35.1
IMI or MERO	526	44.0	1636	38.6	288	34.7
CPM	30	73.3	118	66.9	30	73.3
Klebsiella pneumoniae						
CTR or TAZ	394	76.1	584	70.4	213	70.0
IMI, MERO, or ETP	444	3.8	632	3.8	237	3.4
Acinetobacter baumannii						
IMI or MERO	605	46.3	1209	52.4	113	38.9
Escherichia coli						
CTR or TAZ	193	53.9	274	67.9	343	41.7
IMI, MERO, or ETP	214	3.7	299	3.0	302	4.6
FQs	181	46.4	142	59.9	300	35.0

***AMK*, amikacin; *CPM*, cefepime; *CTR*, ceftriaxone; *ETP*, ertapenem; *FQs*, fluoroquinolones (ciprofloxacin, levofloxacin, moxifloxacin, or ofloxacin); *IMI*, imipenem; *MERO*, meropenem; *OXA*, oxacillin; *PIP*, piperacillin; *PTZ*, piperacillin-tazobactam; *TAZ*, ceftazidime; *VAN*, vancomycin.**

Central line-associated primary bloodstream infections (CLABs), catheter-associated urinary tract infections (CAUTIs), and Ventilator-associated pneumonias (VAPs)

Table 8.2: Features of the participant hospitals and patients

	Hospital							Overall
	A	B	C	D	E	F	G	
No. of ICUs	1	3	4	1	1	1	1	12
Surveillance period	July 2004 to July 2005	Sept. 2004 to May 2006	Apr. 2006 to Mar. 2007	July 2005 to Mar. 2007	Jan. 2005 to Mar. 2006	Jan. 2007 to Mar. 2007	Feb. 2006 to Mar. 2006	July 2004 to Mar. 2007
Experience of the infection control practitioner (years)	17	17	2	2	20	2	1	1–20
No. of patients studied	3052	2655	2151	2032	751	151	43	10,835
Total ICU days	15,302	14,681	8121	10,532	2873	791	218	52,518
Men (%)	84.0	81.3	62.8	66.1	69.2	79.5	74.4	74.6
Mean age (years)	58.2	57.6	55.76	56.37	43.46	49.32	42.7	56.0
Mean ASIS	2.34	2.10	3.15	2.51	3.69	3.99	2.70	2.60
Device use								
Ventilator-days	3632	680	2846	4060	1903	228	132	13,481
Ventilator use, proportion	0.24	0.05	0.35	0.39	0.66	0.29	0.61	0.26
CVC-days	17,960	5030	3054	7081	3196	30	196	36,857
CVC use, proportion	1.17	0.34	0.38	0.67	1.11	0.43	0.90	0.70
Urinary catheter-days	9213	3748	5289	8519	2824	662	209	30,464
Urinary catheter use, proportion	0.60	0.26	0.65	0.81	0.98	0.84	0.96	0.58

ASIS=Average severity of illness score; CVC=Central venous catheter; ICU=Intensive care unit

(Reproduced with permission from Mehta et al (2007)[25].

Indian Scenario

Since 2002, International Nosocomial Infection Control Consortium (INICC) has been measuring healthcare-associated infection rates in 46 developing countries including India. (http://www.inicc.org/english/index.php). Nosocomial control programme is at a nascent stage in Indian hospitals. Unfortunately, many hospitals lack the required infrastructure to establish a central sterilization and supply department (CSSD) and also need to appoint trained staff for initiating infection control program. The research by Global Antibiotic Resistance Partnership (GARP) in India also revealed that 30 per cent of 1.9 lakh neonatal deaths occur due to sepsis. Bacteria involved in the blood stream infection causing sepsis show antibiotic resistance.(http://archive.indianexpress.com/news/1183-per-cent-hospitalacquired-infections-in-india-report/856112/)

A study carried out by Pawa et al in 1997 in the Pediatrics division of New Delhi hospitals revealed multidrug resistant *Klebsiella* sp. causing nosocomial septicemia and pneumonia followed by *Pseudomonas aeruginosa*[23]. Various risk factors detected to be significantly associated with infection were identified as 1. birth weight < 1500 g, 2. assisted ventilation >72 h and 3. very low birth weight (VLBW) neonates. The risk of infection and death of neonates increase further especially those undergoing interventions such as mechanical ventilation. Therefore, strict protocol for asepsis must be adhered to when handling these high risk infants[23].

One study was carried out simultaneously in 12 ICUs in seven hospitals of Indian cities from July 2004 to March 2007. Each hospital had an infection control team (ICT) with a physician and an infection control practitioner (ICP) with at least one year's experience in infection control and had a microbiology laboratory to provide in-vitro susceptibility testing of clinical isolates using standardized methods. Using Centers for Disease Control and Prevention–National Nosocomial Infections Surveillance (CDC-NNIS) definitions, rates of central venous catheter-associated bloodstream infection (CVC-BSI), catheter-associated urinary tract infection (CAUTI) and ventilator-associated pneumonia (VAP) were determined and represented in Table 8.2.

Data generated from Health care associated infection studies in various hospitals of India represented overall 87.5 percent of all *Staphylococcus aureus* HCAIs, caused by meticillin-resistant strains, 71.4 percent of Enterobacteriaceae were resistant to ceftriaxone and 26.1 percent to piperacillin–tazobactam; 28.6 percent of the *Pseudomonas aeruginosa* strains were resistant to ciprofloxacin, 64.9 percent to ceftazidime and 42.0 percent to imipenem. The data may not reflect the clinical setting of the hospitals throughout the country. However, infection control programs including surveillance and antibiotic policies should be dealt on a priority basis in India.

REFERENCES

1. Koenig SM, Truwit JD (2006) Ventilator-Associated Pneumonia: Diagnosis, Treatment, and Prevention. *Clin Microbiol Rev.*19: 637–657.
2. Chastre J, Fagon JY (2002). "Ventilator-associated pneumonia". *Am. J. Respir. Crit. Care Med.* 165 : 867–903.
3. Grammatikos AP, Siempos II, Michalopoulos A, Falagas ME (2008) Optimal duration of the antimicrobial treatment of ventilator-acquired pneumonia. *Expert Rev Anti Infect Ther 6:* 861–866.
4. Gaynes R, Edwards JR (2005) Overview of nosocomial infections caused by gram-negative bacilli. *Clin Infect Dis.* 41:848–854.
5. Hidron AI, Edwards JR, Patel J, et al (2008) NHSN annual update: antimicrobial-resistant pathogens associated with healthcare-associated infections: annual summary of data reported to the National Healthcare Safety Network at the Centers for Disease Control and Prevention, 2006–2007. *Infect Control Hosp Epidemiol.* 29:996–1011.

6. Nordmann P, Cuzon G, Naas T (2009) The real threat of *Klebsiella pneumoniae* carbapenemase producing bacteria. *Lancet Infect Dis.* 9:228–236.
7. Souli M, Galani I, Giamarellou H (2008) Emergence of extensively drug-resistant and pandrug-resistant Gram-negative bacilli in Europe. *Euro Surveill.* 13:19045
8. Yokoe DS, Mermel LA, Anderson DJ, et al (2008) A compendium of strategies to prevent healthcare associated infections in acute care hospitals. *Infect Control Hosp Epidemiol.* 29 Suppl 1:S12–S21.
9. Tenke P, Kovacs B, Bjerklund Johansen TE, Matsumoto T, Tambyah PA, Naber KG (2008) European and Asian guidelines on management and prevention of catheter-associated urinary tract infections. *Int J Antimicrob Agents* 31 Suppl 1:S68–S78.
10. Canto´n, R., and Coque TM (2006) The CTX-M ß-lactamase pandemic. *Curr. Opin. Microbiol.* 9:466–475.
11. Nicolas-Chanoine MH, Blanco J, Leflon-Guibout V, et al. (2008) Intercontinental emergence of *Escherichia coli* clone O25:H4-ST131 producing CTX-M-15. *J Antimicrob Chemother.* 61 : 273–281.
12. Peleg AY and Hooper DC (2010) Hospital acquired infections due to gram-negative bacteria, *N Engl J Med.* 362: 1804–1813.
13. Amine, A. E. K., Helal, M. O. M., & Bakr, W. M. K. (2014). Evaluation of an intervention program to prevent hospital-acquired catheter-associated urinary tract infections in an ICU in a rural Egypt hospital. *GMS Hygiene and Infection Control, 9(2),* Doc15. doi:10.3205/dgkh000235
14. Cooper BS, Stone SP, Kibbler CC, Cookson BD, Roberts JA, Medley GF, et al (2004) Isolation measures in the hospital management of methicillin resistant *Staphylococcus aureus* (MRSA): systematic review of the literature. *BMJ* 329:533–41.
15. Tacconelli, E; De Angelis, G; Cataldo, MA; Pozzi, E; Cauda, R (2008) Does antibiotic exposure increase the risk of methicillin-resistant Staphylococcus aureus (MRSA) isolation? A systematic review and meta-analysis. *J. Antimicrob. Chemother.* 61: 26–38.
16. Toyokawa M, Asari S, Nishi I, Horikawa M, Tsukamoto H, Sunada A, et al (2003) In vitro combined effects of cefozopran/teicoplanin and cefozopran/vancomycin on methicillin-resistant *Staphylococcus aureus. J Chemother* 15:31–6.
17. Boyle-Vavra, S; Daum, RS (2007). “Community-acquired methicillin-resistant Staphylococcus aureus: the role of Panton-Valentine leukocidin”. *Lab. Invest.* 87 (1): 3–9
18. Hanaki H, Yamaguchi Y, Yanagisawa C, Uehara K, Matsui H, Hososaka Y, et al (2005) Investigation of β-lactam antibiotic-induced vancomycin-resistant MRSA (BIVR). *J Infect Chemother* 11:104–6.
19. Haraga I, Nomura S, Fukumachi S, Ohjimi H, Hanaki H, Hiramatsu K, et al.(2002) Emergence of vancomycin resistance during therapy against methicillin-resistant *Staphylococcus aureus* in a burn patient – importance of low-level resistance to vancomycin. Int *J Infect Dis* 6:302–8.
20. Uttley AH, Collins CH, Naidoo J, George RC (1988) Vancomycin-resistant enterococci *Lancet* 1 (8575-6): 57–8.
21. Leclercq R, Derlot E, Duval J, Courvalin P (1988) Plasmid-mediated resistance to vancomycin and teicoplanin in *Enterococcus faecium. N. Engl. J. Med.* 319:157–61.
22. Papanicolaou GA, Meyers BR, Meyers J, Mendelson MH, Lou W, Emre S, Sheiner P, Miller C (1996) Nosocomial infections with vancomycin-resistant *Enterococcus faecium* in liver transplant recipients: risk factors for acquisition and mortality. *Clin Infect Dis.* 23:760-6.
23. Pawa AK,Ramji S, Prakash K and Thirupuram S (1997) Neonatal nosocomial infection : Profile and Risk factors. *Ind.Paediatrics* 34:297-302.
24. Rosenthal VD, Maki DG, Jamulitrat S,et al… ; INICC Members. (2010) International Nosocomial Infection Control Consortium (INICC) report, data summary for 2003-2008, issued June 2009. Am J Infect Control. 38:95-104.
25. Mehta A, Rosenthal VD, Mehta Y, Chakravarthy M, Todi SK, Sen N, Sahu S et al (2007) Device-associated nosocomial infection rates in intensive care units of seven Indian cities. Findings of the International Nosocomial Infection Control Consortium (INICC) *Jr Hospital Infect.* 67: 168-174.

Index

A

A beta-hemolytic streptococcus (GAS) 92
Acanthamoeba 314
Acquisition of Disease 12
Acute otitis media (AOM) 99
Aedes aegypti 29, 239
Aedes albopictus 237
Aflatoxin Poisoning 168
African Sleeping Sickness 308
Airborne 16
Albendazole 273
Amebic liver abscess 176
Amebic Meningoencephalitis 313
American trypanosomiasis 9
Amoebiasis 172
Ancylostoma duodenale 170
Antibiotic associated diarrhea (AAD) 156
ARDS 320
Ascaris lumbricoides (roundworm) 170
Ascaris (round worm) 7
Aspergillus flavus 168
Astrovirus 164
Athlete's Foot (Tinea Pedis) 82
Atopic Dermatitis (AD) 62, 64
Avian Influenza 43

B

Babesiosis 265
Bacillus anthracis 38
Bacterial Endocarditis 221
Bacterial Pneumonia 103
Bacterial vaginosis (BV) 198, 217
Bartonella henselae 5
Basic reproduction number 32
BCG 116
Bioterrorism 37
Biowarfare 37
BIVR 326
Blood-brain barrier (BBB) 306
Blood-flukes 265
Blood poisoning 222
Bloodstream Infection 323
BMC Infectious Diseases 52
Bordetella Pertussis 110
Borrelia burgdorferi 4, 68
Borrelioses 68
Botulinum toxin 293
Botulism 292
Bronchiolitis 88
Brucellosis 227
Bubonic Plague 230
Bundibugyo ebolavirus (BDBV) 244
Burkholderia mallei 38
Burkitt's Lymphoma 234

C

CagA 144
C. albicans 215
Calicivirus 164
Caliciviruses 167
Campylobacter 150
Campylobacter jejuni 4, 146
CA-MRSA 326
Candidiasis 198
Carbapenems 323
Carbuncle 63
Carrier 13
Catheter-associated urinary tract infection (CAUTI) 329
CDC 39
Central venous catheter-associated bloodstream infection (CVC-BSI) 329
Cephalosporins 323
Chagas Disease 9, 246
Chikungunya 29
Chikungunya Fever 236
Chlamydia 198
Chlamydia trachomatis 198
Cholera Toxin 160
Claviceps purpurea 168
Clostridial hemolysins 156
Clostridium botulinum 38, 293
Clostridium Perfringens 156
Clostridium tetani 289

C. neoformans-derived microvesicles (CnMVs) 306
C. neoformans var. gattii isolates (serotype B) 307
C. neoformans var. grubii isolates (serotype A) 307
C. neoformans var. neoformans strain (serotype D) 307
CNS 29
CNS infection 29
Coccidioides immitis 122
Coccidioidomycosis 122
Coccidioidomycosis (Valley Fever) 122
Coevolution of a Host and a Parasite 17
Common cold 88
Coronaviruses 100
Corynebacterium diphtheriae 12
Cpe plasmid 157
Croup 88
Cryptococcal Meningoencephalitis 305
Cryptococcus neoformans 305
Cryptosporidiosis 182
Cryptosporidium 182
Cryptosporidium parvum 4
CSF 29, 306
Cutaneous HPV 72
Cutaneous leishmaniasis (CL) 261
Cyclospara cayetanensis 4, 181
Cyclosporiosis 180

D

DDC - WHO/AFRO 47
DEET 239
Dengue 10
Dengue Virus 239
Dental caries 88
Dental Plaque 134
DF 240
DHF 240
Diarrheagenic Escherichia coli (DEC) 157
Diarrheal diseases 3
Diethylcarbamazine (DEC) 273
Diphtheria 94
Diphtheria Toxin 96
DoD 42
DoD-GEIS 42
DOT test 115
DPT 97
Dracunculiasis 9
Dracunculus medinensis 9
DTaP and DT 291
Duffy antigens 257

E

Ebola Virus 4, 243
Ehrlichia chaffeensis 4
EID Weekly Updates 53
EIEC 157
Elephantiasis 7
Emerging and Re-emerging diseases 20, 22
Emerging Infectious Diseases (CDC) 53
Encephalitis 80, 295
Encephalitozoon cuniculi 5
Encephalitozoon hellem 5
Endemic Steady State 34
Endogenous infection 13
Entamoeba histolytica 172
Enteric adenovirus 164
Enteric Adenoviruses 166
Enteric fever 147
Enteritis 147
Enterocytozoon bieneusi 4
Enterohemorrhagic E. coli (EHEC) 157
Enteropathogenic E.coli (EPEC) 157
Enterotoxigenic E.coli 146
Enterotoxigenic E. coli (ETEC) 157
Enterotoxin B 38
Epimastigote 310
Epsilon toxin 38
Ergot poisoning 168
Escherichia Coli 4
Eurosurveillance Weekly 53
EVD 243
EWGLI 118
Extended spectrum beta-lactamases (ESBLs) 325

F

FASTATS 50
Filamentous hemagglutinin" (FHA) 111
Filaroid life cycle 270
First-line drugs 115
Flat warts 72
Flaviviridae 239
Fluconazole 307
Folliculitis 63
Foodborne 16
Food botulism 294
Francisella tularensis 38, 225
Furuncle 63

G

Gardnerella vaginalis 217
Genital Herpes 198, 204

Genital Tract Infections 198
Genital Warts 73, 198, 202
Giardia duodenalis 179
Giardia lamblia 176
Glomerulonephritis 93
Gonorrhoea 198, 200
Granulomatous Amebic Encephalitis (GAE) 314
Guanarito virus 5
Guinea worm 9

H

Haemophilus influenzae 98
Hansen's disease 285
Hantaan virus 4
Hantaviruses 38
Hardin MD 42
Hartmannella vermiformis 118
Hearing Loss 81
Helicobacter pylori 4, 141
Hepatitis B 3
Hepatitis C 43
Hepatitis C virus 4
Hepatitis E virus 4
Herd Immunity 33
Herpes Simplex Virus 74
HHV-6 76
Histoplasma capsulatum 122, 124, 125
Histoplasmosis 122, 124
HIV 198, 211
HIV/AIDS 3, 21
HIV-associated cryptococcal infection 306
HIV protease 212
hMPV 120
Hookworm 170
Hospital Acquired Infections 319
Hospital acquired Urinary Tract Infection (HAUTI) 324
HSV-1 204
HSV-2 204
Human African trypanosomiasis 9
Human herpesvirus 6 (HHV-6) 4
Human herpesvirus 8 (HHV-8) 5
Human herpesviruses (HHV) 74
Human metapneumovirus 120
Human Papilloma Virus (HPV) 71, 202
Human rabies 301
Human T-Lymphotrompic virus 1(HTLV1) 4

I

IBIS 105
ICAM-1 100
Inactivated poliovirus vaccine (IPV) 301
Infant botulism 293
Infectious cycle 17
Influenza 25
Influenza viruses 27
INH 115
INSPEAR 49
Inter-Cellular Adhesion Molecule 100
Itraconazole 307
Ivermectin 273
Ixodes (Ixodes ricinus) 68

J

Japanese encephalitis virus (JEV) 296
Jock Itch (Tinea Cruris) 83

K

Kidney Infections 197
Klebsiella pneumonia 106

L

Legionella pneumophila 4, 118
Legionellosis 117
Leishmaniasis 8, 261
Leprosy 9, 285
Leshmania donavani 261
Life Cycle 262, 266, 310
Listeria monocytogenes 147, 152
Listeriosis 152
Lower respiratory infections 3
L. tropica and L. braziliensis 261
Lyme borreliosis (LB) 70
Lyme Disease 43, 68
Lymphatic Filariasis (LF) 7, 269

M

Malaria 3, 24, 254
MALT 144
Marburg virus (MARV) 28
Mass Vaccination 34
MDR TB 116
Measles 3
Medscape Resource Center 42
Meningitis 3, 80, 281
Meningococcaemia 282
Meningococcal disease 281
Metronidazole 176
Metronidazole (Flagyl) 180
Microbial Threats 51
Microfilaremia 272
Miliaria 62
MMR 29
M-nontypeable (MNT) isolates 68

Model System 19
Mode of Transmission 15
Mononucleosis 235
Moraxella catarrhalis 98
Mucocutaneous leishmaniasis 261
Mucous membrane portal 14
Multibacillary 286
Multidrug resistant (MDR) 320
Multidrug-resistant Mycobacterium tuberculosis 38
Multidrug therapy (MDT) 288
Mumps Virus (MuV) 29, 79
MuV 29
Mycobacterium Leprae 285
Mycobacterium ulcerans 10
Mycoplasma Pneumonia 109
Myocarditis 81

N

Naegleria fowleri 313
Necator americanicus (hookworms) 170
Necatoror Ancylostoma (hookworm) 7
NEDDS 50
Neisseria gonorrhoeae 200
Neisseria meningitidis 281
Nipah virus 38
Norovirus 44

O

Onchocerca volvulus 8
Onchocerciasis 8
Opportunistic diseases associated with HIV infection 213
Oral poliovirus 300
Orchitis and oophoritis 81
Otitis media (OM) 88, 89, 98

P

PAHO 47
Pancreatitis 81
Paramyxovirus virions 140
Parenteral portal 14
Parvovirus 4
Paucibacillary 286
Pelvic inflammatory disease (PID) 198
Peridontitis/Gingivitis 135
Periungual warts 72
Pertussis 3
Pertussis Caused 110
P. falciparum 254
Pharyngitis 88
Phlebotomus spp. 263
Pinworm (Enterobius vermicularis) 170
Pityriasis alba 62
Plague 229
Plantar warts 72
Plasmodium 254
P. malariae 254
Pneumocystis jirovecii 126
Pneumocystis Pneumonia (PCP) 126
Poliomyletis 297
Poliovirus (PV) 297
Porphyromonas gingivalis 135
Portals of entry 13
Portals of Entry to Nervous System 16
Portals of exit 14
P. ovale 254
Primary 208
ProMED 41
Propionibacterium 59
Pseudomonas 63
Pseudomonas aeruginosa 329
P. vivax 254
Pyoderma 67
Pyuria 195
PZA 115

Q

Q fever 38

R

Rabies virus (RV) 302
RADT 92
RAM 115
Repellents 239
Reservoir 13
Respiratory Syncytial Virus-(RSV) 119
Reston ebolavirus (RESTV) 244
Rheumatic fever 94
Rhinoviruses 100
Ringworm (Tinea Corporis) 83
Risk Factors 321
River blindness 8
RNTCP 115
Rotavirus 4, 164
Roundworm (Ascaris lumbricoides) 170
Rubulavirus 79

S

Sabia virus 5
Salmonella serovar 147
Salmonella typhi 147
Sandflies 263
SAPNA 105
SARS 27
S. aureus 63

S. boydii 155
Scabies 62
Scalded skin syndrome (SSS) 66
Scarlet fever 93
Schistosoma 265
Schistosomiasis 7
S. dysenteriae 155
Seborrheic dermatitis 62
Secondary 208
Second-line drugs 115
Selection dynamics 18
Septicemia 147, 222
Sexually Transmitted Infections (STI) 198
S. flexneri 155
Shigellosis 155
Sin Nambre virus 5
Sinusitis 98
Skin portal 14
Snail fever 7
Soil-transmitted helminthes (STH) 6, 170
S. paratyphi 147
Sporadic diarrhea (SD) 156
S. sonnei 155
Staphylococcal scalded skin syndrome (SSSS) 66
Staphylococcus aureus 4, 64, 222
Staphylococcus epidermidis 64, 222
STH 170
Stomatits 88
Streptococcal Pharyngitis 91, 93
Streptococcus mutans 134
Streptococcus Pneumonia 103
Streptococcus pneumoniae 98, 222
Streptococcus pyogenes 12, 67
Streptococcus viridans 222
Strongyloidiasis (Thread worm) 170
STSS 92
Subungual 72
Sudan ebolavirus (SUDV) 244
Sympathetic over activity (SOA) 289
Syphilis 3, 198

T

Taï Forest ebolavirus (TAFV) 244
Tapeworm (Taenia solium) 170
T. brucei brucei 309
T. brucei gambiense 309
T. brucei rhodesiense 309
TEM and SHV genes 325
Tertiary 208
Tetanus 3, 289
Tetanus toxin 290
Thyroiditis 81
Tickborne Encephalitis viruses 38
Tickborne hemorrhagic fever viruses 38
Tick Relapsing Fever 44
Tinea Versicolor 83
Tinidazole 176
Topical corticosteroids 65
Toxoplasma gondii 250
Toxoplasmosis 250
Trachoma 8
Treponema pallidum 208
Triatomine bug (Triatoma infestans) 247
Trichiuris trichiuria (whipworm) 170
Trichomonas 198, 205
Trichuris (whipworm) 7
Tropical diseases 3
Tropical pulmonary eosinophilia(TPE) 271
Trypanosoma brucei 308
Trypanosoma brucei gambiense 9
Trypanosoma brucei rhodesiense 9
Trypanosoma cruzi 246
Trypanosomiasis caused 246
Trypansoma cruzi 9
Trypomastigote 310
Tsetse fly 9
Tuberculosis (TB) 3, 25
Tularemia 225

U

Urticaria 62
USAMRIID 49
UTIs 195

V

VacA 144
Vaccine 97
Vaccine derived polioviruses 300
Vancomycin Resistant Enterococci 326
Vancomycin resistant Enterococcus faecium (VREF) 326
Varicella (Chicken pox) 77
varicella zoster virus (VZV) 77
Variola major 38
V cholerae O139 160
Veillonella 134
Ventilator-associated pneumonia (VAP) 320
Vibrio Cholerae 160
Vibrio Cholerae O139 5
Vibrio cholerae O1 or O139 162
Vir typing 68
Visceral leishmaniasis 261
Vulvovaginal candidiasis (VVC) 215

W

Waterborne 16
Weekly Epidemiological Record (WHO) 53
West Nile Virus 28
Whipworm (Trichuris trichiura) 171
WHO-APCRI 304
Whooping cough 88, 110
Wild polioviruses 300
WNV 28
World Health Organization 48, 49
Wound botulism 294
Wuchereria bancrofti 269

Y

Yellow fever virus 38
Yersinia 229
Yersinia enterocolitica 151
Yersinia pestis 38, 230

Z

Zaire ebolavirus (EBOV) 244
Zoonosis 13